Mechanical Design of Process Systems

Volume 2

Shell-and-Tube Heat Exchangers
Rotating Equipment
Bins, Silos, Stacks

Volume 1: Piping and Pressure Vessels

1. Piping Fluid Mechanics
2. The Engineering Mechanics of Piping
3. Heat Transfer in Piping and Equipment
4. Engineering Mechanics of Pressure Vessels

Appendix A. Partial Volumes and Pressure Vessel Calculations
Appendix B. National Wind Design Standards
Appendix C. Properties of Pipe
Appendix D. Conversion Factors

Volume 2: Shell-and-Tube Heat Exchangers; Rotating Equipment; Bins, Silos and Stacks

5. The Engineering Mechanics of Bins, Silos and Stacks
6. Rotating Equipment
7. The Mechanical Design of Shell-and-Tube Heat Exchangers
8. External Loadings on Shell Structures

Appendix A. Partial Volumes and Pressure Vessel Calculations
Appendix B. National Wind Design Standards
Appendix C. Properties of Pipe
Appendix D. Conversion Factors

Mechanical Design of Process Systems

Volume 2

Shell-and-Tube Heat Exchangers
Rotating Equipment
Bins, Silos, Stacks

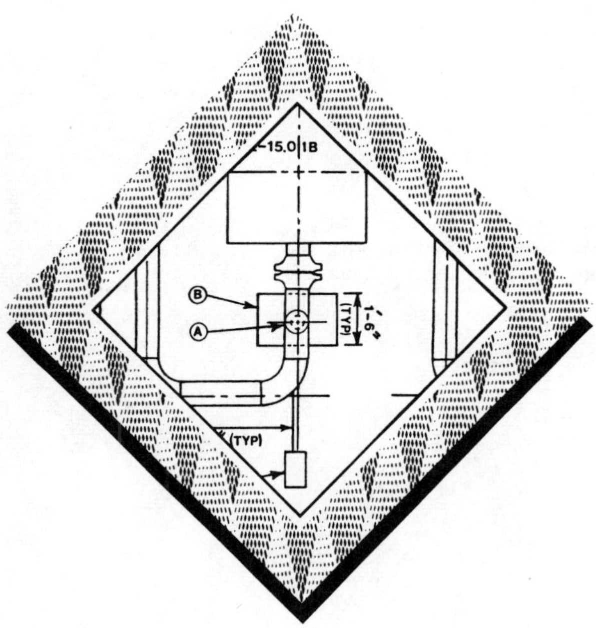

A. Keith Escoe

Gulf Publishing Company
Book Division
Houston, London, Paris, Tokyo

Mechanical Design of Process Systems
Volume 2
Shell-and-Tube Heat Exchangers • Rotating Equipment • Bins, Silos, Stacks

Library of Congress Cataloging-in-Publication Data

Escoe, A. Keith.
Mechanical design of process systems.

Includes bibliographies and indexes.
Contents: v. 1. Piping and pressure vessels—v. 2. Shell-and-tube heat exchangers; rotating equipment; bins, silos, stacks.
1. Chemical plants—Design and construction.
I. Title.
TP155.5.E83 1986 660.2 ′ 81 85-22005

ISBN 0-87201-562-9 (v. 1)
ISBN 0-87201-565-3 (v. 2)

Contents

Foreword

The engineer who understands the impact of process design decisions on mechanical design details is in a position to save his client or his company a lot of money.

That is because the test of any process design is in how cost-effectively it yields the desired product, and how "cost" generally translates to "equipment": How much will the process require? How long will it last? How much energy will it consume per unit of product?

In this two-volume work on *Mechanical Design of Process Systems,* A. K. Escoe has performed a monumental service for mechanical design engineers and chemical process engineers alike. The information is presented in such a manner that even the neophyte engineer can grasp its full value. The author has produced an in-depth review of the way in which process design specifications are interpreted into precise equipment designs. Perhaps most valuable of all are the extensive *worked examples* throughout the text, of actual designs that have been successfully executed in the field.

The piping system is the central nervous system of a fluid flow process, and the author has treated this with proper respect in two excellent chapters on fluid mechanics and the engineering mechanics of piping (Volume 1).

The chapter on heat transfer in vessels and piping illustrates lucidly the interrelationship between process and mechanical design. Every engineer working with industrial process systems will benefit from reading this chapter.

Although the author has made a herculean effort in covering the mechanical design of pressure vessels, heat exchangers, rotating equipment, and bins, silos and stacks (Volume 2), it is true that there are omissions. It is hoped that, as the author hints in his preface, a future volume might be added covering multiphase flow, specific cogeneration processes, turbines, and detailed piping dynamics.

Still, at this writing these two volumes comprise an outstanding practical reference for chemical and mechanical engineers and a detailed instructional manual for students.

I recommend these volumes highly for each design engineer's professional library.

John J. McKetta, Ph.D., P.E.
Joe C. Walter Professor of Chemical Engineering
University of Texas, Austin

Dedication

To the memory of my beloved parents, Aubrey H. Escoe and Odessa Davies Escoe; and to the dedicated engineer, Dr. Judith Arlene Resnik, U.S. astronaut aboard the ill-fated space shuttle *Challenger* (Flight 51-L).

Preface to Volume 2

This book's purpose is to show how to apply mechanical engineering concepts to process system design. Process systems are common to a wide variety of industries including petrochemical processing, food processing and pharmaceuticals, power generation (including cogeneration), ship building, and the aerospace industry. The book is based on years of proven, successful practice, and almost all of the examples described are from process systems now in operation.

While practicality is probably its key asset, this second volume contains a unique collection of valuable information, such as a practical approach to bin and silo design as well as practical methods of controlling wind vibrations of stacks using vortex strakes; new information on nozzle loadings on compressors and turbines; comprehensive discussions and examples on sizing pumps and compressors for various process applications; expanded tube count tables for shell-and-tube heat exchangers; a practical approach to design against tube bundle vibration; and a comparative synopsis of the various national wind codes.

Topics included in the text are considered to be those typically encountered in engineering practice. For reasons of time and space the dynamic analyses of seismic response spectra and an extensive discussion on pulsation response spectra in piping induced by acoustic pulsation are not discussed. However, a short discussion is given on pulsation response spectra induced by acoustic pulsations. Single-phase flow is much more common in mechanical systems than two-phase flow, so because of time and space two-phase flow is not discussed.

This book is not intended to be a substitute or a replacement of any accepted code or standard. The reader is strongly encouraged to consult and be knowledgeable of any accepted standard or code that may govern. It is felt that this book is a valuable supplement to any standard or code used.

The book is slanted toward the practices of the ASME vessel and piping codes and the TEMA standard for shell-and-tube heat exchangers. The intent is not to be heavily prejudiced toward any standard, but to discuss the issue—engineering. If one feels that a certain standard or code should be mentioned, please remember that there are others who may be using different standards and it is impossible to discuss all of them.

The reader's academic level is assumed to be a bachelor of science degree in mechanical engineering, but engineers with bachelor of science degrees in civil, chemical, electrical, or other engineering disciplines should have little difficulty with the book, provided, of course, that they have received adequate academic training or experience.

Junior or senior undergraduate engineering students should find the book a useful introduction to the application of mechanical engineering to process systems. Professors should find the book a helpful reference (and a source for potential exam problems), as well as a practical textbook for junior-, senior-, or graduate-level courses in the mechanical, civil, or chemical engineering fields. The book can also be used to supplement an introductory level textbook.

The French philosopher Voltaire once said, "Common sense is not very common," and unfortunately, this is sometimes the case in engineering. Common sense is often the by-product of experience, and while both are essential to sound engineering practice, neither can be learned from books alone. It is one of this book's goals to unite these three elements of "book learning," common sense, and experience to give the novice a better grasp of engineering principles and procedures, and serve as a practical design reference for the veteran engineer.

Finally, I wish to thank Dr. John J. McKetta, professor of chemical engineering at the University of Texas at Austin, who had many helpful comments, suggestions, and words of encouragement; other engineering faculty members at the University of Texas at Austin for their comments; Albert T. Taube, P.E., who was so kind to offer helpful and useful comments while reviewing Chapter 6; and John D. Guenther, P.E., for his helpful critique of Chapter 7. Last, but certainly not least, I wish to express gratitude to William J. Lowe and Timothy W. Calk of Gulf Publishing Company whose hard work and patience made this book possible.

A. Keith Escoe, P.E.

5

The Engineering Mechanics of Bins, Silos, and Stacks

The engineering mechanics of bins and silos differ from the mechanics of pressure vessels because solids behave differently from liquids and gases, both in storage and in flow conditions. The mechanics of stacks are almost identical to those of towers, but are somewhat simpler. An engineer has more flexibility and approaches for solving vortex shedding around stacks than around towers, because stacks rarely have as many attached structures.

SILO AND BIN DESIGN

The mechanics of solid flow theory is a fairly complicated subject. The proper design of silos and bins is more than meets the untrained eye, and involves every aspect of engineering mechanics. This chapter only "sketches" methods of approaching this complex phenomenon, and refers the interested reader to literature on this specialty.

The field of solids handling has been augmented the past twenty years by two researchers—Jenike and Johanson [1]. The methods presented in this chapter are largely influenced by their work.

Bins and silos appear to be very simple devices, but what goes on inside is not so simple. To design an efficient bin the design engineer must understand why solids in bins do not flow (Figure 5-1):

1. Development of a rathole or stable arch that ceases flow.
2. Erratic flow—transient arches form within the solid resulting in variance of the bulk density such that flow becomes unstable.
3. Flushing—the fluidization and flushing of powders creates erratic flow.
4. Dead storage—residual build-up of solids caused by the inability to exit bin.
5. Segregation—a heterogenous solid of varying specific gravity in which the lighter particles exit the bin first, leaving behind the heavier particles.
6. Degradation—the chemical change of solids caused by remaining in storge too long. Spoilage, caking, and oxidation are some examples.

Solids behave differently from gases or liquids because they can transfer shear stresses without movement, and because of their cohesive strength, they can retain their shape under load. The shear stress transferred between the solid and the channel walls is a function of the normal pressure, w. The relationship between the two is as follows:

$$\mu = \tan \phi' = \frac{S}{w} \tag{5-1}$$

where ϕ' = kinematic angle of friction between the solid and the bin wall
μ = coefficient of friction between the bulk solid and the bin wall

Typical values of ϕ' are given in Table 5-1 for various solids and bin materials. This table can be used in applications where the bulk solid properties are not known (as is commonly the case). The value of ϕ' is required by the methods presented to be a constant value so that using the table will produce a conservative design.

There are two flow conditions that can occur—mass flow and funnel flow. Mass flow is a flow pattern in which all the material in the hopper or bin is in motion and the flow occurs along the bin walls. Funnel flow is a flow pattern in which the material flows primarily in the center region of the bin.

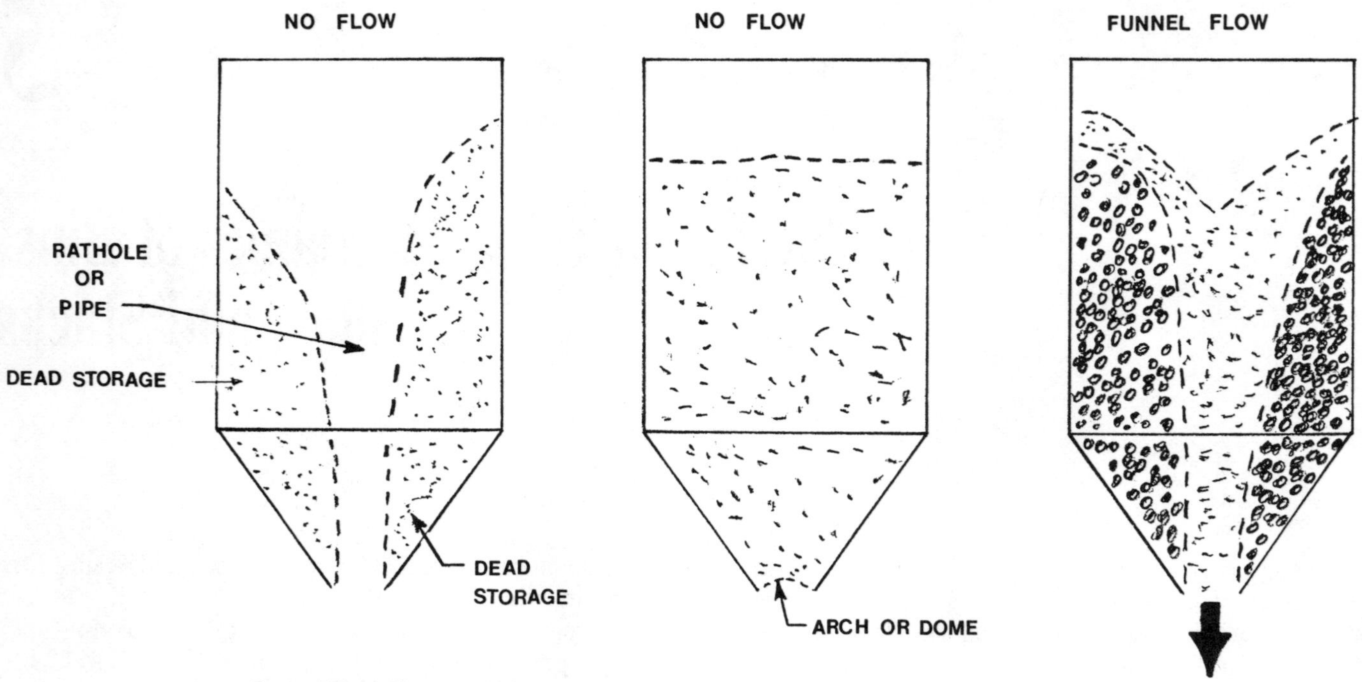

Funnel Flow Characteristics

1. Material segregates and ratholes may be formed.
2. Flow may be erratic.
3. Low headroom.
4. Powders will flush.

Figure 5-1A. Problems of flow of solids.

Table 5-1
Properties of the Materials Used in the Stacking-Out Bins [1]

Material	Hopper Opening	δ°	ϕ_1°	ϕ'°	γ lb/ft³	σ_1' lb/ft²	ff	ff_a	T, sec	q_1 calculated, lb/sec
Iron ore concentrate	Rec.*	39	33	23	190	585	1.25	5.50	5.7	16,700
(H_2O, 1.5%)	Cir.**	39	33	23	190	467	1.25	4.65	5.7	18,200
Iron ore (H_2O, 10.0%)	Rec.	63	47	25	141	343	0.97	1.50	9.5	8,450
	Cir.	63	47	25	141	395	1.05	1.50	9.5	9,250
Dolomite—Michigan	Rec.	46	40	25	105	286	1.14	3.20	8.1	8,150
(H_2O, 4.2%)	Cir.	46	40	25	105	229	1.14	2.80	8.1	8,600
Dolomite—(Moller)	Rec.	55	39	26	100	263	1.05	1.70	11.9	6,220
(H_2O, 8.2%)	Cir.	55	39	26	100	210	1.05	1.60	11.9	6,660

** Rectangular opening, 4 by 2.5 ft.*
*** Circular opening, 4 ft diam.*

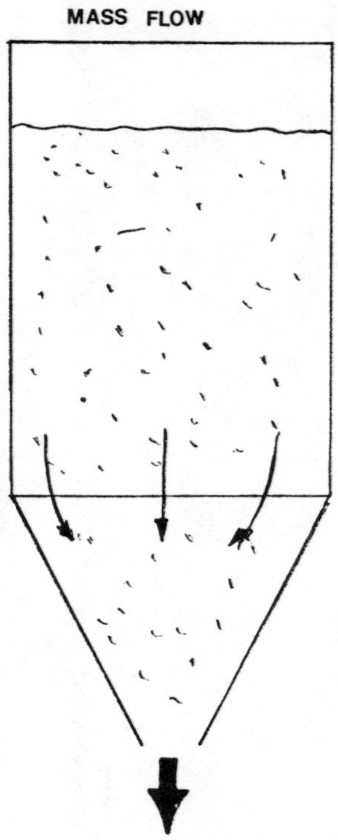

Figure 5-1B. Ideal flow of solids—mass flow.

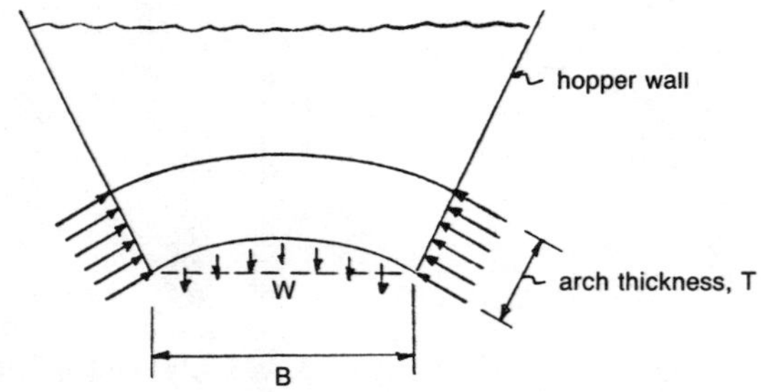

Figure 5-2. Formation of an arch.

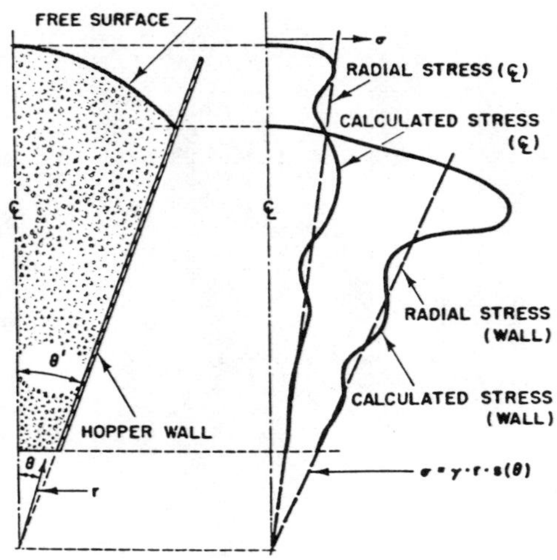

Figure 5-3. Stress distributions along hopper wall [1].

The strength of the solid material is the criterion for flow behavior in bins. Failure conditions of the solid particles can result in arching, no flow, piping (a hole formed in the solid formation), or limited flow. Figure 5-2 illustrates an arch formed by a solid in a hopper. The failure of the arch will occur when the major compressive stress, P, equals the unconfined yield strength, f_c. To prevent arching, the critical dimension, B, of the hopper opening must be

$$B > \frac{f_c}{\gamma(1 + m)} \tag{5-2}$$

where m = 0 for slot opening of width B
m = 1 for circular opening of diameter B
γ = bulk density of the solid, lb/ft^3

The calculated stress and radial stresses are shown in Figure 5-3. When the stresses induced between the solid particles and the hopper wall are not compatible with radial stress, a flow pattern will not develop along the hopper wall. When the hopper angle is less than 30°, the limits of radial stresses will occur in conical hoppers, as shown in Figure 5-4.

Even though the hopper opening is large enough to prevent arching, mass flow piping will occur. The critical diameter at which the pipe is unstable is given by the following:

$$D \geqslant 4\frac{f_c}{\gamma}\Phi \tag{5-3}$$

Figure 5-5 shows a plot of the piping factor, Φ, against the angle of internal friction, ϕ. The limiting relations for arching and piping in Equations 5-2 and 5-3 are functions of the material yield strength, f_c. This parameter can be determined empirically only if the consolidating pressure σ_1 for steady flow is known. This pressure is denoted by

$$\sigma_1 = \gamma BQ \tag{5-4}$$

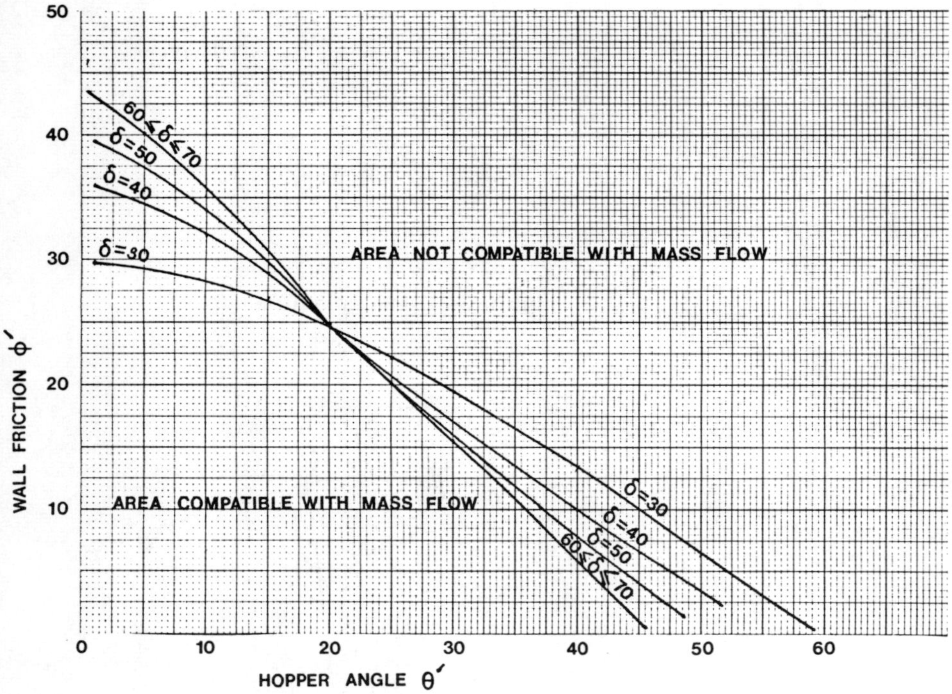

Figure 5-4. The criteria for mass flow when $\theta' < 30°$.

$$\text{where } Q = \frac{\sigma(1 + \sin \delta)}{2 \sin \theta} \tag{5-5}$$

θ = angle of hopper slope
σ = computed stress function along the wall

Combining Equations 5-2 and 5-5 we obtain

$$\frac{\sigma_1}{f_c} \geqslant (1 + m)Q \tag{5-6}$$

where σ_1/f_c = flow factor of solid

The critical flow factor for arching in channels (ff) is represented by

$$ff = \left(\frac{\sigma_1}{f_c}\right)_{critical} = (1 + m)Q \tag{5-7}$$

Figures 5-6–5-9 show the values of ff for straight-walled converging bins with various material properties and wall slopes. These factors are presented as straight lines in the f_c vs. σ_1 graph in Figure 5-10.

The consolidating pressure σ_1 that the flowing solid particles exert in a vertical cylindrical channel is

$$\sigma_1 = D\gamma G \tag{5-8}$$

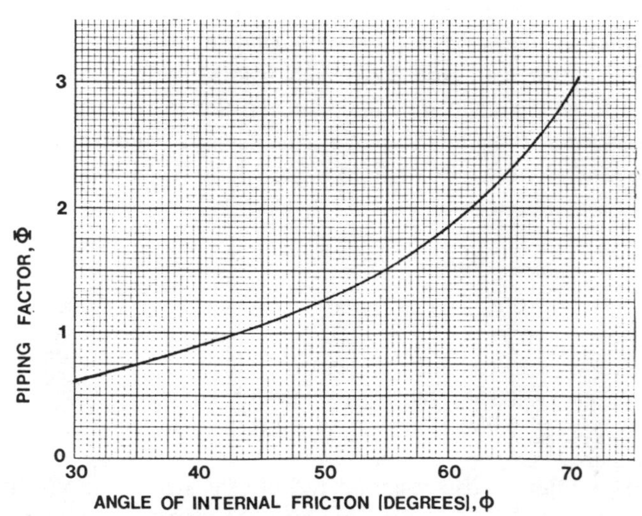

Figure 5-5. Piping factor, Φ, versus angle of internal friction, ϕ.

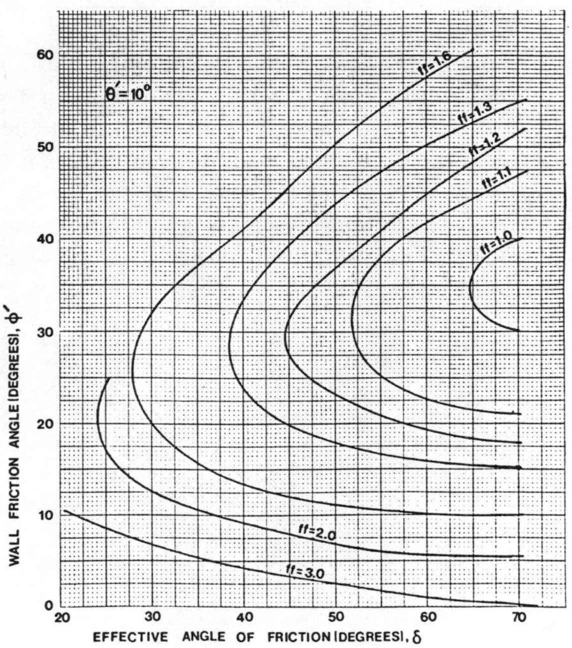

Figure 5-6. Wall friction angle, ϕ', versus effective angle of friction, δ.

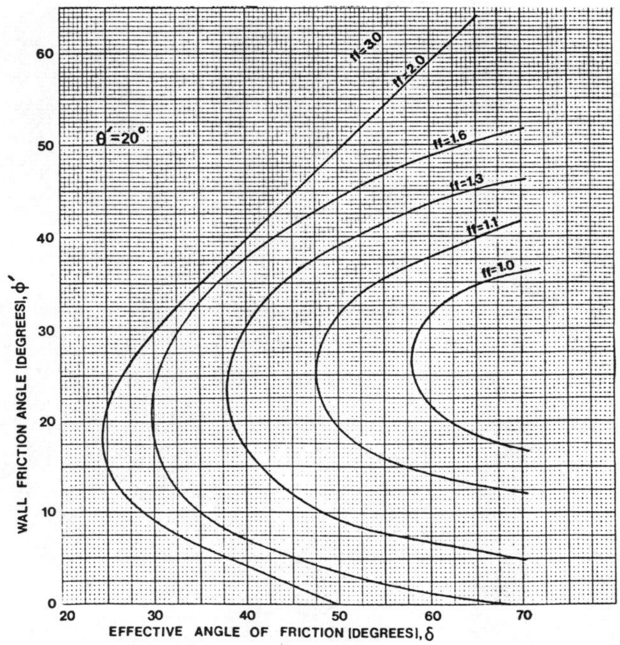

Figure 5-7. Wall friction angle, ϕ', versus effective angle of friction, δ.

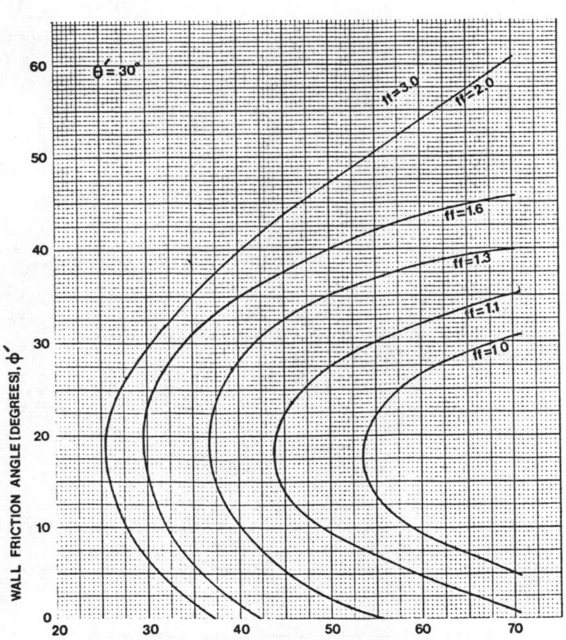

Figure 5-8. Wall friction angle, ϕ', versus effective angle of friction, δ.

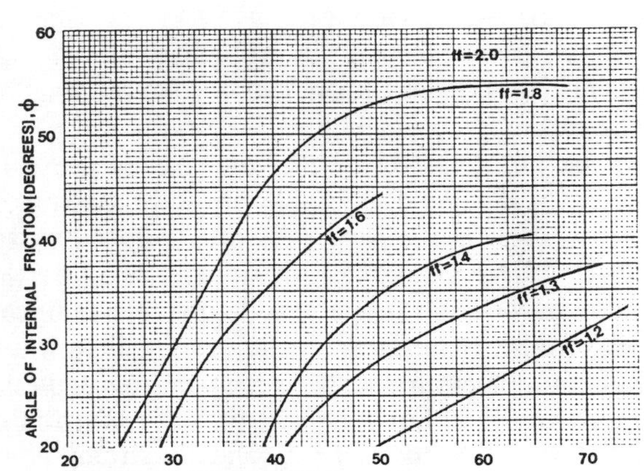

Figure 5-9. Wall friction angle, ϕ', versus effective angle of friction, δ.

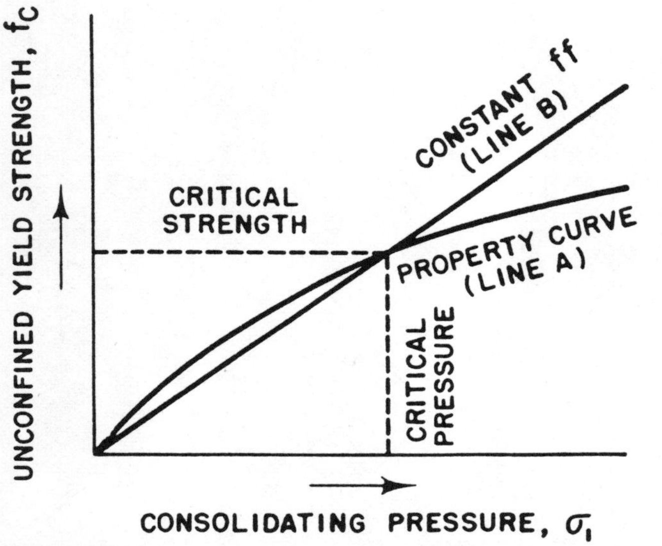

Figure 5-10. Critical values of σ_1 and f_c. Line A represents strength properties and Line B the constant flow factor [1].

where G is a function of the effective angle of friction, δ, and the internal angle of friction, ϕ. This consolidating pressure, σ_1, provides the strength of the material that forms the pipe in the bin. Combining Equation 5-3 with 5-8 we have

$$\left(\frac{\sigma_1}{f_c}\right) \geqslant 4\Phi G$$

$$\text{or } ff = \left(\frac{\sigma_1}{f_c}\right)_{critical} = 4\Phi G \tag{5-9}$$

The value of ff is plotted against δ and ϕ in Figure 5-11.

Figure 5-12 shows flow properties of a typical bulk solid, which are quite useful in problem solutions. Table 5-2 lists critical hopper dimensions for the material with flow properties given in Figure 5-12.

Once the problems of arching and piping are solved and the bin is designed to handle the solid mixture, the next step is to examine flow pressures induced by solid particle flow. As mentioned previously, solid particles suspended in vertical storage bins do not behave linearly, such as liquids. To a much greater extent than liquids, solids manifest shear forces between particles and on bin walls. Figure 5-13 shows typical pressure distributions for mass flow and funnel flow, and illustrates how in mass flow the pressure is maximum at the bin-hopper junction point. The geometric discontinuity causes an increase in flow pressure because of change in momentum of the flow of solid particles. This pressure is reduced internally somewhat because as the solid particles descend through the hopper, a vacuum in the void between particles develops and produces a negative gauge pressure. As the particles approach the outlet, atmospheric pressure is obtained.

While the wall pressure is maximum at the bin-hopper tangent line in mass flow, it is only a fraction of a hydrostatic pressure for a liquid head equivalent to the height of the solid in the bin. Thus, designing solid bins for hydrostatic loads results in overdesign of the bins. As a guideline, the maximum hoop pressure at the bin-hopper tangent point is about seven times that of the pressure of the solid induced by gravity. That is,

$$P_{max} = 7\,(\gamma)\,\frac{16}{ft^3}\,(H)\text{ ft} \tag{5-10}$$

where γ = bulk density of the solid, lb/ft³
H = height that solid is stored in bin, ft

Table 5-2
Critical Hopper Dimensions for Material With Flow Properties Shown in Figure 5-12 [1]

Type	Critical width of a slot opening for arching, ft	
	Freshly stored	**Stored for 24 hr**
Flat bottom or nonmass flow bins	0.2	1.0
Stainless lined hopper ($\theta' = 30°$, $\phi' = 27°$)	0*	0.4
Mild steel hopper ($\theta' = 30°$, $\phi' = 35°$)	0*	0.6
	Critical diameter of a circular opening for arching, ft	
Flat bottom or nonmass flow bins	0.4	2.0
Stainless lined conical hopper ($\theta' = 15°$, $\phi' = 27°$)	0*	0.9
Mild steel conical hopper ($\theta' = 15°$, $\phi' = 35°$)	0.4**	2.0**
	Critical dimensions for piping, ft	
For flat bottom or nonmass flow bins	5.6	7.7

* *Dictated only by particle size or dynamic conditions.*
** *These values are the same as the flat bottom bin values because the mild steel conical hopper when $\phi' = 35°$ is too rough to provide flow along the walls of the cone when $\theta' = 15°$*

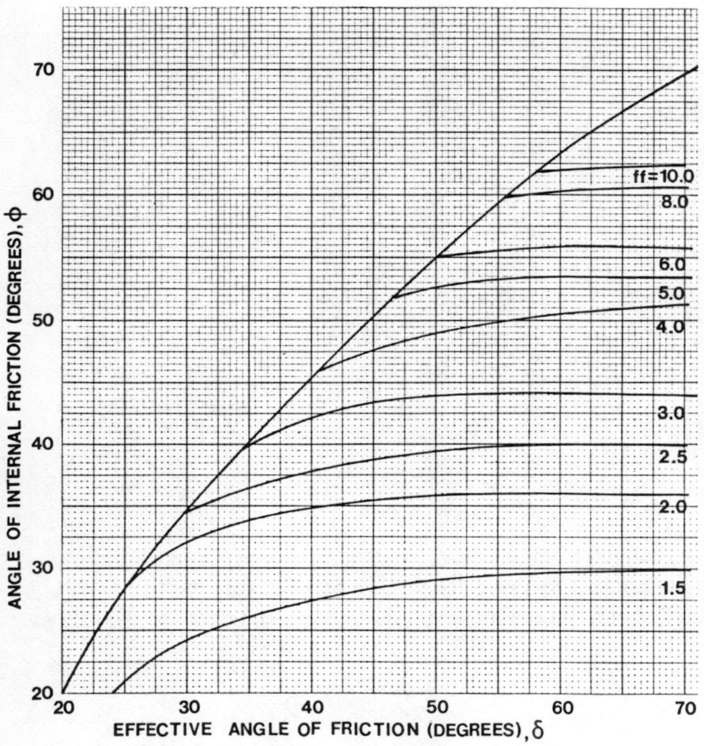

Figure 5-11. Critical flow factor for piping.

The internal pressure in Equation 5-10 can be increased by the use of pneumatic air supplied to the bin. In the case of bins where funnel flow exists or for small bins with cohesive solids, supplying forced air through ducts in the bin is desirable to prevent the formation of arches and pipes within the solid itself. To compensate for the additional internal pressure, Equation 5-9 becomes

$$P_{max} = 7\gamma H + P_{air} \tag{5-11}$$

where P_{air} = air pressure, psig

The use of pneumatic air in bins is often desirable and in the situations where air cannot be used because of chemical interaction with the solids in a closed system, nitrogen is commonly used.

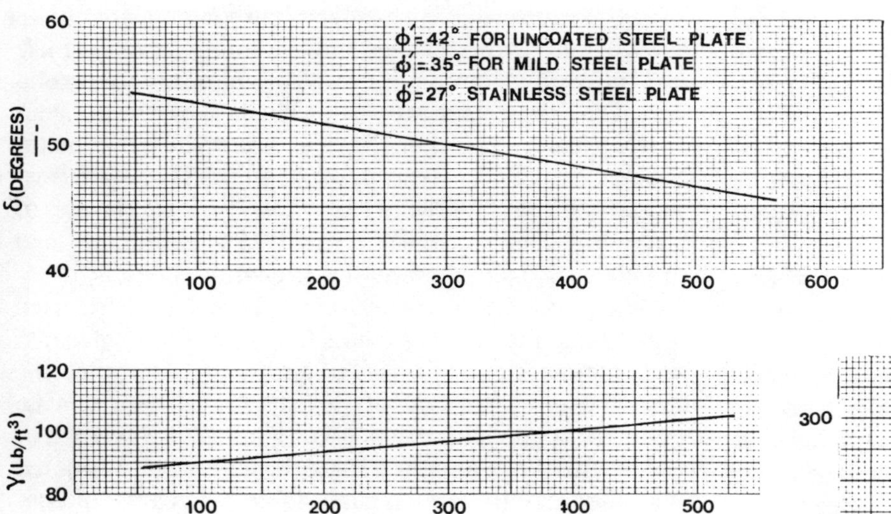

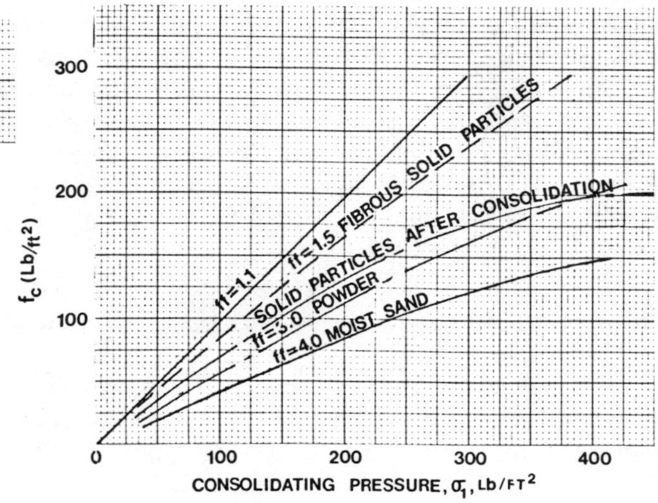

Figure 5-12. Typical bulk solid flow properties used to determine critical dimensions for piping and arching.

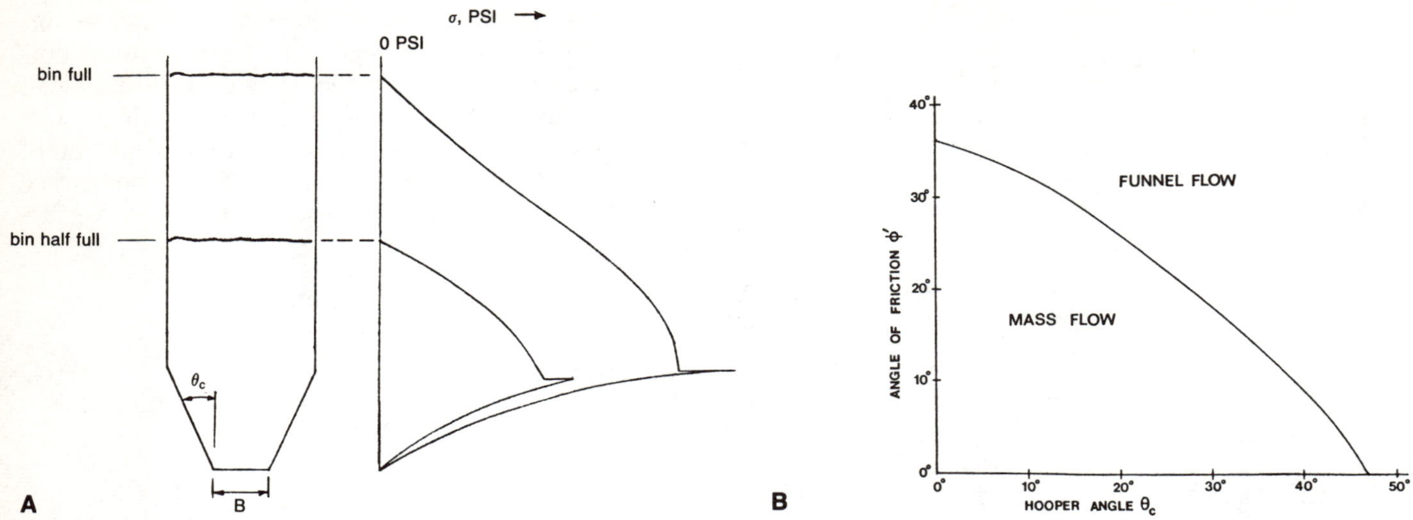

Figure 5-13. (A) Pressure distribution for solid flow is maximum at cylinder-cone intersection primarily because of discontinuity stresses; (B) The relationship between mass flow and funnel flow for conical sections. The angle of kinematic friction, ϕ', is a function of the coefficient of friction between the solid and bin material and the compression the solid is subjected to in storage.

STACK DESIGN

The analyses of stacks subjected to wind and seismic response spectra are identical to those methods used for process towers discussed in Chapter 4. The differences in the two types of equipment are twofold: (1) stacks have different values for logarithmic decrement and dynamic magnification factor, and (2) the solution to problems induced by vortex shedding are different. Both of these factors are a result of stacks having simpler geometries.

The simpler geometry of the stack works for and against the engineer. The positive aspect comes as a result of the methods used to break vortex shedding—vortex breakers are much easier and more practical to install on stacks than on process towers. The negative aspect of stacks is that they do not have connected piping and structures to break up vortices and to damp wind-induced vibrations. Thus, we will focus our discussion on those aspects of wind design that are peculiar to stacks, remembering that the fundamental basis of design is the same for stacks and towers.

Vortex Shedding and Frequency Response

As explained in Chapter 4, only the fundamental mode of vibration is considered for process towers and stacks. Consequently, the Rayleigh method is applied to obtain the vibration characteristics of the stack.

In stacks, lining is often used where high temperatures are encountered and carbon structural steel is the stack material. Lining must be used for temperatures in excess of 800°F because of the danger of carbon precipitation in the steel. To avoid this and not use lining, one must use hot-rolled, high-strength low-alloy steels that have good elevated-temperature properties. Such steels are not generally pressure vessel quality and require heat treatment, such as the Cr-Mo steels described in ASTM specifications A-387 and A-542. These low-alloy steels are of structural quality, contain 0.75–1.25% chromium, and are cheaper than pressure-vessel-quality alloys.

When common carbon structural steel is to be used with lining, the effect of gunite lining must be considered with the mass and stiffness to accurately determine the fundamental frequency of the stack. An approximate value of the modulus elasticity of gunite is 1.3×10^6 psi.

The effect of lining in a stack must also be considered with the flexibility of the base. Table 5-3 is a list of conservative values of the logarithmic decrement and dynamic magnification factors for various soil conditions for lined and unlined stacks. For explanation and use of these values the reader is referred to Chapter 4.

Ovaling

When slender stacks, i.e., rings in which the thickness is small in comparison to the radius, are subjected to vortex shedding caused by air currents, the elastic strain energy of the cylinder is distributed in such a manner as to induce flexural and torsional modes of vibration. The ring is subjected to the following modes:

1. Extensional (axial elongation and contraction about the ring's own axis).
2. Torsional (twisting of the ring about its own axis).
3. In-plane flexural (inextensional vibrations in the plane of the ring).
4. Out-of-plane flexural (inextensional displacements in the plane of the ring).

The flexural modes are generally the only modes of practical significance since the fundamental natural frequencies of the torsional and extensional modes are much greater than the fundamental natural frequencies of the flexural modes. Figure 5-14 shows these various modes.

The flexural modes, in-plane and out-of-plane, are used in determining the resonance frequency of the stack caused by ovaling. Since out-of-plane flexural vibrations are coupled to torsional vibrations, it is the out-of-plane frequency used to describe the vibration of the stack; however, the natural frequencies of the flexure modes in and out of the plane of the ring vary only slightly for circular cross sections. The natural frequency of the ring is given as

$$f_r = \frac{1}{2\pi}\left[\frac{EIn^2(n^2 - 1)^2}{\rho Ar^4(n^2 + 1 + \nu)}\right]^{0.5} \qquad (5\text{-}12)$$

The lowest flexural mode exists when n = 2 and Equation 5-12 reduces to

$$f_r = \frac{4.409t}{r^2}\sqrt{E} \qquad (5\text{-}13)$$

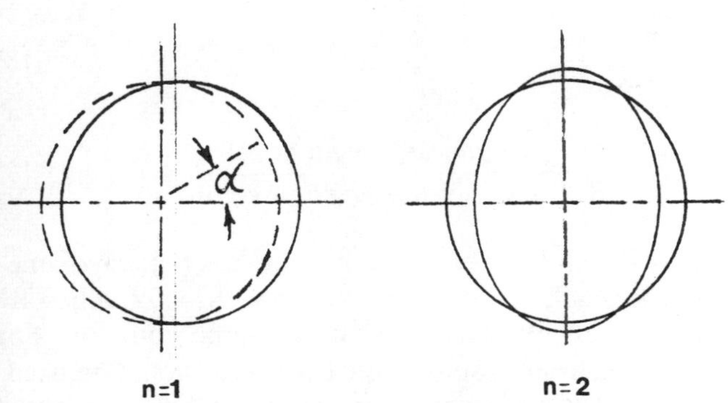

Figure 5-14. Stack mode shapes.

These relationships were formulated by the great pioneers Michell and Love during the nineteenth century. The reader is referred to Example 5-4 for further clarification of units.

In practical stack design, because vortices form alternately on either side of the stack, the flexural frequency (ovaling frequency) given in Equation 5-13 is taken to be twice that of the vortex shedding frequency. The vortex shedding frequency is given by Equation 4-101 as

$$f_v = \frac{0.2V}{D} \qquad (4\text{-}101)$$

Now since f = 2f, we solve for V and obtain

$$V_c = \frac{60f_rD}{2s} \qquad (5\text{-}14)$$

in which s = the Strouhal number (is equal to 0.2 for a wide range of Reynolds numbers). The value of V_c is the critical wind velocity in which ovaling occurs.

Both the vortex shedding and flexural frequencies should be evaluated at each elevation if ovaling rings are to be used. Normally, the upper third of the stack is all that is required to be investigated, based on various wind tunnel tests.

Now we come to the most practical aspect of stack design—how to alleviate flexural excitation. This can be done in two ways—ovaling rings or vortex strakes.

Ovaling rings are used to increase the mass distributed along the tower to dampen flexural vibrations. When the flexural frequency equals twice the vortex shedding frequency, i.e., if the design wind speed range includes the critical wind velocity, V_c, stiffeners are added at those sections where f = 2f. The section modulus of the stiffeners is given by

$$S_m = \frac{(7 \times 10^{-7})V_c^2D^2H_s}{\sigma_t} \qquad (5\text{-}15)$$

where V_c = critical wind velocity (Equation 5-14), fpm
D = internal stack diameter at elevation under investigation, ft
H_s = stiffening ring spacing, ft
σ_t = allowable tensile stress of stack material, psi

Ovaling rings provide a redistribution of the mass of the stack, resulting in localized stiffening that tends to offset flexural frequency modes. This is particularly desirable with stacks of several diameters. However, with stacks of constant or tapering cross section the use of vortex strakes is becoming increasingly popular.

Helical Vortex Breaker Strakes

For critical wind velocities less than 35 mph, dynamic stresses should be investigated. One optimum solution for such stresses in stacks has been found in wind tunnel tests and in practice to be helical vortex strakes.

The application of helical vortex strakes to vertical cylindrical towers has shown remarkable results. The strakes' function is to break up vortices such that flexural frequency modes are quickly dampened. It is significant to note that adding the strakes increases drag and thus wind loading. These strakes are shown in Figure 5-15.

To minimize the flow-induced drag and optimize the vortex-breaking effect, the strake height, W(ft), should be in the following range:

$$0.09D \leqslant W \leqslant 0.10D$$

where D = OD of stack, ft

Figure 5-15 shows a helix generated on a cylinder by taking a template πD long by L high and wrapping it around a cylinder. The length, L, of the helix is the top 1/3 of the stack. Wind tunnel tests have shown that vortex breaking devices are most effective on the upper third of a stack. The helix angle, ϕ, should fall into the following range:

$$54° \leqslant \phi \leqslant 58°$$

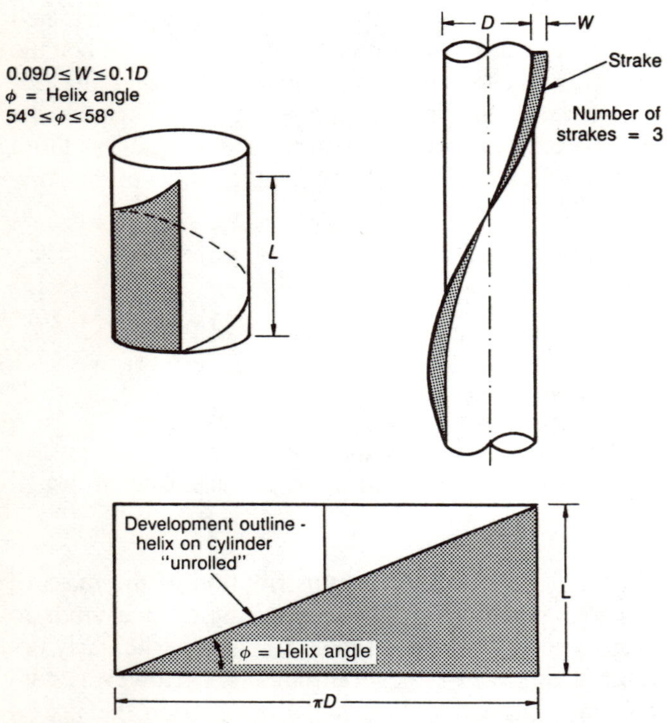

Figure 5-15. Cylindrical strake helix geometry.

There are always three strakes per stack to counter the alternate formation of vortices on either side of the stack.

Strakes can be fabricated from a flat piece of metal, normally 3/16-in. or 5 mm thick. Each strake is divided up into a certain number of strips, usually five to twenty segments, depending on the length of the stack. The overall length of the individual strakes that is divided up is determined by

$$S = [(\pi D)^2 + L^2]^{0.5} \tag{5-16}$$

where D = OD of stack, ft
L = height of tower portion straked (1/3 of total stack height), ft

The number "S" is divided into individual strips that are cut from a larger piece of plate shown in Figure 5-16. The strips must be cut to a radius of curvature, r, that is determined as follows:

$$r = \frac{a^2\omega^2 + b^2}{a\omega^2} \tag{5-17}$$

where $a = \frac{D}{2}$, ft

$$b = \frac{L}{2\pi w}$$

ω = number of revolutions around stack cylinder made by helical strake (usually $\omega = 1$)

An alternative formula, developed by Dr. Frank Morgan, and two to three percent in error of Equation 5-17, is

$$r = \frac{\lambda W}{1 - \lambda} \tag{5-18}$$

$$\text{where } \lambda = \frac{S_i}{S_o} = \frac{\text{interior arc length of helix}}{\text{exterior arc length of helix}} \tag{5-19}$$

The value S_i is determined by using the outside diameter of the stack in Equation 5-15, and S_o is obtained by using D + 2W in place of D in the same equation. For the most accurate results, Equation 5-16 should be used, as it is the exact radius of curvature of a helix projected on a cylinder [3].

Strips are laid out, as shown in Figure 5-16, with an inner radius of curvature determined by Equation 5-17 and outer radius of curvature of r = r + W. It is desired that the helix be perpendicular to the centerline of the cylinder along the entire length of the helical strake shown in Figure 5-15. To obtain this each metal strip is placed in a rig shown in Figure 5-17. The rig is composed of two clamps, each 45° from the plane perpendicular to the table, or 90° offset from each other. Once the metal strip is clamped-in, a hot torch is run up and down the length of the metal strip hot-forming it to the shape formed by the clamps. The strip should not be heated any longer than necessary to hot-form.

The metal strips should be the same material as the stack. The effectiveness of the system is not impaired by a gap of 0.005D between the helical strake and cylinder.

This method leads to ease and quickness in fabricating helical vortex strakes.

EXAMPLE 5-1: GRANULE BIN DESIGN FOR ROOFING PLANT

Twelve granule bins are to be designed to provide granules for the manufacture of roofing shingles of Example 3-6. Each bin is to contain 10.02 tons of granules, yielding 120.24 tons total capacity for all twelve bins. The client desires to use an existing steel frame that limits the bin to a rectangular shape with an off-centered opening as shown in Figure 5-18. From this figure we consider the first criterion in bin design—to satisfy flow conditions such that the granules will move.

As seen in Figure 5-13b, the minimum hopper angle for mass flow is

$\theta' = 37.747°$

From Figure 5-4, $\phi' = 10$

From Figure 5-5, $\Phi \cong 0$, which implies that we will not have piping forming in the bin

$\delta = 70°$

For a circular opening, m = 1

$$Q = \frac{s'(1 + \sin\delta)}{2\sin\theta'} \qquad (5\text{-}5)$$

From Figure 5-6, ff = 1.6

$$ff = (1 + m)Q \qquad (5\text{-}7)$$

$$Q = \frac{1.6}{2} = 0.80$$

$\sigma_1 = \gamma BQ$

$\gamma = 90\ \text{lb/ft}^3$

$B = 0.667\ \text{ft}$

$\sigma_1 = (90)(0.667)(0.80) = 48.02\ \text{lb/ft}^2$

From Figure 5-12,

$f_c = 50\ \text{lb/ft}^2$

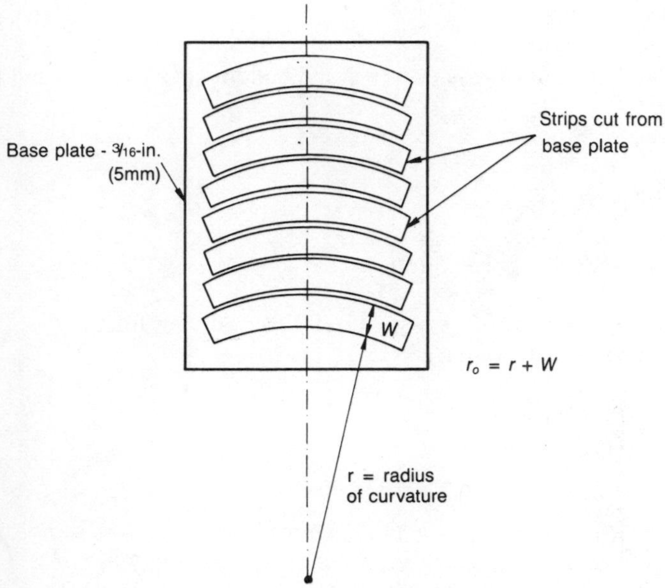

Figure 5-16. Strake fabrication detail.

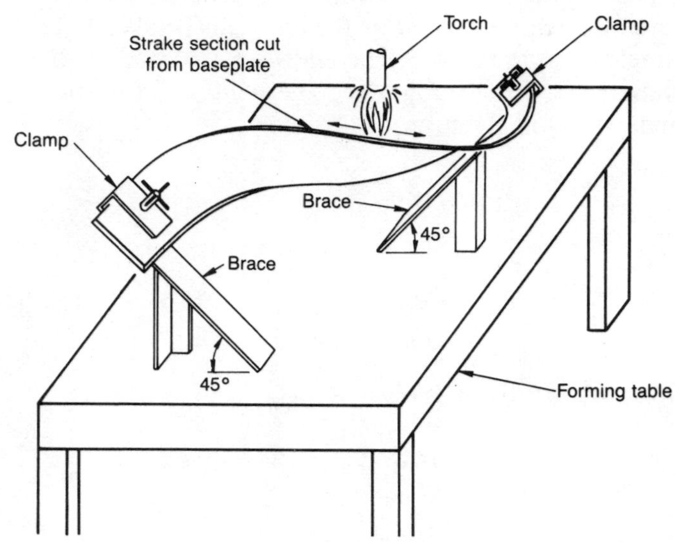

Figure 5-17. Clamping each strip on 45 degree offsets and hot forming with torch obtains desired geometry.

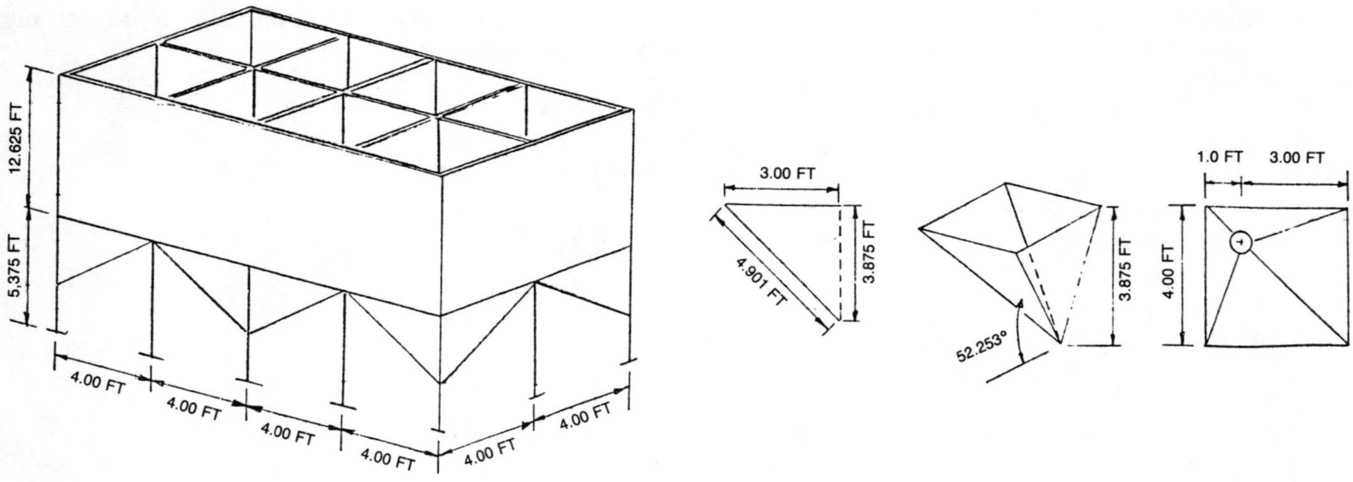

Figure 5-18. Granule bin silo.

Thus, the critical arching dimension is

$$B = \frac{f_c}{\gamma(1 + m)} = \frac{50}{(90)(2)} = 0.278 \text{ ft} \quad (5\text{-}2)$$

Since 0.278 ft < 0.667 ft = 8 in. arching will not form in the bin

After flow criteria have been met, we proceed to the structural design of the bin. The allowable stress used in the case of bin design is the ASME allowable, since the granule weight forms a pressure distribution, thus making the bin walls pressurized components.

For simplicity and ease in calculations, the solid pressure distribution exerted on the bin walls is taken to be a simple hydrostatic load. The bin walls are fixed on three ends and free on the top edge. The solution for the maximum stress is given by

$$\sigma_b = \frac{\Psi_1 P b^2}{t^2} \quad \text{and } F = \Omega_1 P b \text{ at } x = 0,\ z = 0 \quad (5\text{-}20)$$

$$\sigma_a = \frac{\Psi_2 P b^2}{t^2} \quad \text{and } F = \Omega_2 P b \text{ at } x = \pm\frac{a}{2},\ z = b \text{ if } a > b \text{ or } z = 0.4b \text{ if } a \leqslant b \quad (5\text{-}21)$$

where Ψ_1, Ψ_2, Ω_1, and Ω_2 are in Figure 5-19

F = reaction force exerted on the plate edge normal to the plate surface, lb/in.

P = load per unit area, psi

t = plate thickness, in.

In this problem, a = 12.625 ft and b = 4.00 ft. The pressure at the bottom of the plate is

$$P = 90\,\frac{\text{lb}}{\text{ft}^3}(12.625)\text{ft}\left(\frac{1 \text{ ft}^2}{144 \text{ in.}^2}\right) = 7{,}891 \text{ psi}$$

$$\frac{a}{b} = \frac{4.000}{12.625} = 0.317$$

From Figure 5-19 we obtain the following:

$$\Psi_1 = 0.030$$

$$\Psi_2 = 0.032$$

The maximum stress occurs at the bottom side at x = 0 and z = 0

$$\sigma_{max} = \frac{\Psi_1 P b^2}{t^2} \quad (5\text{-}20)$$

b = 12.625 ft = 151.50 in.

For SA-516 Gr. 55, σ_{all} = 13,700 psi. Solving for t in Equation 5-20 we have

$$t = \left(\frac{\Psi_1 P b^2}{\sigma}\right)^{0.5}$$

$$t = \left[\frac{(0.030)(7.891)\,\frac{\text{lb}}{\text{in.}^2}\,(151.50)^2\text{in.}^2}{13{,}800\,\frac{\text{lb}}{\text{in.}^2}}\right]^{0.5} = 0.627 \text{ in.}$$

The stress at mid-plane is

$$\sigma_a = \frac{\Psi_2 P b^2}{t^2}$$

z = 0.4b, P = 4.734 psi

$$t_{min} = \left[\frac{(0.032)(4.734)(151.50)^2}{13,800}\right]^{0.5} = 0.502 \text{ in.}$$

Selecting SA-516 Gr. 70, $\sigma_{all} = 17,500$

$$t_{min} = \left[\frac{(0.030)(7.891)(151.50)^2}{17,500}\right]^{0.5} = 0.557 \text{ in. at bottom edge}$$

$$t_{min} = \left[\frac{(0.032)(4.734)(151.50)^2}{17,500}\right]^{0.5} = 0.446 \text{ in. at } z = 5.050 \text{ ft}$$

Deflections of bin plate,

$$\frac{b}{a} = \frac{12.625}{4.0} = 3.156$$

At x = 0, z = b = 12.625 ft

$$w = \frac{(0.00020)(7.891)\ \dfrac{\text{lb}}{\text{in.}^2}\ (48.0)^2 \text{ in.}^2}{D\ \dfrac{\text{lb}}{\text{in.}}}$$

$$D = \text{flexural rigidity of plate} = \frac{Et^3}{12(1 - \nu^2)}$$

$$t = \frac{9}{16} \text{ in.} = 0.562 \text{ in.}$$

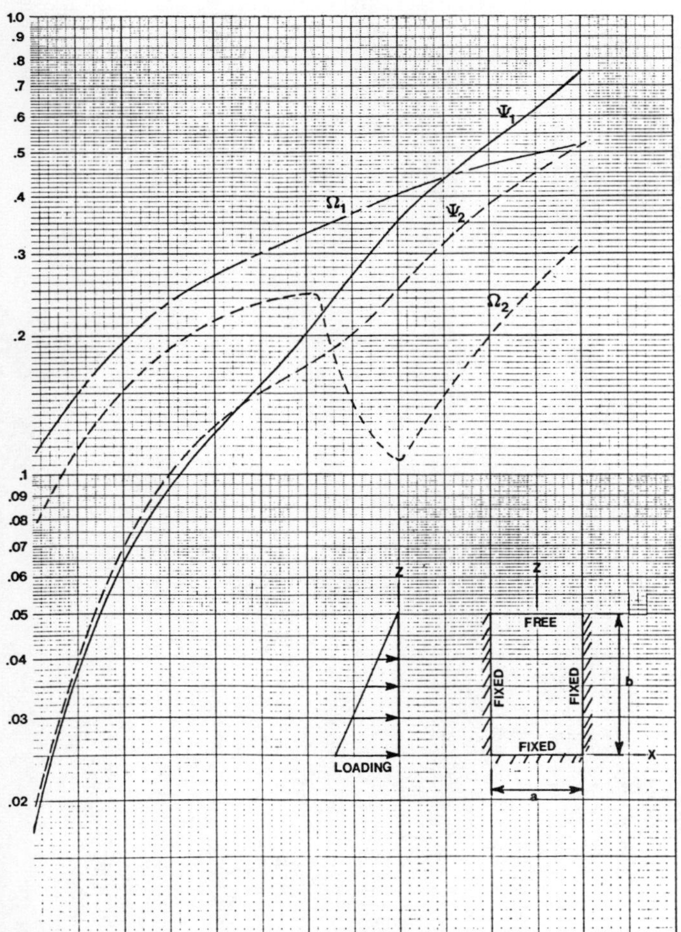

Figure 5-19. Rectangular flat plate solutions.

$$D = \frac{(30.0 \times 10^6)(0.562)^3}{12(1 - 0.3^2)} = 487{,}649.253$$

in which,

$w = 7.4565 \times 10^{-6}$ in.

At $x = 0$, $z = \frac{b}{2}$,

$$w = \frac{(0.004)(7.891)(48.0)^2}{487{,}649.253}$$

$w = 1.4913 \times 10^{-4}$ in.

For a 9/16-in. plate deflections are negligible and no stiffeners are required for this plate thickness.

Bin Stiffener Design

To reduce bin plate thickness, stiffeners can be used with thinner plate. A thinner bin plate makes fabrication simpler because a thinner plate is easier to weld and is cheaper.

With stiffeners, each enclosed area is analyzed as a flat plate with three edges fixed and one edge simply supported. The stress in the plate is given by the following:

$$\sigma_b = \frac{\Psi_1 P b^2}{t^2} \text{ and } F = \Omega_1 P b \text{ at } x = 0,\ z = 0 \quad (5\text{-}22)$$

$$\sigma_a = \frac{\Psi_2 P b^2}{t^2} \text{ and } F = \Omega_2 P b \text{ at } x = \pm\frac{a}{2},\ z = 0.4b \quad (5\text{-}23)$$

where Ψ_1, Ψ_2, Ω_1, and Ω_2 are shown in Figure 5-20

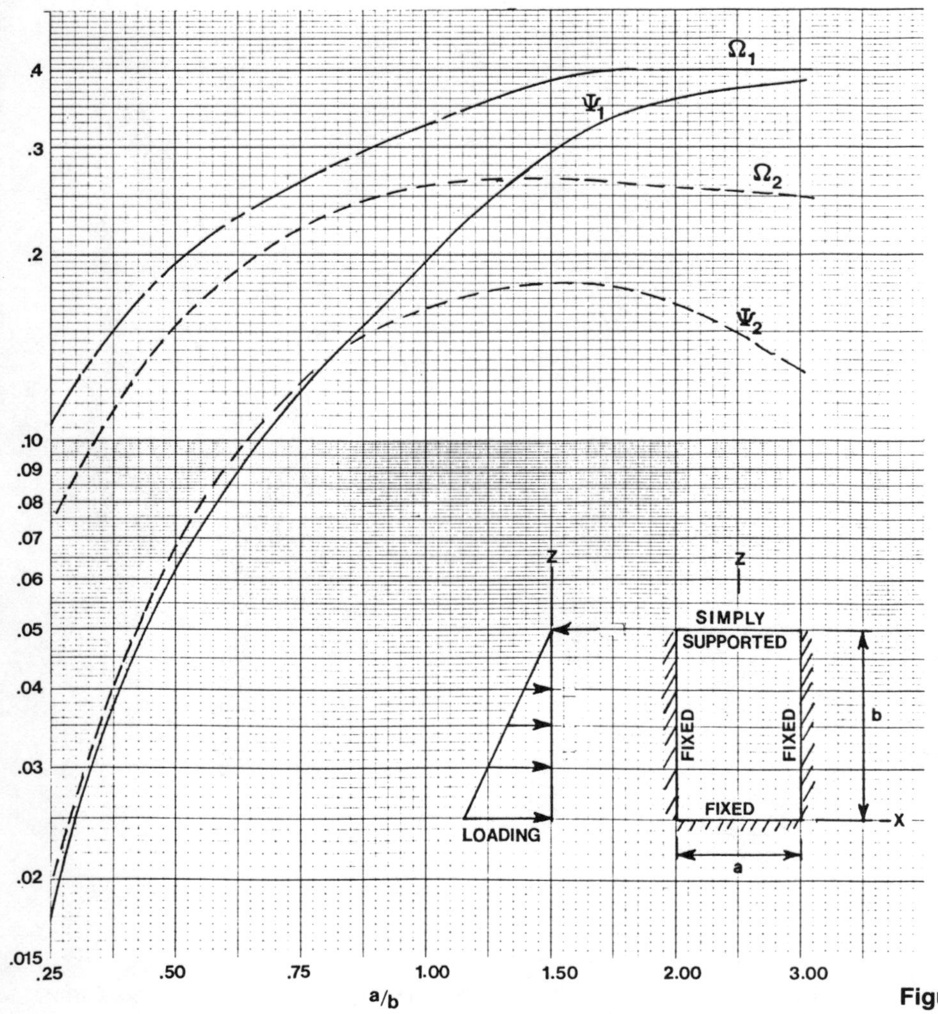

Figure 5-20. Rectangular flat plate solutions.

F = reaction force exerted on the plate edge normal to the plate surface, lb/in.
P = load per unit area, psi
t = plate thickness, in.

First Stiffener

Consider b = 8.0 ft, a/b = 0.50. From Figure 5-20 we obtain $\Psi_1 = 0.064$. Thus, from Equation 5-22 we have

$$\sigma_{max} = \frac{(0.064)(7.891)(96)^2}{(0.141)} = 33{,}009.228 \text{ psi} > \text{allowable}$$

Consider b = 4.0 ft, a/b = 1.0. From Figure 5-20 we have $\Psi_1 = 0.192$ and from Equation 5-22,

$$\sigma_{max} = 24{,}756.921 \text{ psi} > \text{allowable}$$

Similarly, considering b = 2.0 ft,

$$\sigma_{max} = 11{,}475.865 \text{ psi} < \text{allowable} = 17{,}500 \text{ psi}$$

By a process of iteration we obtain a value of b = 2 ft 8 in., in which

$$\sigma_{max} = 17{,}364.229 \text{ psi} < 17{,}500 \text{ psi allowable}$$

Thus, we place the first stiffener 2 ft 8 in. above the bottom seam.

Second Stiffener

At 2 ft 8 in. the maximum pressure exerted on the bin wall is

$$P = 90\,\frac{\text{lb}}{\text{ft}^3}\,(12.625 - 2.667)\text{ ft}\left(\frac{1 \text{ ft}^2}{144 \text{ in.}^2}\right) = 6.224 \text{ psi}$$

Consider b = 4.0 ft. a/b = 1.0 in which $\Psi_1 = 0.192$ from Figure 5-19. Thus,

$$\sigma_{max} = \frac{(0.192)(6.224)(48)^2}{(0.141)} = 19{,}526.9 \text{ psi} > 17{,}500 \text{ psi}$$

By a process of iteration we arrive at b = 3 ft 1 in. in which $\sigma_{max} = 17{,}502$ psi

Third Stiffener

At the new elevation, 6.167 ft above the bottom seam, we obtain the maximum pressure exerted on the wall.

$$P = 90\frac{\text{lb}}{\text{ft}^3}\Big[12.625 - (2.667 + 3.50)\Big]\text{ft}\left(\frac{1 \text{ ft}^2}{144 \text{ in.}^2}\right)$$

$$= 4.036 \text{ psi}$$

The top portion of the bin is now a plate with three sides fixed and the top edge free. Thus, Equations 5-20 and 5-21 hold, using Figure 5-18. By iteration we obtain

b = 6.458 ft, P = 4.036 psi, a/b = 0.619, $\Psi_1 = 0.091$ and $\sigma_{max} = 15{,}643$ psi $\bullet$ 17,500 psi

Since the maximum stress is less than the allowable for the top portion, no third stiffener is required.

First Stiffener Design

$$a = 4 \text{ ft-0 in.};\ b = 2 \text{ ft-8 in.}\quad a/b = \frac{4.00}{2.667} = 1.500$$

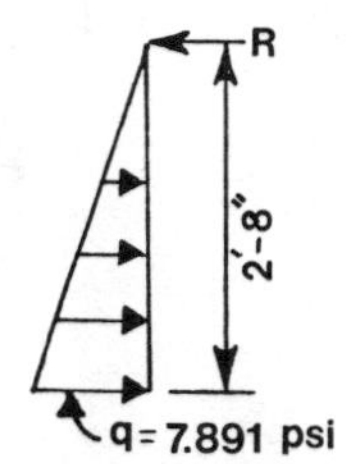

$\gamma_1 = 0.383$

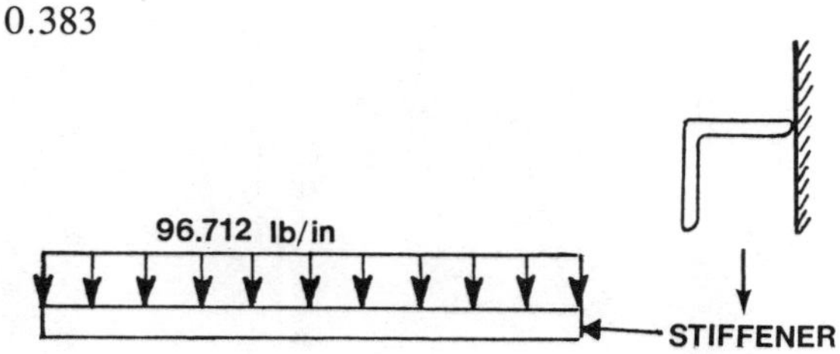

$$R = \gamma_1 P b = (.383)(7.891)(32.0) = 96.712 \text{ lb/in.} = w$$

With plate pushing uniformly on stiffener, the latter will be analyzed as a fixed end beam with uniform loading.

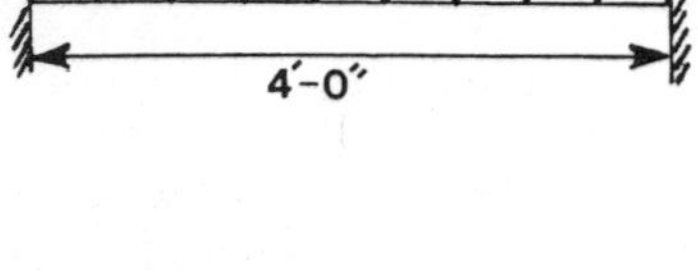

$$M_{max} = \frac{W\ell}{24};\ W = w\ell$$

$$W = (96.712)(48) = 4{,}642.18 \text{ lb}$$

$$M_{max} = \frac{(4{,}642.181)(48.0)}{24} = 9{,}284.361 \text{ in.-lb}$$

$$= 773.697 \text{ ft-lb}$$

$$\sigma = \frac{Mc}{I}$$

For design purposes select a design stress of $\sigma = 17{,}000$ psi. With a factor of safety of 2. This would give a yield stress of 34,000 psi, which is conservative.

$$I = \frac{Mc}{\sigma}$$

Select a 3-in. × 2-in. $\measuredangle$ 1/4-in. thick,

$$I_{min} = \frac{(9{,}284.361) \text{ in-lb}(.49) \text{ in.}}{17{,}000 \text{ lb/in.}^2} = 0.268 \text{ in.}^4$$

I of $\measuredangle$ = 0.39 in.^{4}

Therefore, 3-in. × 2-in. × 1/4-in. $\measuredangle$ is sufficient

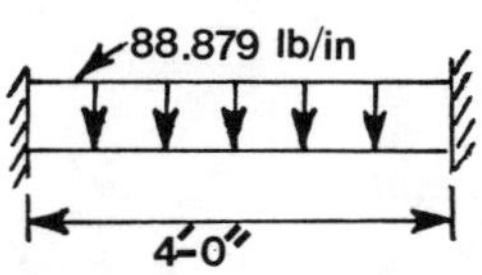

Second Stiffener Design

P = 6.224 psi

$$a = 4 \text{ ft-0 in.};\ b = 3 \text{ ft-6 in.};\quad a/b = \frac{4.0}{3.5} = 1.143$$

By linear interpolation, $\gamma_1 = 0.340$

$$R = (0.340)(6.224) \text{ lb/in.}^2(42.0) \text{ in.} = 88.879 \text{ lb/in.}$$

$$M_{max} = \frac{w\ell}{24};\ W = (88.879) \text{ lb/in.}(48.0) \text{ in.}$$

$$W = 4{,}266.192 \text{ lb}$$

$$M_{max} = \frac{(4{,}266.192)(48.0)}{24} = 8{,}532.384 \text{ in.-lb}$$

$$I = \frac{Mc}{\sigma}$$

Select a 2 1/2-in. × 2-in. × 1/4-in. $\measuredangle$

$$I_{min} = \frac{(8{,}532.384) \text{ in.-lb } (0.54) \text{ in.}}{17{,}000 \text{ lb/in.}^2} = 0.271 \text{ in.}^4$$

I = 0.37 in.^{4}

Therefore, 2 1/2-in. × 2-in. × 1/4-in. $\measuredangle$ is sufficient

Stiffener at Junction Point of Bin

$$\text{Hoop Force} = \frac{P}{\gamma D}$$

From data provided by the client, P ≈ 400 lb/ft^2 at junction point. Using a factor of 7 we have

$$P = 7(400) = 2{,}800 \text{ lb/ft}^2$$

$$P = 2{,}800 \text{ lb/ft}^2 \left(\frac{1 \text{ ft}^2}{144 \text{ in.}^2}\right) = 19.444 \text{ psi}$$

Use P ≈ 20 psi

For bottom plate, a = 4 ft-0 in.; b = 2 ft-8 in., a/b = 1.500

$$\gamma_1 = 0.383$$

$$R = (0.383)(20.0)(32.0) = 245.120 \text{ lb/in.}$$

$$M_{max} = \frac{W\ell}{24};\quad W = (245.120)(48.0) = 11{,}765.760 \text{ lb}$$

$$M_{max} = \frac{(11{,}765.760)(48)}{24} = 23{,}531.520 \text{ in-lb}$$

Select a 3 1/2 in. × 3 in. × 1/4 in. $\measuredangle$

$$I_{min} = \frac{(23{,}531.520)(0.79)}{17{,}000} = 1.094 \text{ in.}^4$$

I = 1.3 in.^{4} for section

Therefore, 3 1/2-in. × 3-in. × 1/4-in. $\measuredangle$ is sufficient with long side facing bin

Bottom Portion of Bin

Bottom portion of bin will be approximated with four triangular plates welded together, as shown in Figure 5-18.

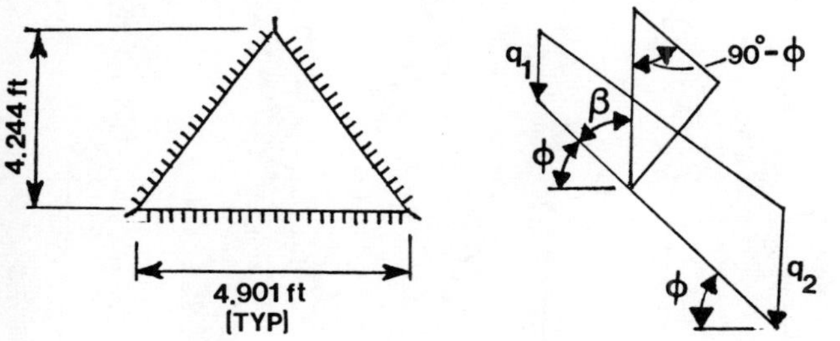

$P_1 = 7.891$ lb/in.^{2}

$$P_2 = 90\ \text{lb/ft}^3(16.50\ \text{ft})\left[\frac{1\ \text{ft}^2}{144\ \text{in.}^2}\right] = 10.313\ \text{psi}$$

At an angle of $90°\text{-}\phi = 37.747°$,

$P = 10.313 \sin 37.747° = 6.313$ psi

By linear interpolation, $\beta_1 = 0.3650$

$$\beta = \frac{\beta_1}{2} = 0.1825$$

$$\sigma = \frac{0.1825Pa^2}{t^2};\quad \sigma_{all} = 17,500\ \text{psi}$$

$$t = \sqrt{\frac{(0.1825)(6.313)(50.928)^2}{17,500}} = 0.413\ \text{in.}$$

with t = 3/8 in.,

$$\sigma = \frac{0.1825(6.313)(50.928)^2}{(0.375)^2} = 21,249.532\ \text{psi}$$

For SA-516 Gr. 70, minimum yield = 38,000 psi

$$\%\ \text{yield} = \frac{21249.532}{38,000} = 55.92\%\ \text{of minimum yield}$$

$$\%\ \text{yield} = \frac{21249.532}{70,000} = 30.36\%\ \text{of ultimate yield}$$

with 7/16 in. ℄ ,

$$\sigma = \frac{0.1825(6.313)(50.928)^2}{(0.438)^2} = 15,576.278\ \text{psi}$$

Therefore, use 7/16 in. ℄ for bottom plates

Bending Stress in Bottom Portion

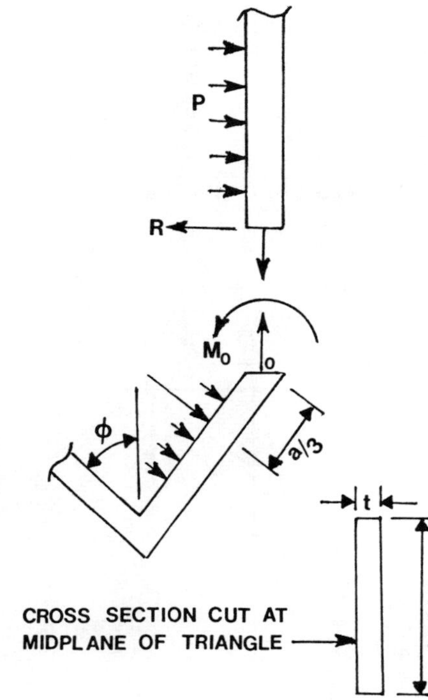

From previous information,

P = 6.313 psi on triangular plate

$$A = \text{area of triangle} = \left(\frac{4.901}{2}\right)(4.244) = 10.40\ \text{ft}^2$$

or A = 1,497.589 in.^{2}

⇒F = (6.313) lb/in.^{2}(1,497.589) in.^{2} = 9,454.279 lb

a/3 = (4.244)(12)/3 = 16.976 in.

$M_o = F(a/3) = 160,495.84$ in.-lb

$$\sigma = \frac{Mc}{I}$$

$$I = \frac{th^3}{12} = \frac{t(50.928)^3}{12} = 11,007.498\ t$$

$$c = a/3 = \frac{50.928}{3} = 16.976 \text{ in.}$$

$$\sigma = \frac{(160{,}495.84)(16.976)}{(11{,}007.498)t}; \quad \sigma_{all} = 17{,}500 \text{ psi}$$

$$t = \frac{(160{,}495.84) \text{ in.-lb } (16.976) \text{ in.}}{(11{,}007.498)(17{,}500) \text{ in.}^3\text{-lb/in.}^2} = 0.0141 \text{ in.}$$

Therefore, 7/16 in. ℄ is sufficient.

Vessel Supports

Consider all trusses as pin connected.

Side Truss

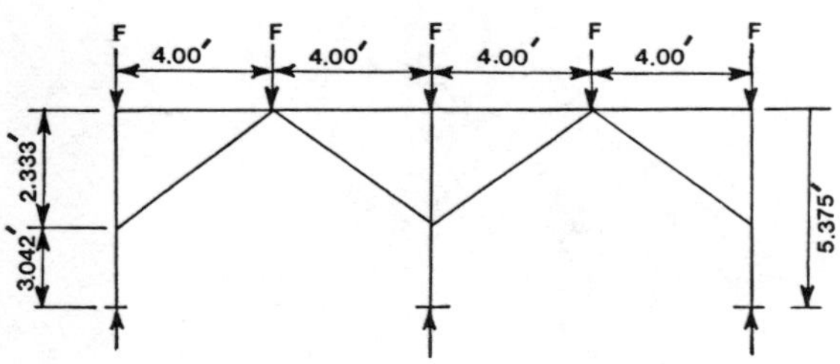

$$\text{Weight of internal load} = (120.24) \text{ tons} \left(\frac{2{,}240 \text{ lb}}{\text{ton}}\right)$$

$$= 269{,}337.60 \text{ lb}$$

End Truss

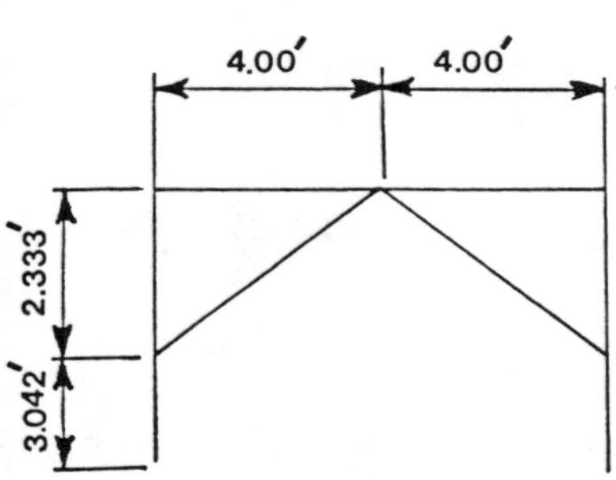

Weight of steel (Wt):

$$(12.625 \text{ ft})(16.0 \text{ ft})(t) = (151.50)(192)(0.563) = 16{,}362.0 \text{ in.}^3$$

$$\gamma_{steel} = 0.283 \text{ lb/in.}^3$$

$$Wt = (16{,}362.0)(0.283) = 4{,}630.446 \text{ lb}$$

For three horizontal plates,

$$(12.625 \text{ ft})(8.0 \text{ ft})t = \frac{4{,}630.446}{2} = 2{,}315.223 \text{ lb}$$

or for three plates,

$$Wt = 6{,}945.669 \text{ lb}$$

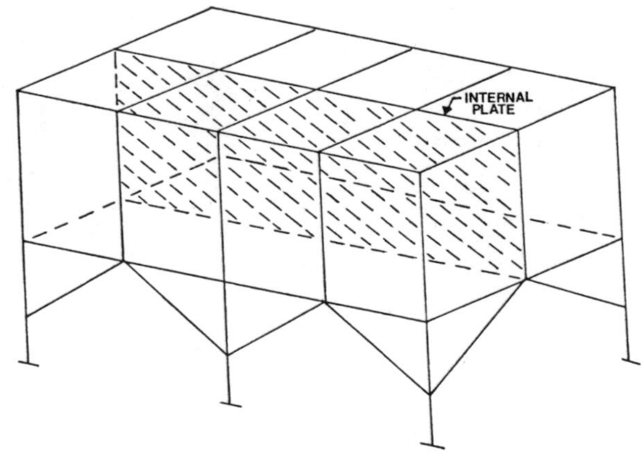

For simplicity and to keep things conservative, let us analyze the internal plate to determine if we need any supports on inside of structure.

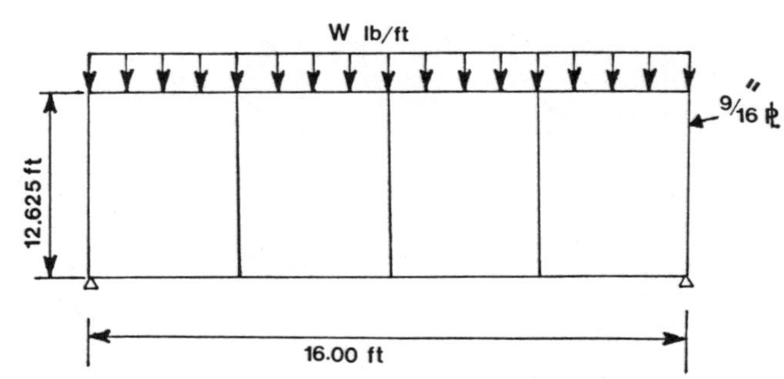

For two outside plates, t = 3/8 in.;

$$Wt = (12.625)(8.0)(0.375)(144)(.283) = 1{,}543.482 \text{ lb}$$

$$Wt_{total} = 3{,}086.964 \text{ lb}$$

For two side plates,

$$Wt_{total} = 2(3{,}086.964) = 6{,}173.928 \text{ lb}$$

Under Bins—4 Triangular Plates

For each bin,

$$A = 4\left(\frac{4.901}{2}\right)(4.244)(144) = 5{,}990.355 \text{ in.}^3 \text{ of metal}$$

$$\text{Wt of each bin} = (5{,}990.355)(.283) = 1{,}695.270 \text{ lb}$$

Number of Bins = 8

Therefore, Wt_{total} = 13,562.164 lb

Empty weight of structure = 4,630.446 lb + 6,945.669 lb + 3,086.964 lb + 6,173.928 lb + 13,562.164 lb
= 34,399.171 lb

Wt of granules = 269,337.60 lb

Total Wt loaded = 303,736.771 lb

Total number of internal plates = 4

Total length = 4.0 ft

$$w = \frac{303{,}736.771}{4.0} = 75{,}934.193 \text{ lb/ft}$$
$$= 911{,}210.313 \text{ lb/in.}$$

Considering the plate in Figure 5-18,

$$M = \frac{w\ell}{8};\ W = (911{,}210.313)\ \frac{\text{lb}}{\text{in.}}\ (192) \text{ in.}$$

$$= 174{,}952{,}380.1 \text{ lb}$$

$$M = \frac{(174{,}952{,}380.1)(192)}{8} = 4{,}198{,}857{,}123 \text{ in.-lb}$$

Therefore, bin must have internal supports under bottom.

$$\text{Number of vertical supports} = 9 \Rightarrow R = \frac{303{,}736.771}{9}$$

$$= 33{,}748.530 \text{ lb}$$

$$\text{Number of joint supports} = \frac{6}{15}$$

$$F = \frac{303{,}736.771}{15} = 20{,}249.118 \text{ lb}$$

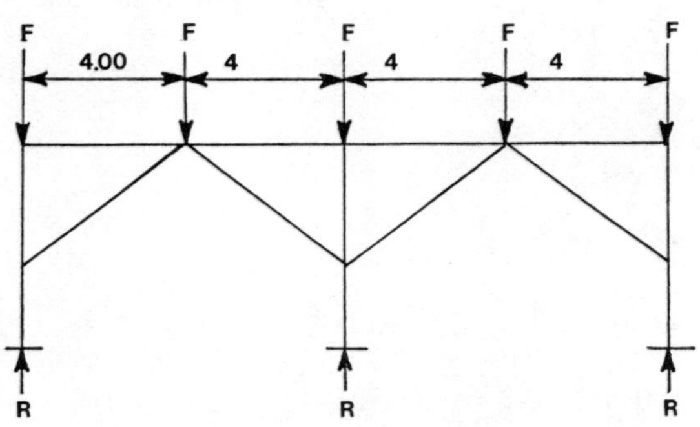

The frame structure shown in Figure 5-18 is analyzed as continuous beams in the longitudinal and lateral directions.

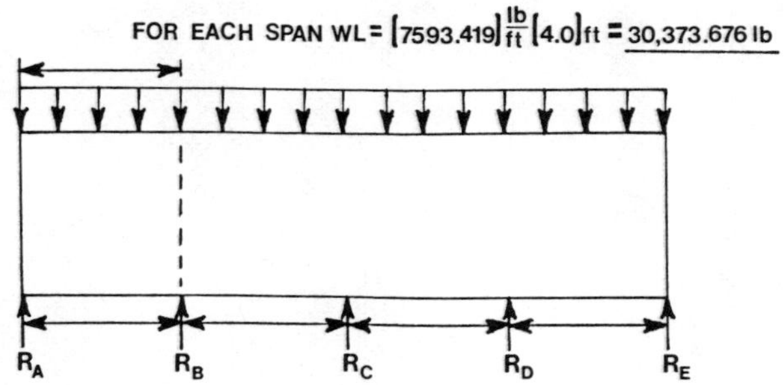

R_A = 0.393 wℓ = 0.393(30,373.676) = 11,936.855 lb

R_B = 1.143 wℓ = 1.143(30,373.676) = 34,717.112 lb

R_C = 0.928 wℓ = 0.928(30,373.676) = 28,186.771 lb

R_D = 1.143 wℓ = 1.143(30,373.676) = 34,717.112 lb

Solving for reactions in lateral plate,

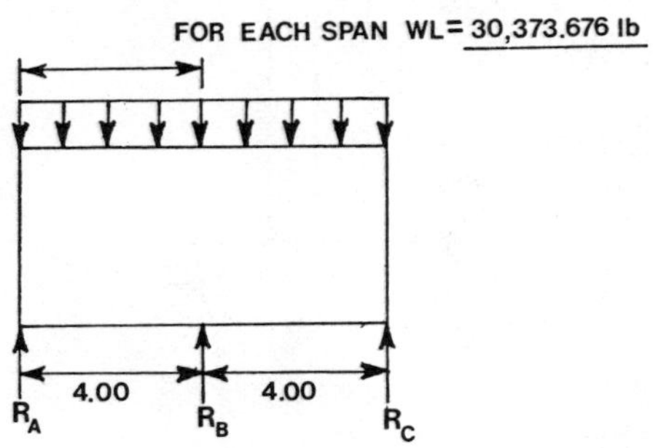

V_{max} = 0.607(30,373.676) lb

V_{max} = 18,436.821 lb

$$R_B = \frac{5}{8}(30{,}373.676)(2) = 37{,}967.095 \text{ lb}$$

R_A = R_C = 11,390.129 lb

Design each support column for 37,967.095 lb ≈ 38,000 lb side structure

The bin structural detail is shown in Figure 5-21.

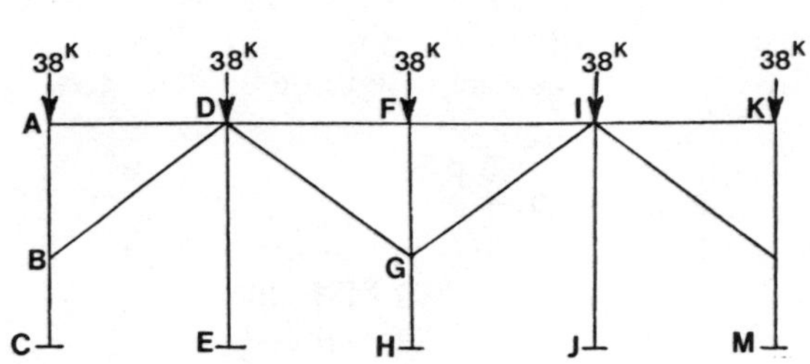

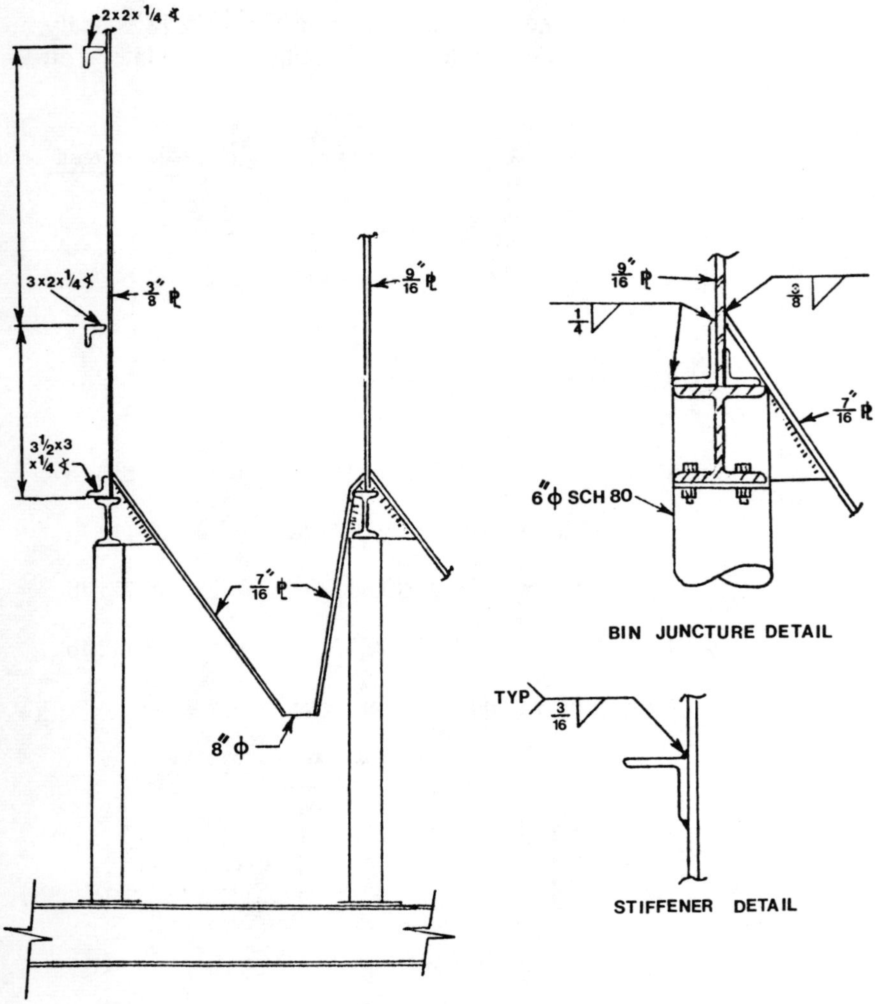

Figure 5-21. Bin structural frame detail.

EXAMPLE 5-2: HIGH-PRESSURE FLARE STACK DESIGN

A high-pressure flare stack shown in Figure 5-22 is to be designed and constructed to the following specifications:

Base diameter = 10 ft
Height from bottom of steel base to tip of flare stack = 200 ft
Gas pressure in stack = 2 psig
Gas temperature = 100°F
Design wind velocity = 100 mph
Maximum gas flow rate = 300 MMscfd
Earthquake design = World Mercali 6–7

Effective Diameters

Add 12 in. for platforms and 12 in. for ladders.
Add 4–2-in. ϕ lines.
2-in. ϕ dia. line = 2.375 in.—Add 1/2 in. insulation
D = (3.375)(4) = 13.50 in.

$$D_{excess} = 2(12) + 13.50 = 37.50 \text{ in.}$$

$$D_A = 42 + 37.50 = 79.50 \text{ in.}$$

$$D_B = 90 + 37.50 = 127.50 \text{ in.}$$

$$D_C = 120 + 37.50 = 157.50 \text{ in.}$$

Height (ft)	Wind Pressure P_z (lb/ft²)	$w = B \times D_e \times P_z$ lb/ft
0–30	26	$= (0.6)\left(\frac{157.50}{12}\right)(26) = 204.75$
30–40	33	$= (0.6)\left(\frac{157.50}{12}\right)(33) = 259.88$
40–74	38	$= (0.6)\left(\frac{157.50}{12}\right)(38) = 299.25$
74–76.5	44	$= (0.6)\left(\frac{157.50}{12}\right)(44) = 346.50$
76.5–125	44	$= (0.6)\left(\frac{127.50}{12}\right)(44) = 280.50$
125–159	48	$= (0.6)\left(\frac{127.50}{12}\right)(48) = 306.00$
159–174	48	$= (0.6)\left(\frac{79.50}{12}\right)(48) = 190.80$
174–200	51	$= (0.6)\left(\frac{79.50}{12}\right)(51) = 202.73$

Wind Load	Moment
5,270.98	$(5{,}270.98)\text{ lb}\left(\frac{26.0}{2}\right)\text{ ft} = 68{,}522.74$ ft-lb
2,862.00	$(5{,}270.98)(13.0 + 15.0) + (2{,}862.00)\left(\frac{15.0}{2}\right) = 169{,}052.44$ ft-lb
10,404.00	$(5{,}270.98)(28.0 + 34.0) + (2{,}862.0) \times (7.5 + 34.0) + (10{,}404.0)\left(\frac{34.0}{2}\right) = 622{,}441.76$ ft-lb
13,604.25	$(5{,}270.98)(62.0 + 48.5) + (2{,}862.0) \times (41.5 + 48.5) + (10{,}404.0)(17.0 + 48.5) + (13{,}604.25)\left(\frac{48.5}{2}\right) = 1{,}851{,}388.35$ ft-lb
866.25	$(5{,}270.98)(110.5 + 2.5) + (2{,}862.0) \times (90.0 + 2.5) + (10{,}404.0)(65.5 + 2.5) + (13{,}604.25)(24.25 + 2.5) + (866.25)\left(\frac{2.5}{2}\right) = 1{,}932{,}824.24$ ft-lb

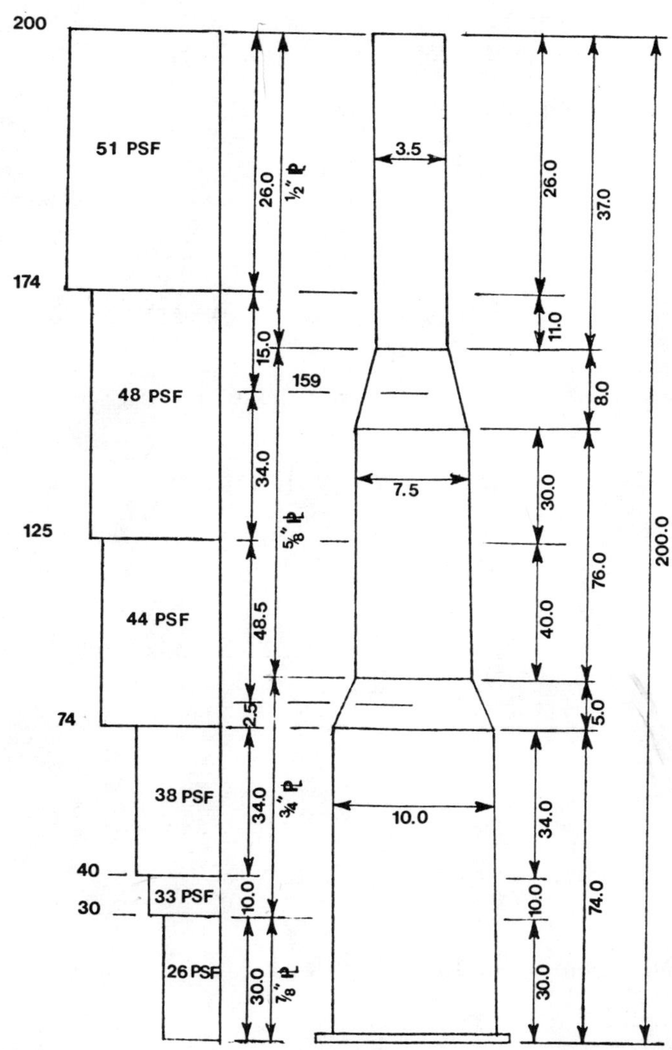

Figure 5-22. High-pressure flare stack; unless otherwise indicated, all dimensions in feet, design wind speed = 100 mph.

Wind Load	Moment
10,174.50	$(5{,}270.98)(113.0 + 34.0) + (2{,}862.0) \times (92.5 + 34.0) + (10{,}404.0)(68.0 + 34.0) + (13{,}604.25)(26.75 + 34.0) + (866.25) \times (1.25 + 34) + (10{,}174.5)\left(\frac{34.0}{2}\right) = 3{,}228{,}045.06$ ft-lb
2,598.80	$(5{,}270.98)(147.0 + 10.0) + (2{,}862.0) \times (126.50 + 10.0) + (10{,}404.0)(102.0 + 10.0) + (13{,}604.25)(60.75 + 10.0) + (866.25) \times (35.25 + 10.0) + (10{,}174.5) \times (17.0 + 10.0) + (2{,}598.80)\left(\frac{10.0}{2}\right) = 3{,}672{,}858.86$
6,142.50	$(5{,}270.98)(157.0 + 30.0) + (2{,}862.0) \times (136.50 + 30.0) + (10{,}404.0)(112.0 + 30.0) + (13{,}604.25)(70.75 + 30.0) + (866.25) \times (45.25 + 30.0) + (10{,}174.5)(27.0 + 30.0) + (2{,}598.80)(5.0 + 30.0) + (6{,}142.50) \left(\frac{30.0}{2}\right) = 5{,}138{,}419.76$ ft-lb

Allowable Shell Buckling Stress

$$\sigma_c = \frac{0.56\ t_a\ E}{d\ (1 + 0.004\ E/y)}\ ;\ E = 29 \times 10^6 \text{ psi};$$

$y = 30{,}000$ psi

For Section D

$$\frac{t_a}{d} = \frac{(0.875 - 0.125)}{120} = 0.006 > 0.00425$$

$$\sigma_c = \frac{(0.56)(0.006)(29.0 \times 10^6)}{[1 + (0.004)(29.0 \times 10^6)/(30{,}000)]}$$

$= 20{,}021.918$ psi

For Section C

$$\frac{t_a}{d} = \frac{(0.750 - 0.125)}{120} = 0.005 > 0.00425$$

$$\sigma_c = \frac{(0.56)(0.005)(29.0 \times 10^6)}{[1 + 0.004(29.0 \times 10^6)/(30{,}000)]}$$

$= 16{,}684.932$ psi

For Section B

$$\frac{t_a}{d} = \frac{(0.625 - 0.125)}{90.00} = 0.006$$

$\sigma_c = 20{,}021.918$ psi

For Section A

$$\frac{t_a}{d} = \frac{(0.500 - 0.125)}{42} = 0.009$$

$\sigma_c = 30{,}032.877$ psi

Section Weights—Uncorroded Weight

Section A

$$Wt = (0.2833)\ \frac{lb}{in.^3}\ (37.0)(12)\ in.\ \pi \left[\left(\frac{42}{2}\right)^2 - \left(\frac{41}{2}\right)^2\right] in.^2$$

$= 8{,}199.69$ lb

Section B

$$Wt = (0.2833)\ \frac{lb}{in.^3}\ (76.0)(12)\ in.\ \pi \left[\left(\frac{90}{2}\right)^2 - \left(\frac{88.75}{2}\right)^2\right] in.^2$$

$= 45{,}340.61$ lb

Section C

$$Wt = (0.2833)\ \frac{\text{lb}}{\text{in.}^3}\ (44.0)(12)\ \text{in.}\ \pi \left[\left(\frac{120}{2}\right)^2 - \left(\frac{118.50}{2}\right)\right]\ \text{in.}^2$$
$$= 42{,}029.09\ \text{lb}$$

Section D

$$Wt = (0.2833)\frac{\text{lb}}{\text{in.}^3}(30.0)(12)\ \text{in.}\ \pi \left[\left(\frac{120}{2}\right)^2 - \left(\frac{118.25}{2}\right)^2\right]\ \text{in.}^2$$

$$= 33{,}397.19\ \text{lb}$$

Total Wt = 128,966.580 lb

Required ⅊ Thickness

$$t_r = \frac{16\ D_o\ M_T}{\pi(D_o^2 + D_i^2)(D_o + D_i)\sigma E} + \frac{W}{\pi(D_o + D_i)\sigma E}$$

Section D

$$t_r = \frac{(16)(120.0)(5{,}138{,}419.76)(12)}{\pi[(120)^2 + (118.25)^2](120.0 + 118.25)(14{,}182.19)(1.0)} + \frac{128{,}966.580}{\pi(120.0 + 118.25)(14{,}182.19)(1.0)}$$

$t_r = 0.381$ in. ⇒ 7/8 in. ⅊, OK for buckling

Section C

$$t_r = \frac{(16)(120.0)(3{,}672{,}858.86)(12)}{\pi[(120)^2 + (118.5)^2](120.0 + 118.5)(16{,}684.932)(1.0)} + \frac{95{,}569.39}{\pi(120.0 + 118.5)(16{,}684.932)(1.0)}$$

$t_r = 0.245$ in. ⇒ 3/4 in. ⅊, OK for buckling

Section B

$$t_r = \frac{(16)(90.0)(1{,}851{,}388.35)(12)}{\pi[(90)^2 + (88.75)^2](90.0 + 88.75)(20{,}021.918)(1.0)} + \frac{53{,}540.300}{\pi(90.0 + 88.75)(20{,}021.918)(1.0)}$$

$t_r = 0.183$ in. ⇒ 5/8-in. ⅊, OK for buckling

Section A

$$t_r = \frac{(16)(42.0)(169{,}052.44)(12)}{\pi[(42)^2 + (41)^2](42.0 + 41.0)(30{,}032{,}877)(1.0)} + \frac{8{,}199.69}{\pi(42.0 + 41.0)(30{,}032.877)(1.0)}$$

$t_r = 0.052$ in. ⇒ 1/2 in. ⅊, OK for buckling

Anchor Bolt Design

Try 24–1¾-in. ϕ anchor bolts ⇒
$d_{BC} = 120 + 2(2.50) = 125.00$

Total tension in each bolt = W_B

$$W_B = \frac{48(5{,}138{,}419.76)}{24(125.00)} - \frac{128{,}966.58}{24} = 76{,}841.109\ \text{lb}$$

$\sigma_B = 40{,}000$ psi

$$A_B = \frac{76{,}841.109}{40{,}000} = 1.921\ \text{in.}^2 < 1.980\ \text{in.}^2$$
$= 1\frac{3}{4}$ in. dia, 8-thread series

Check

$$A_B = \frac{\left[\left(\frac{4M}{d}\right) - W\right]}{N\ \sigma_a}$$

$$= \frac{\left[\frac{(4)(12)(5{,}138{,}419.76)}{(125.50)} - 128{,}966.58\right]}{(24)(40{,}000)}$$

$A_b = 1.913$ in.^{2} < 1.980 in.^{2}

$$\text{Bearing pressure} = P_b = \frac{48M}{\pi\ D_b^2\ j} + \frac{W}{\pi\ D_b\ j}\ ;\ j = 7\tfrac{1}{2}\ \text{in.}$$

$$P_b = \frac{48(5{,}138{,}419.76)}{\pi(125.00)^2(7.50)} + \frac{128{,}966.58}{\pi(125.00)(7.50)}$$

$P_b = 713.734$ psi $< F_b = 1.33(900) = 1{,}197$ psi

T_B = Base ⅊ thickness, T_T = compression ⅊ thickness

$$T_B = e\left(\frac{3\ P_b}{20{,}000}\right)^{1/2}\ ;\ e = B + C = 2\tfrac{3}{4}\ \text{in.} + 2\tfrac{3}{4}\ \text{in.} = 5.50$$

$$T_B = (5.50)\left[\frac{3(713.734)}{20{,}000}\right]^{1/2} = 1.800 \text{ in.}$$

$$T_T = \left[\frac{3(W_B)\,6}{4(20{,}000)e}\right]^{1/2} = \left[\frac{3(76{,}841.109)(6)}{4(20{,}000)(5.5)}\right]^{1/2}$$

$$= 1.773 \text{ in.}$$

Brownell and Young Base ⅊ Method

Bolt circle $\phi = 125.00$ in.

Base P :

$$d_o = 125.00 + 2(2.50) = 130.00 \text{ in.}$$
$$d_i = 130.00 - 2(7.00) = 116.00 \text{ in.}$$
$$t_3 = \frac{130.00 - 116.00}{2} = 7.00 \text{ in.}$$

with $f_{C(BC)} = 1{,}000$; $K = 0.333$

$$f_{C(BC)} = (1{,}200)\left[\frac{2(0.333)(125.00)}{2(0.333)(125.00) + 7.00}\right]$$

$$= 1{,}106.925 \text{ psi}$$

For $K = 0.333$; $c_c = 1.588$; $C_t = 2.376$; $z = 0.431$; $j = 0.782$

$$F_t = \frac{(5{,}138{,}419.76) - (128{,}966.58)(0.431)\left(\frac{125.00}{12}\right)}{(0.782)\left(\frac{125.00}{12}\right)}$$

$$= 559{,}723.403$$

$$t_1 = \frac{A}{\pi\, d} = \frac{1.980 \text{ in.}^2\,(12)}{\pi(125.00)} = 0.061 \text{ in.}$$

$$f_S = \frac{F_t}{t_1 r c_t} = \frac{(559{,}723.403)}{(0.061)\left(\frac{125.00}{2}\right)(2.376)} = 61{,}789.855$$

$$F_C = 559{,}723.403 + 128{,}966.58 = 688{,}689.983$$

$$t_2 = 7.00 - 0.061 = 6.939 \text{ in.}$$

$$f_C = \frac{688{,}689.983}{[6.939 - (10)(0.061)]\left(\frac{125.00}{2}\right)(1.588)} = 1{,}096.373$$

After one iteration,

$$K = \frac{1}{1 + \dfrac{(61{,}789.855)}{(10)(1{,}096.373)}} = 0.151$$

After six iterations,

$$K = 0.178$$

$$f_S = n\,\frac{E_S}{E_C} = 10(1{,}096.373) = 10{,}963.73$$

$$f_{C(\text{max induced})} = (1{,}096.373)\left[\frac{2(0.178)(125.0) + 7.00}{2(0.178)(125.00)}\right]$$

$$= 1{,}268.836 \text{ psi}$$

$$t_4 = A\left(\frac{3f_C}{f_{\text{allow}}}\right)^{1/2} = (5.0)\left[\frac{3(1{,}268.836)}{20{,}000}\right]^{1/2}$$

$$= 2.181 \text{ in. (without gussets)}$$

Using 24 gusset ⅊'s, gusset spacing is

$$b = \frac{\pi(125.0)}{12} = 32.725 \text{ in.},\ l = A = 5.00 \text{ in.}$$

$$\frac{\eta}{b} = \frac{5.00}{32.725} = 0.153$$

From Table 4-8, using linear interpolations,

$$M_y = -0.467 f_C \eta^2$$

$$M_y = -0.467(1{,}268.836)(5.00)^2 = 14{,}813.660 \text{ in.-lb}$$

$$t_4 = \left[\frac{(6)(14{,}813.660)}{20{,}000}\right]^{1/2} = 2.108 \text{ in.}$$

⅊ ⇒ use $2\frac{1}{8}$ in base

Brownell and Young External Chair Design

$$a = \frac{\eta}{2} = \frac{5.00}{2} = 2.50 \text{ in.}$$

b = gusset spacing = 32.725 in.

For $1\frac{3}{4}$-in.ϕ bolts, $e = \dfrac{2.75}{2} = 1.375$ in.

P_B = max. bolt load on upwind side
$= f_S A_B = (10{,}963.73)(1.980) = 21{,}708.185$ lb

$$M_y = \frac{21{,}708.185}{4\pi}\left[(1 + 0.30)\ \ln\left[\frac{2(5.00)}{\pi(1.375)}\right] + 1\right]$$

$$= 3{,}612.549 \text{ in.-lb}$$

$$t_5 = \left[\frac{(6)(3612.549)}{15{,}000}\right]^{1/2} = 1.202$$

$\approx 1\frac{1}{4}$ in. in. ℙ for compression ring

Calculation of Gusset ℙ Thickness for Compression Rings

$$I = \eta t_6 r^2 \Rightarrow r^2 = \frac{t_6^2}{12}, \quad \text{with } t_6 = 3/8\ (1.250) = 0.469$$

$$r = \left[\frac{(0.469)^2}{12}\right] = 0.135 \text{ in.}$$

$h = G + H + 2\frac{1}{8}$ in. $= 9 + 1\frac{1}{4}$ in. $+ 2\frac{1}{4}$ in. $= 12\frac{1}{2}$ in.

$$\frac{h}{r} = \frac{12.500}{0.135} = 92.376 \Rightarrow \frac{P}{a} = \frac{\text{Bolt Load}}{\eta t_6}$$

$$= \frac{18{,}000}{1 + \left(\dfrac{h^2}{1{,}500\ t_6^2}\right)}$$

$$18{,}000\eta t_6^3 - P t_6^2 - \frac{h^2 P}{1{,}500} = 0$$

$$18{,}000(5.00)t_6^3 - (21{,}708.185)t_6^2 - \frac{(12{,}500)^2(21{,}708.185)}{1{,}500} = 0$$

$$t_6^3 - 0.241 t_6^2 - 0.025 = 0$$

$t_6 \cong 0.40$ in. $\Rightarrow$ 1/2-in. ℙ is OK

Skirt-to-Base Ring Weld

$$\xi = \left(\frac{4M}{\pi D^2_{SK}}\right) + \left(\frac{w}{\pi D_{SK}}\right) \text{ lb/lin in.}$$

$$\xi = \left[\frac{(4)(5{,}138{,}419.76)(12)}{\pi(120.0)^2}\right] + \frac{128{,}966.58}{\pi(120.00)} = 5{,}794.127$$

$f_W = 1.33\gamma_A(0.55)$, for wind or earthquake

$f_W = 1.33(20{,}000)(0.55) = 14{,}630.00$

$$\text{Weld size} = \frac{5{,}794.121}{14{,}630.00} = 0.396$$

or

$$w_1 = \frac{0.396}{2} = 0.198 \Rightarrow 1/4 \text{ in. minimum weld each side}$$

Cantilever Vibration

$$D_r = \left(\frac{41}{100}\right)(3.5) + \left(\frac{59}{100}\right)(7.5) = 5.860 \text{ ft}$$

Corroded Stack Weight

$$Wt_A = (395.166)\left[\left(\frac{42}{2}\right)^2 - \left(\frac{41.25}{2}\right)^2\right] = 6{,}168.294 \text{ lb}$$

$$Wt_B = (811.692)\left[\left(\frac{90}{2}\right)^2 - \left(\frac{89}{2}\right)^2\right] = 36{,}323.217 \text{ lb}$$

$$Wt_C = (469.927)\left[\left(\frac{120}{2}\right)^2 - \left(\frac{118.75}{2}\right)^2\right] = 35{,}060.960 \text{ lb}$$

$$Wt_D = (320.405)\left[\left(\frac{120}{2}\right)^2 - \left(\frac{118.75}{2}\right)^2\right] = 23{,}905.217 \text{ lb}$$

101,457.688 lb

$L_C = 8.00 + 5.00 = 13.0$ ft

$$\frac{L_C}{L} = \frac{13.00}{200} = 0.065 < 0.5$$

$$\frac{W}{L D_r^2} = \frac{101{,}457.688}{(200)(5.860)^2} = 14.773 < 20$$

Therefore, vibration analysis *must* be performed.

$$W_\delta = 101{,}457.688 \text{ lb}, \ L_e = 200 - 13.0 + \frac{13.0}{2}$$

$$= 193.50 \text{ ft}$$

$$T = \frac{1.648\ L_e^2}{D_r(E)^{0.5}} = \frac{1.648(193.50)^2}{(5.860)(29.0 \times 10^6)^{0.5}} = 1.955 \text{ seconds}$$

$$f = \frac{1}{1.955} = 0.511 \text{ cps}$$

$V_C = 3fD_r = 3(0.511)(5.860) = 8.983$ mph

$V_{30} = 100$ mph

$$V_w = (100)\left(\frac{200}{30}\right)^{0.143} = 131.165 \text{ mph}$$

Maximum gust velocity = 131.165(1.3) = 170.515 mph

$$K = \frac{0.0077 D_r{}^5 E}{L_e{}^3 W_S} = \frac{(0.0077)(5.860)^5(29.0 \times 10^6)}{(193.50)^3(101{,}457.688)}$$

$$= 0.002 < \frac{1}{15}$$

Therefore, the stack is free from cantilever vibration.

Static Deflection

$$P_C = C_\ell \rho \frac{(1.467 V_C)^2}{2}$$

$$= \frac{(1.0)(0.00238)(1.467)^2(8.983)^2}{2} = 0.207 \text{ psf}$$

$$t = \left(\frac{41}{100}\right)(0.50 - 0.125) + \left(\frac{59}{100}\right)(0.625) = 0.523 \text{ in.}$$

$$r_o = \left(\frac{41}{100}\right)\left(\frac{42}{2} + 0.125\right) + \left(\frac{59}{100}\right)\left(\frac{90}{2} + 0.125\right)$$
$$= 35.285 \text{ in.}$$

$$r_i = \left(\frac{41}{100}\right)\left(\frac{41 + 0.25}{2}\right) + \left(\frac{59}{100}\right)\left(\frac{88.75 + 0.25}{2}\right)$$
$$= 34.711 \text{ in.}$$

$$I = \frac{\pi}{4}[(35.285)^4 - (34.711)^4] \text{ in.}^4 = 77{,}307.326 \text{ in.}^4$$

$$D_S = \frac{P_C D_r (L_e)^4 (12)^3}{8EI}$$

$$= \frac{(0.207)(5.860)(193.50)^4(12)^3}{8(29.0 \times 10^6)(77{,}307.326)} = 0.164 \text{ in.}$$

Dynamic Deflection

Using a magnification factor of 30,

$\delta = 0.164\ (30) = 4.915$ in., which is permissible

Ovaling Vibration

Natural frequency of free ring = f_r

$$f_r = \frac{7.58 t_r (E)^{0.5}}{60 D^2}$$

At 42-in. dia = 3.50 ft,

$$f_r = \frac{7.58(0.375)(29.0 \times 10^6)^{0.5}}{60(3.50)^2} = 20.826 \text{ cps}$$

f_v = vortex shedding frequency

$$= \frac{0.2V}{D} = \frac{0.2(66)}{(3.5)} = 3.771 \text{ cps}$$

$2f_v = 7.543 < f_r$

At 90-in. dia = 7.5 ft,

$$f_v = \frac{7.58(0.5)(29.0 \times 10^6)^{0.5}}{60(7.5)^2} = 6.047 \text{ cps}$$

$$f_v = \frac{0.2(66)}{7.5} = 1.760$$

$2f_v = 2\ (1.760) = 3.520 < 6.047$ cps

At 120-in. dia = 10 ft,

$$f_v = \frac{7.58(0.625)(29.0 \times 10^6)^{0.5}}{60(10.0)^2} = 4.252 \text{ cps}$$

$$f_v = \frac{0.2(66)}{10} = 1.320 \text{ cps}$$

$2f_v = 2.640 < 4.252$

At bottom section,

$$f_v = \frac{7.58(0.75)(29.0 \times 10^6)^{0.5}}{60(10.0)^2} = 5.102 \text{ cps}$$

$$f_v = \frac{0.2(66)}{10} = 1.320 \text{ cps}$$

$2f_v = 2.640 < 5.102$

Therefore, stack is free from ovaling vibration.

ANCHOR BOLT TORQUE

Anchor bolt torque on stack bolts is handled exactly like tower anchor bolts as discussed in Chapter 4. Using Equation 4-66 and considering lubricated bolts we have

$$T = CDF_i \tag{4-66}$$

where the uplift load on each bolt, F, is

$$F_i = \frac{4(5{,}138{,}419.76)(12)}{(24)(125.00)} - \frac{101{,}457.688}{24} = 77{,}987.312 \text{ lb}$$

which results in a required bolt torque of

T = (0.15) (1.75)(77,987.312)
= 20,471.67 in.-lb = 1,706 ft-lb

Use 1,706 ft-lb torque with lubricant grease Fel-Pro C-5A, or equivalent.

The skirt base and anchor bolt detail for the stack is shown in Figure 5-23.

Design Summary

Static wind shear at base = 22,355.110 lb
Static wind moment at base = 1,299,115.509 ft-lb
Dynamic wind shear at base = 22,844.841 lb
Dynamic wind moment at base = 1,308,916.974 ft-lb
Total deflection at top of original tower = 4.418 in.
Total deflection at top of modified tower = 5.898 in.
Base plate thickness = $2^1/_8$-in. plate
Compression plate = $1^1/_4$-in. plate
(16) $1^3/_4$-in. anchor bolts
Required anchor bolt torque = 1,710 ft-lb
Total operating weight = 128,966.580 lb

EXAMPLE 5-3: STACK VORTEX STRAKE DESIGN

An exhaust stack 126 ft tall is to be provided with helical vortex strakes. The length of the stack to be straked is the top portion 31 ft 6 in. long. Compute the radius of curvature of the strake to be cut from flat plate. Referring to Figure 5-15 we have the following:

D = OD of stack = 7 ft 4 in.

L = 31 ft 6 in.

$$a = \frac{D}{2} = \frac{7.333}{2} = 3.667$$

$$\omega = 1$$

$$b = \frac{L}{2\pi\omega} = \frac{31.5}{2\pi(1)} = 5.013$$

Now,

$$r = \frac{a^2\omega^2 + b^2}{a\omega^2} \qquad (5\text{-}17)$$

$$r = \frac{(3.667)^2(1)^2 + (5.013)^2}{(3.667)(1)^2}$$

$$r = 10.521 \text{ ft}$$

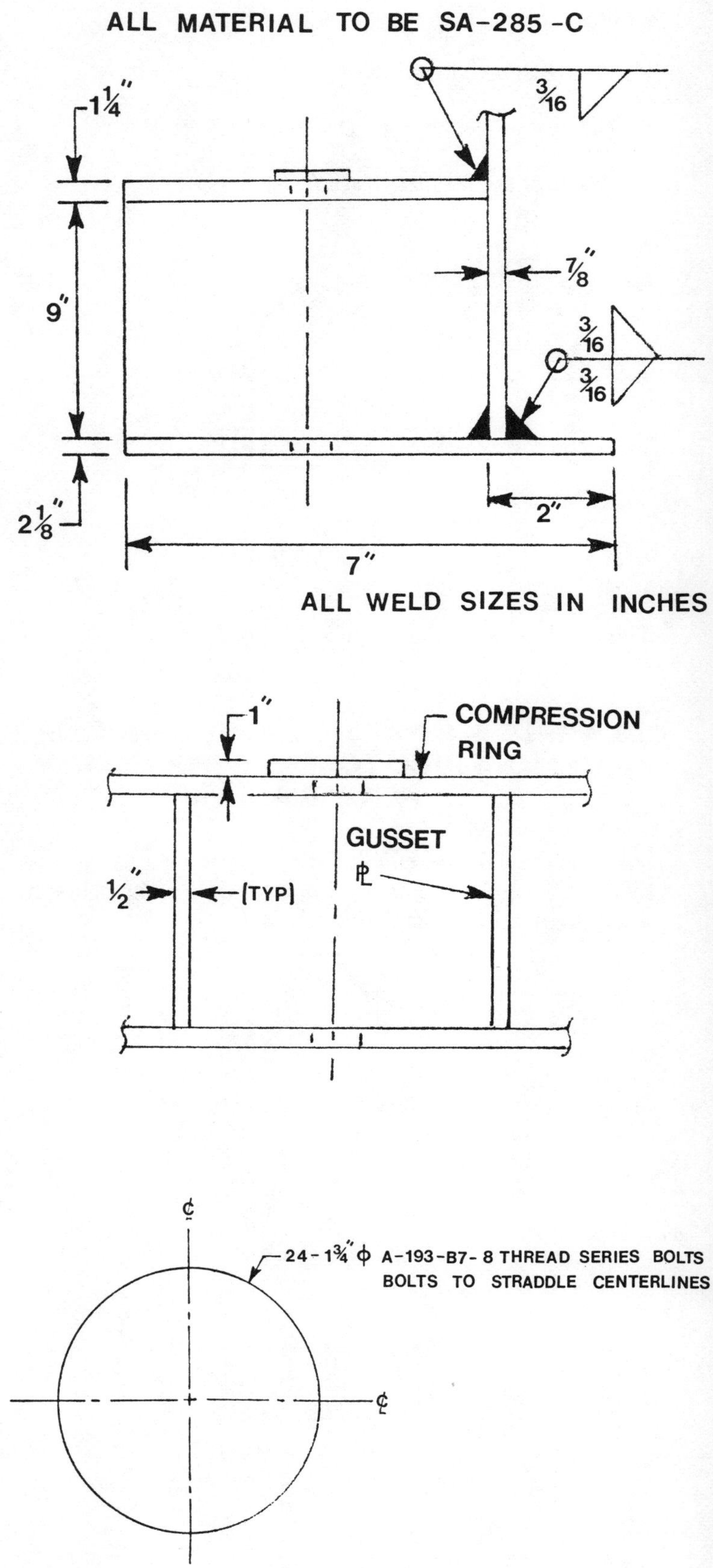

Figure 5-23. High-pressure flare stack base support detail.

Check

Using the approximate Morgan equation we have,

S_i = interior arc length = $[(\pi D_o)^2 + L^2]^{0.5}$ = 39.025 ft

S_2 = exterior arc length = $[[\pi(8.667)]^2 + (31.5)^2]^{0.5}$
= 41.637 ft

$$\lambda = \frac{S_i}{S_o} = \frac{39.025}{41.637} = 0.937 \qquad (5\text{-}19)$$

$$r = \frac{\lambda W}{1 - \lambda} \qquad (5\text{-}18)$$

$$r = \frac{(0.937)(0.667)}{1 - 0.937} = 9.966 \text{ ft} = 9 \text{ ft } 11.594 \text{ in.}$$

$$\% \text{ error} = \frac{10.521 - 9.966}{10.521} = 5.276\% \text{ error}$$

The final product is shown in Figure 5-24.

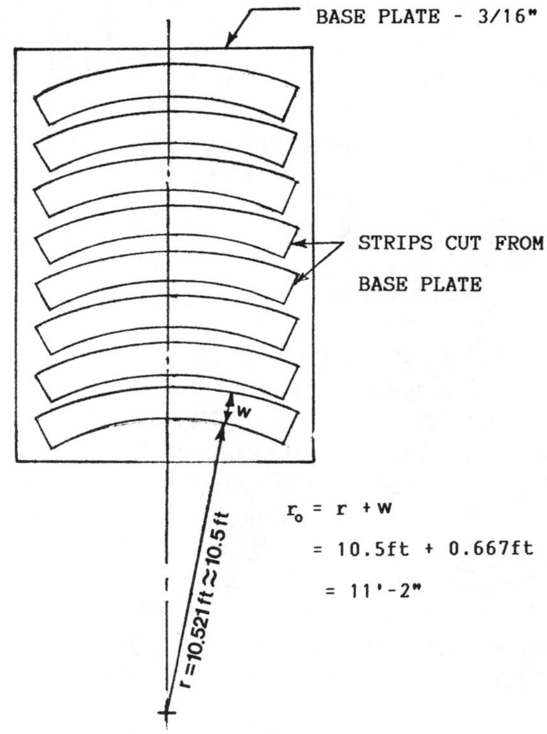

EXAMPLE 5-4: NATURAL FREQUENCY OF OVALING RING FORMULA (MICHELL FORMULA)

To use the Michell equation (5-12) dimensional analysis must be applied to obtain Equation 5-13. The original Michell equation is as follows:

$$f_r = \frac{1}{2\pi}\sqrt{\frac{E\,I\,n^2\,(n^2 - 1)^2}{\rho\,A\,r^4\,(n^2 + 1 + \nu)}} \qquad (5\text{-}12)$$

where ρ = 0.283 lb/in.^{3} for steel
A = (t) in. × (1) in.
r^4 = in.^{4}
E = lb/in.^{2}
$I = \frac{t^3}{12}$, per unit length of ring, in.^{3}
$n = 2$
ν = 1/3 for steel

$$f_r = \frac{1}{2\pi}\sqrt{\frac{E\,\frac{lb_f}{in.^2}\left(\frac{386\ lb_m\text{-}in.}{\frac{sec^2}{lb_f}}\right)(36)\ I\ (in.^4)}{(0.283)\,\frac{lb_m}{in.^3}\ t\ in.^2\ r^4\ in.^4\ (5.333)}}$$

$$f_r = \frac{4.409t}{r^2}\sqrt{E} \qquad (5\text{-}13)$$

Figure 5-24. Manufactured strake elements.

NOTATION

A = cross-sectional area of stack, in.^{2}
A_B = anchor bolt area, in.^{2}
a = stack radius = D/2, ft
B = critical arching parameter, dimensionless
D = critical diameter at which piping is unstable, dimensionless; internal stack diameter (Equation 5-15), ft; outside diameter of stack (Equation 5-16), ft; dynamic magnification factor (Table 5-3)
E = modulus of elasticity, psi
f_c = material yield strength, psi
ff = critical flow factor for arching in channels, dimensionless
f_r = natural frequency of a ring, Hz
f_v = stack vortex shedding frequency, Hz
G = consolidation particle parameter (Equation 5-8), dimensionless
H = height that solid is stored in bin, ft
H_s = stiffening ring spacing, ft
I = moment of inertia of stack cross section, in.^{4}
L = height of tower portion straked, ft
m = geometric parameter for arching (Equation 5-2), dimensionless
n = flexural mode (Equation 5-12), dimensionless
P_{air} = air pressure (Equation 5-11), psig
P_{max} = maximum hoop pressure at bin-hopper tangent point, psi
r = outside radius of stack (Equation 5-12), ft; radius of curvature of vortex strake (Equation 5-17), ft
S = over-all length of vortex strake (Equation 5-16), ft
S_i = interior arc length of helix (Equation 5-18), ft
S_o = exterior arc length of helix (Equation 5-18), ft
S_m = section modulus of stiffeners (Equation 5-15), ft^3
t = shell thickness of stack, in.
V = wind velocity, ft/min
V_c = critical wind velocity in which ovaling occurs (Equation 5-14), fpm
w = width of strake, ft; normal pressure applied on bin walls by solid (Equation 5-1), psi
$\tilde{x}$, $\tilde{y}$, $\tilde{z}$ = mode shapes relating translational displacements about the x, y and z axes, respectively

Greek Symbols

γ = bulk density of solid, lb/ft^3
δ = logarithmic decrement, dimensionless
Φ = piping factor, dimensionless
θ' = angle of hopper slope, degrees
$\tilde{\theta}$ = modal shape relating to rotation about an axis perpendicular to stack centerline (Figure 5-14), dimensionless
μ = coefficient of friction between the bulk solid and the bin wall (Equation 5-1), dimensionless
ϕ' = kinematic angle of friction between the solid and the bin wall, degrees
σ_1 = consolidating pressure for steady flow (Equation 5-4), lb$_f$/ft^2
σ_t = allowable tensile stress of stack material, psi
ω = number of revolutions around stack made by a helical strake, dimensionless

Table 5-3
Conservative Values for Logarithmic Decrement and Dynamic Magnification Factor for Various Stacks

	Low Damping		Average Damping		High Damping	
	δ	D	δ	D	δ	D
Unlined Stacks	0.035	90	0.052	60	0.105	30
Lined Stacks						
2″ gunite lining	0.070	45	0.100	31	0.300	10
4″ gunite lining	0.117	27	0.125	25	0.360	9

Low Damping = rocky, very stiff soil; low-stressed pile support, or structural frame support.
Average Damping = moderately stiff soil; normal spread footing or pile support
High Damping = soft soil; foundation on highly stressed friction piles

REFERENCES

1. Jenike, A. W., Johanson, J. R., and Carson, J. W., *Storage and Flow of Solids,* American Institute of Chemical Engineers, New York, New York, 1981.
2. Blevins, R. D., *Formulas For Natural Frequency and Mode Shape,* Van Nostrand Reinhold Company, New York, NY, 1979
3. Thomas, G. B., *Calculus and Analytic Geometry,* Addison-Wesley Publishing Co., Inc., Third Edition, 1960.

6

Rotating Equipment

Fluid movers and their use are vital to the process industries. This chapter focuses on two basic types—pumps and compressors. The sizing of these units and their interaction with the other components of a process system are discussed. This chapter does not address the detailed mechanical design of sophisticated equipment, such as turbine blade design and gas dynamics in a turbine. That type of material is a separate field of study and lies outside this text's objective of examining how to select and apply rotary equipment to process systems. For further reading, see the bibliography at the end of the book.

PUMPS

As the primary movers of liquids, pumps come in many types and an understanding of the various kinds is essential in successfully applying them to process systems.

Pumps are used to transfer liquids from one point to another. They basically fall under two categories—centrifugal and positive-displacement. The centrifugal pump gets its name from the fact that the pump's impeller imparts kinetic energy to the liquid with centrifugal force acquired by the impeller's rotation. This simple mechanism allows the centrifugal pump to be practical for high capacity, at low to medium heads. The aspect of low to medium heads will be discussed shortly. Typical centrifugal pumps include mixed flow, propeller, peripheral, and turbine.

Positive-displacement (PD) pumps trap a quantity of liquid and force it out of the cavity against the pressure of the discharge by means of rotary or reciprocating action. Ideally, a PD pump will produce whatever head is impressed on it by the system restrictions to the flow. Not all PD pumps are purely rotary or reciprocating, but we will focus our attention on these types. PD pumps, by definition, deliver fluids at a rate proportional to the speed of the pump action and this rate is independent of the pressure differential across the pump. For this reason means must be provided to limit the discharge pressure and this will be discussed under the section of positive-displacement pumps. Typical rotary positive-displacement pumps include screw, gear, vane, cam, and lobe. Reciprocating positive-displacement pumps include piston, plunger, and diaphragm.

Selecting the type of pump to use is a function of the service to be handled. Sometimes, the selection is obvious; for example, if you wanted to pump molasses, you would choose a positive-displacement pump. In the situation where neither a standard type of pump is used for the service, nor is it obvious what type to use, a centrifugal pump is always considered first. The reason for considering a centrifugal pump initially is because of its low initial cost, economical cost of maintenance, wide range of materials of construction, and relatively large clearances. Factors to be considered in selecting a pump are as follows:

1. Efficiency
2. Net positive suction head (NPSH) required by pump
3. Operating costs
4. Shaft speed
5. Magnitude of clearances
6. Materials of construction
7. Fluid service to be handled
8. Availability and delivery time of pump

The type of pump to be used for a specified service or duty can be selected from Figure 6-1. This figure clearly indicates how different pumps have overlapping charac-

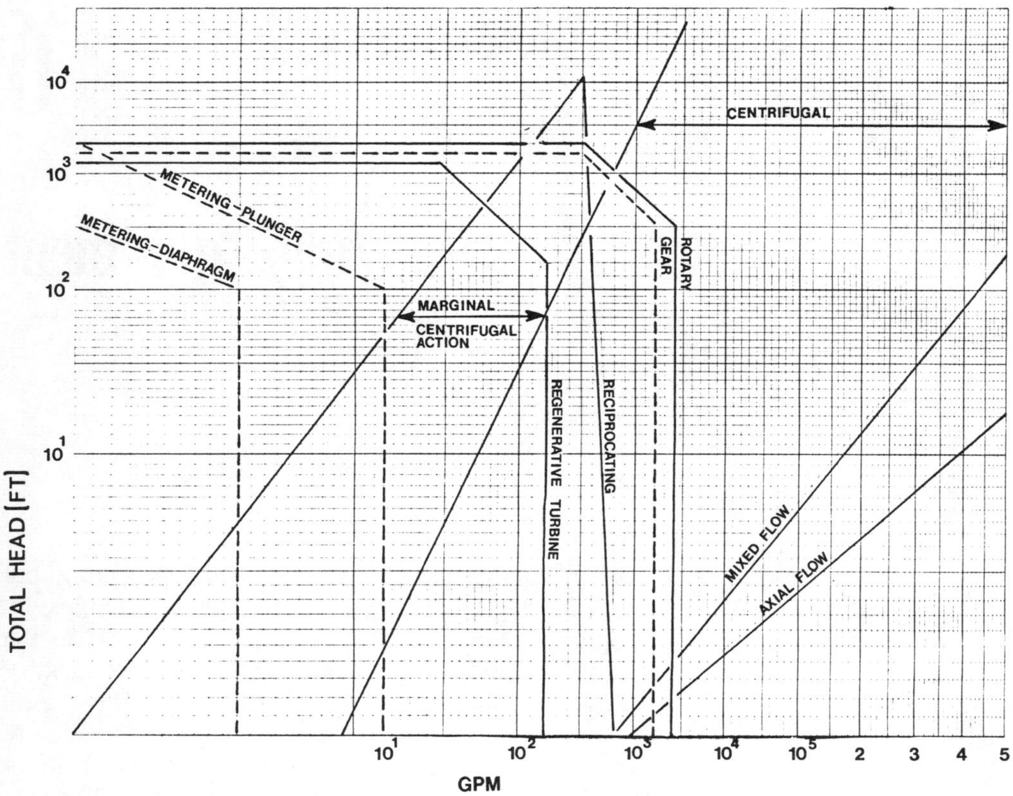

Figure 6-1. Pump selection guide.

teristics. Depending on the relative importance of the previously cited criteria, a certain type of pump will be selected. Figure 6-1 will help the reader determine from a quick glance what type(s) of pump(s) will be required.

Centrifugal Pumps

Centrifugal pumps are the most widely used because of their wide operating range and the reasons previously cited. These pumps come in a variety of types, depending on the type of impeller, casing, stuffing box, and bearings. These components are shown in Figure 6-2.

The radial type impeller is by far the most common centrifugal pump in the process industries. The flow is directed by the impeller imparting motion on the fluid, driving the fluid to the periphery of the impeller. This allows the velocity head to be converted mostly to pressure head in the volute.

The mixed flow pump impeller consists of vanes doubly curved or screw-shaped so that the impeller moves the fluid by both centrifugal and pushing action. The result is a discharge of axial and radial flows.

The axial flow pump impeller develops head by a lifting or pushing hydrodynamic action that results in totally axial flow on discharge.

The impeller is hydrodynamically balanced to ensure minimal vibration. The casings can come in a variety of designs, but are either vertically or horizontally split. A vertical-split casing implies that the casing is bolted together along a vertical plane. Similarly, a horizontally split casing is bolted or connected along a horizontal plane. The advantage of the vertical split casing is that the pump is supported along the shaft allowing for thermal movements without causing shaft misalignment.

Packing and seals on the shaft are the most common source of failure for a pump. In low-pressure applications, soft or metallic packing will suffice in a stuffing box. In most low-pressure applications, a single seal will usually suffice. When pressures exceed about 50 psig and there can be no tolerance for leakage, a double seal is utilized. These seals come in various configurations—tandem, bellows, and face-to-face.

When process conditions get severe enough, a double inside-outside seal, where part of the seal is outside the stuffing box, is used. The disadvantage of this type of seal is that not all stuffing box arrangements allow such a configuration.

For proper cooling and lubrication the seal must be supplied with a fluid, called a seal flush. Figure 6-3 shows such a system.

Group I Standard Pump

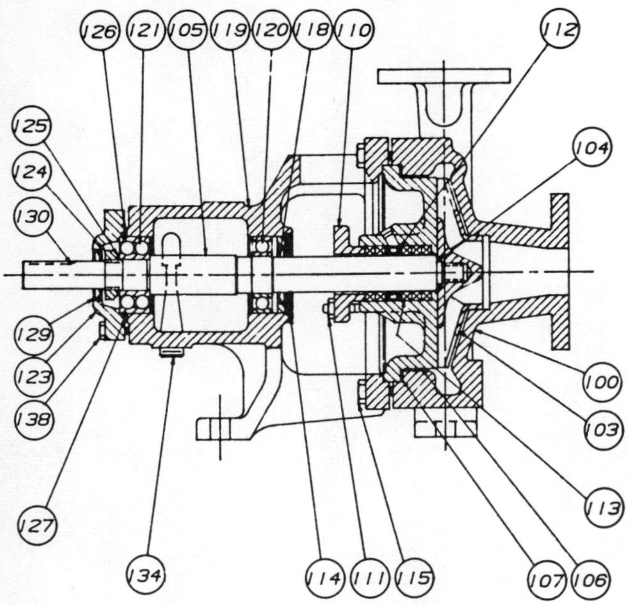

Group II and III Standard Pumps

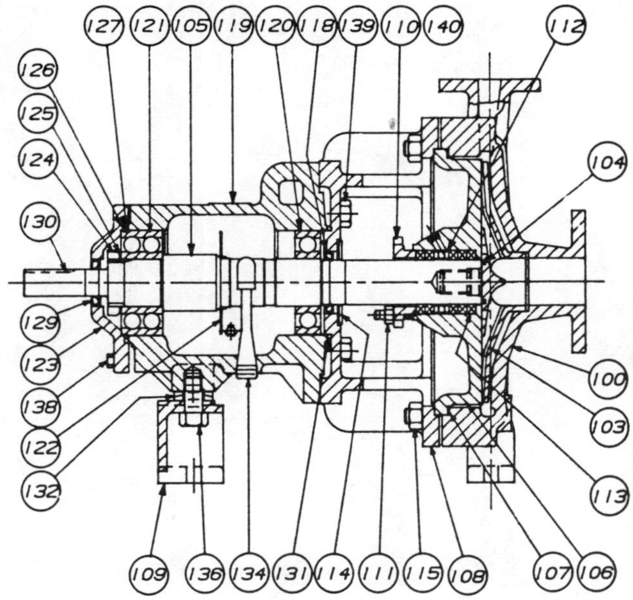

Materials Common to all Alloys Unless Otherwise Noted

Part No.	Part	Material
104	Impeller Gasket*	TFR(F)
107	Rear Cover Plate Gasket*	Durabla
108	Bearing Housing Adapter	Cast Iron
109	Bearing Housing Foot	Cast Iron
111	Gland Studs or Flange Studs with Hex Nuts	304 S.S./303 S.S.
112	Seal Cage*(E)	PTFE
113	Molded Ring Packing*(E)	Kevlar(H)
114	Inboard Deflector	PTFE
115	Casing Studs/Hex Nuts	304 S.S./316 S.S.(G)
118	Inboard Oil Seal*	TFSR
119	Bearing Housing	Cast Iron
120	Inboard Bearing*	Steel
121	Outboard Bearing*	Steel
122	Oil Slinger	Steel
123	Bearing Cover	Cast Iron
124	Bearing LockNut	Steel
125	Bearing LockWasher	Steel
126	Bearing Cover Gasket	Cork
127	Bearing Shim*	Steel
129	Outboard Oil Seal*	TFSR
130	Shaft Coupling Key	Steel
131	Bearing Housing Adapter "O" Ring	SBR
132	Spherical Washer for Foot	Steel
133	Trico Oiler (not shown)	Steel-Plastic
134	Bearing Housing Vented Drain Plug	Plastic
136	Cap Screw for Foot	Steel
138	Cap Screws for Bearing Cover	Steel
139	Machine Bolts for Bearing Housing	Steel
140	Cap Screws for Adapter to Cover	Steel

*Parts normally stocked by customer for emergency repairs
**Tradename of International Nickel Company
(A) Not available in Recessed Impeller pumps.
(B) Not available in Self-Priming pumps
(C) Not available on 4x3 US-10H, 4x3 US-13, or 6x4 US-13A (cast steel substituted)
(D) Jacketed cover plates are carbon steel
(E) Used in packed pumps only
(F) Titanium pumps have Grafoil impeller gaskets
Grafoil is a registered trade name of Union Carbide Corporation
(G) Alloy is B7 Sr on Ductile Iron and Carbon Steel pumps.
(H) Tradename of E.I. Dupont deNemours & Company, Inc.

Figure 6-2. Centrifugal pump components. (Courtesy of the Duriron Company.)

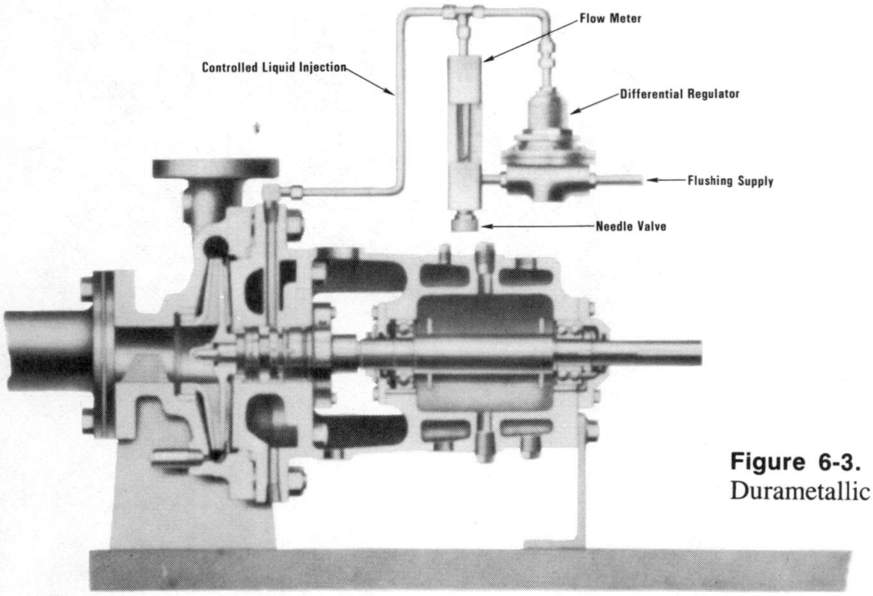

Figure 6-3. A seal flush configuration. (Courtesy of the Durametallic Corporation.)

The various types of seals are shown in Figure 6-4. The pump manufacturer should be relied upon for the choice of seals. Sealing technology is a subject vast enough to encompass this book and the reader is referred to Buchter [1] for additional sources.

Bearings, like seals, are for the most part the main responsibility of the pump manufacturer. In all situations, the bearings should be of the outboard type (not subjected to the process fluid), unless situations prevent this type of arrangement.

Hydraulic Requirements of Centrifugal Pumps

In this section the reader will find it advantageous to refer to Chapter 1. The most important hydraulic parameter in pump selection is the net positive suction head (NPSH), which is the total pressure at the pump suction point minus the vapor pressure of the liquid at the pumping temperature. NPSH is the energy that forces the liquid into the pump, and is expressed in foot-pounds of energy per pound of mass (normally referred to as feet of head) or pounds per square inch of *absolute* pressure. When values of pressure are expressed in feet of liquid, the theoretical height to which a liquid can be lifted at any temperature is the difference between the atmospheric pressure and the vapor pressure of the liquid at that temperature. Figure 6-5 helps simplify the calculation of the NPSH.

In selecting a pump the engineer must refer to the performance curves the pump manufacturer prepares for each model of pump. Most performance curves are plots of flow capacity (gpm) of water versus break horsepower or total dynamic head in feet. Such a curve is shown in the examples that follow. As seen, the efficiency curves are plotted with various lines indicating impeller size and the NPSH required at various points. In reading the performance curves, it is emphasized that the extreme right side of the curve should be avoided, because the capacity and head change abruptly. Pumps are normally selected to operate in the area of high efficiency. The danger in selecting a pump on the extreme left is that at low flows the pump horsepower overheats the liquid. If low rates cannot be avoided, a by-pass may be required to prevent vaporization and subsequent pump damage. Thus, vaporization of the pumped liquid can occur two ways: (1) the NPSH required is not being met and cavitation occurs in the liquid causing vapor bubbles that can severely damage the impeller or (2) the pump horsepower overheats the pumped liquid, forming vapor bubbles that can (and normally will) damage the pump.

Excess heat resulting in pumping a fluid can be avoided by determining the minimum flow required to allow proper heat dissipation. At low flow rates or shut-off conditions, heat is transferred to the liquid contained in the pump casing at a rate representing the power losses of the pump. The power loss is the difference between the brake horsepower consumed and the water horsepower developed. The remnant energy in the pump bearings and that lost to convection to the outside atmo-

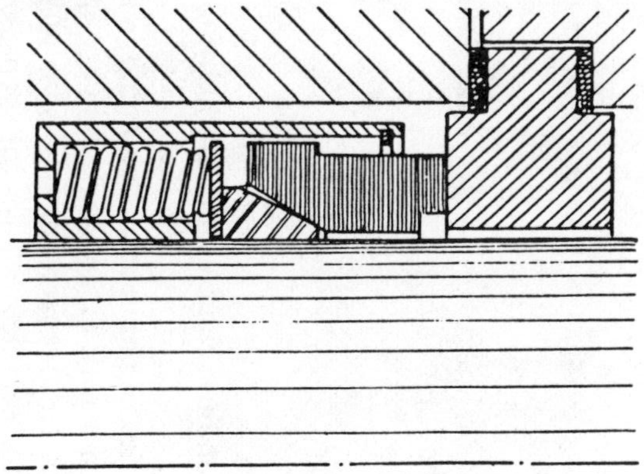

Single seal, mounted internally.

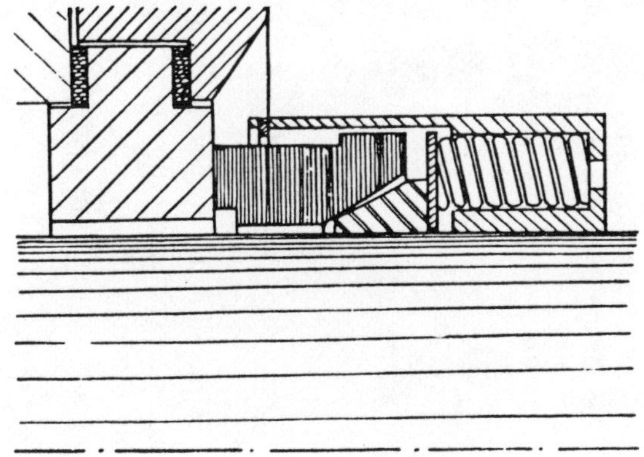

Single seal, outside mounted.

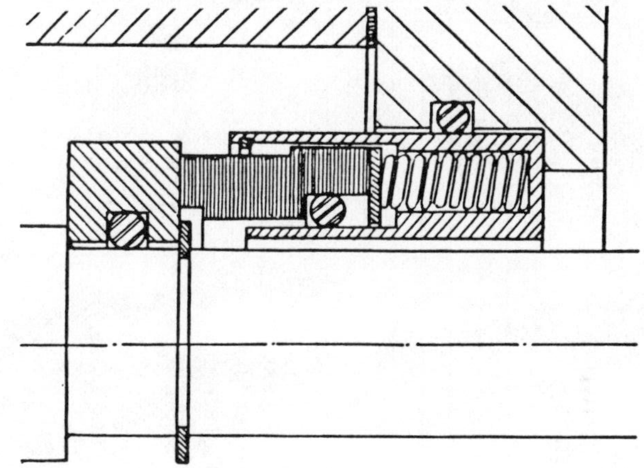

Inside seal, head stationary.

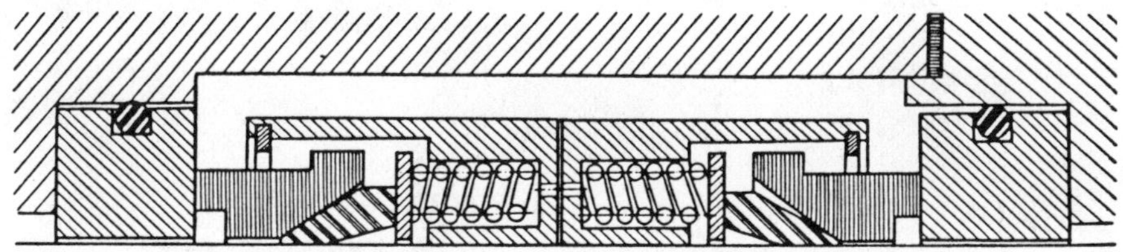

Double seal arrangement, mounted internally.

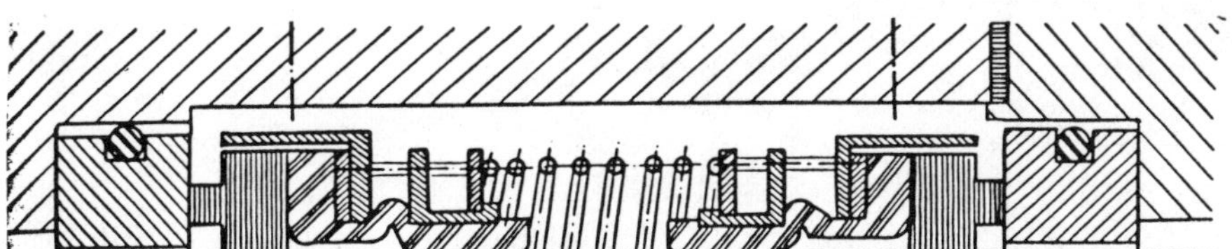

Double seal with elastomeric bellows, internally mounted.

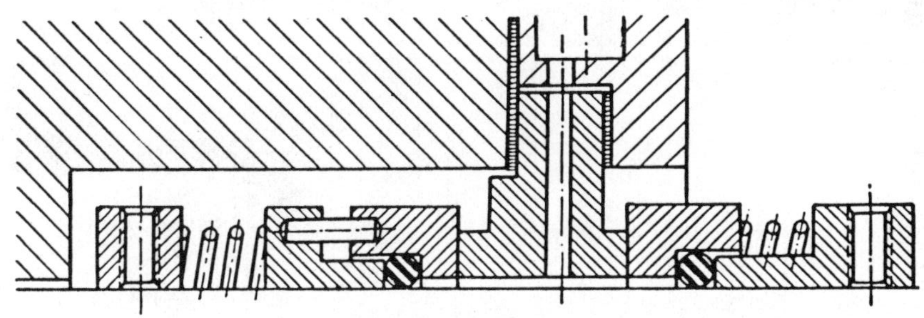

Double inside-outside arrangement.

Figure 6-4. Various types of seals used in centrifugal pumps [1]. (Courtesy of Buchter.)

Pump Hydraulic Design Calculation Sheet

Liquid ____________________
Viscosity at P.T. (Pumping Temp.) ____________________ cp
Vapor pressure at P.T. ____________________ psia
Sp. gr. (γ) at P.T. ____________________
Flow at ambient temp. ____________________ gpm
Operating flow at P.T. ____________________ gpm
Design flow at P.T. ____________________ gpm

Suction				Discharge			
Source* pressure	=	________	psia	Terminal pressure	=	________	psia
Static (+ head)(− lift)	=	________	psi	Static (head)(lift)	=	________	psi
− ΔP_f line loss	=	________	psi	− ΔP_f discharge			
Suction pressure	=	________	psia	Piping system	=	________	psi
− Vapor pressure	=	________	psia	Other	=	________	psi
NPSH avail	=	________	psia	Discharge press.	=	________	psia
NPSH avail	=	________	ft	− Suction press.	=	________	psia
NPSH req'd	=	________	ft	TDH	=	________	psia
				TDH	=	________	feet

Min NPSH avail ⩾ NPSH req'd + 2 ft
*Initial press., e.g., ATM press. or press. pad

① bhp at Duty Condition

$$bhp_D = \frac{(gpm)(TDH)(\gamma)}{(3{,}960)(\eta)} =$$

② bhp at Maximum Capacity Condition

$$bhp_{MC} = \frac{(gpm)(TDH)(\gamma)}{(3{,}960)(\eta)}$$

TDH = total dynamic head
TDH = discharge press. − suction press.
η = pump efficiency, %
P.T. = pumping temperature

Figure 6-5. Pump hydraulic design calculation sheet.

sphere is negligible. The temperature rise per minute is computed by the following relation:

$$\Delta t = \frac{42.2(bhp_{so})}{W_w C_p} \tag{6-1}$$

where Δt = temperature rise per minute, °F/min
bhp_{so} = brake horsepower at shut-off
W_w = weight of liquid in pump, lb
C_p = specific heat of liquid in pump

The break horsepower of the pump is given by

$$bhp = \frac{QH\gamma}{3{,}960\eta} \tag{6-2}$$

where Q = flow rate, gpm
H = total head, ft
γ = specific gravity
η = pump efficiency (fraction)

The water horsepower is given by

$$whp = \frac{QH\gamma}{3{,}960} \tag{6-3}$$

which is the power required if the desired head at the required capacity could be produced with zero losses.

For water flowing through the pump, conditions become stabilized and the temperature rise is determined by the following:

$$\Delta t = \frac{(bhp - whp)\ 2{,}545}{\dot{m}} \tag{6-4}$$

where 2,545 = Btu equivalent of 1 hp-hr
$\dot{m}$ = mass flow rate, lb/hr

Another variant of Equation 6-4 that relates the temperature rise to the total head is

$$\Delta t = \frac{H}{718}\left(\frac{1}{\eta} - 1\right) \tag{6-5}$$

In Equations 6-4 and 6-5 the compressibility of water is neglected.

To prevent overheating of the pumped liquid, a bypass piping arrangement is used to have the pump operating at full capacity. Such an arrangement is shown in Figure 6-6. It is always desirable to pass the bypass liquid through an intercooler to cool the fluid before it enters the pump suction port. Under no circumstances should the bypass line connect directly from the pump discharge to the pump suction.

So far, we have not considered the pumping of viscous liquids. For a liquid that has viscosity greater than about 10 cp, a viscosity correction must be made, because the pump motor must work harder to pump the fluid.

All pump manufacturers' pump performance curves are based on pumping water. To correct for the pumped liquid's viscosity, Figures 6-7 and 6-8 are used to approximate the equivalent water performance. The figures, developed by the Hydraulic Institute, are used by entering the bottom with the viscous flow rate (gpm), moving vertically upward to the desired viscous head (head per stage for multistage pumps), then moving horizontally to the left or right to the viscosity line, and proceeding vertically upward to the correction-factor curves for the head and capacity. The equivalent water-performance values are then obtained by dividing the viscous-performance values by the correction values. Thus, the pump selection can be made on those ratings established for water. The efficiency of the viscous liquid pumping conditions can be calculated using the efficiency correction factor multiplied by the pump efficiency for water. In this manner the viscous performance of the pump can be determined using the manufacturers' performance curves, which are always based on pumping water. This procedure is illustrated in the examples later in this chapter.

Positive Displacement (PD) Pumps

Positive displacement (PD) pumps are usually selected after it has been determined that a centrifugal design cannot meet the requirements. Thus, PD pumps are used where centrifugals cannot operate—under low NPSH requirements or handling a highly viscous liquid. There are several types of PD pumps, as previously mentioned, and their positive attributes are that they

1. Operate at relatively high efficiencies when pumping viscous liquids.
2. Operate under low NPSH conditions and produce high suction lifts.
3. Operate with high heads at a wide range of capacities.
4. Have a wide speed range, which is limited by the liquid's viscosity.
5. Are inherently self-priming.

Selecting the type of rotary pump is primarily a function of cost and the particular requirements that are to be met.

1. *Vane* pumps—normally have a capacity up to about 380 gpm and operate by trapping liquid within vane differential pressures, usually at around 50 psig. The practical limit on viscosity is approximately 100,000 SSU. Vane pumps are subject to wear and should not be used with a liquid that has poor lubricating qualities.
2. *Gear* pumps—normally are used up to about 1,000 gpm and can handle liquids with viscosities up to 5 $\times 10^6$ SSU. These pumps operate at approximately 1,200 rpm with liquids of 10 to 500 SSU viscosity (see Figure 6-9). It is desirable to have internal timing gears and bearings since only one shaft sealing area is required. A variant of a gear pump is shown in Figure 6-10.

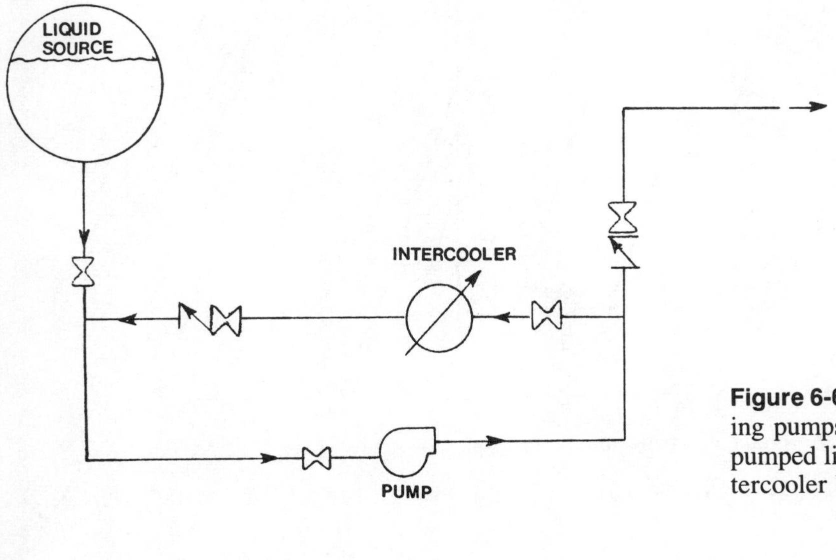

Figure 6-6. Excessive heat build-up is often caused by operating pumps at reduced flow rates. To prevent overheating the pumped liquid, it is advisable to pass the liquid through an intercooler before it enters the pump suction port.

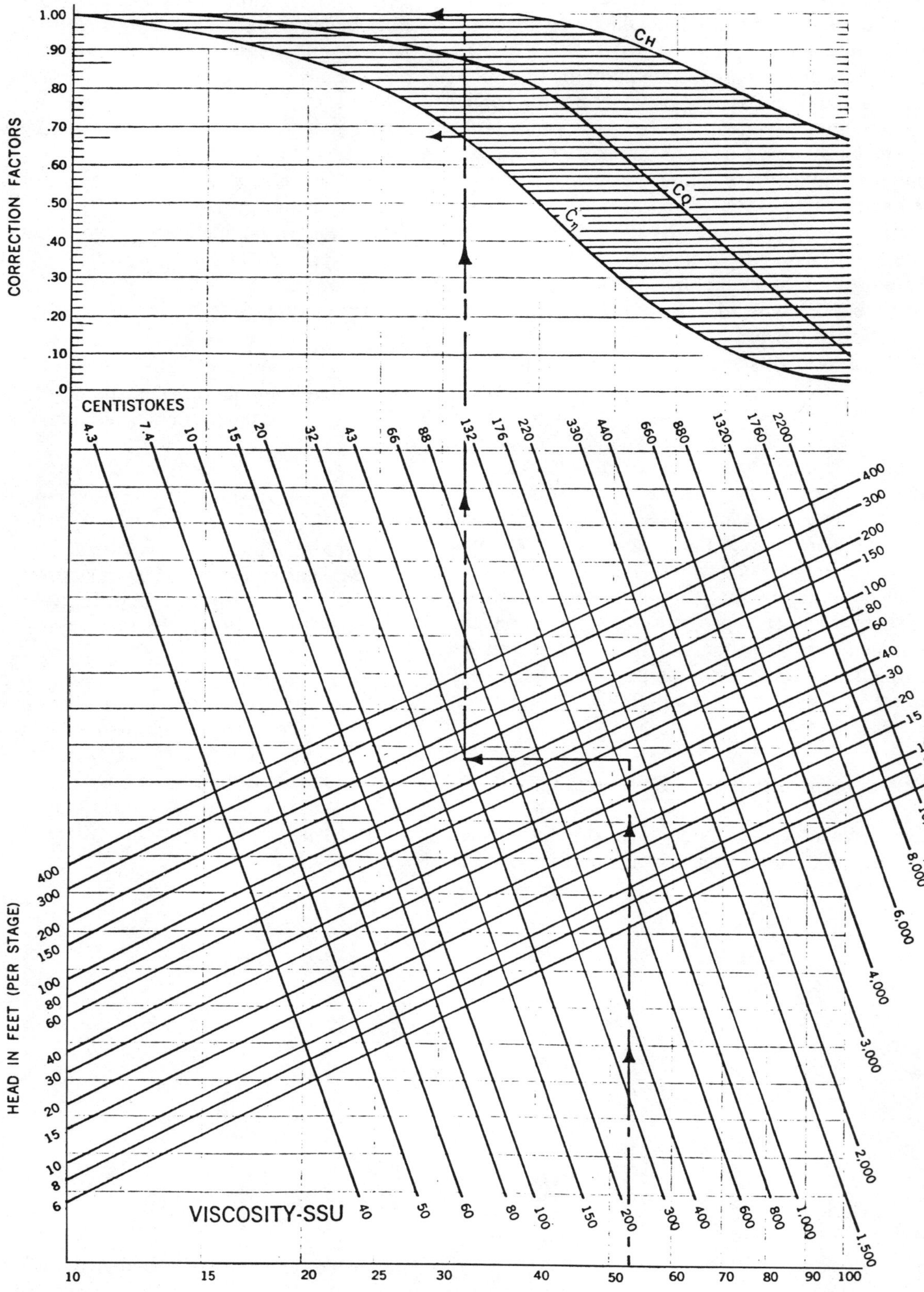

Figure 6-7. Performance correction chart for viscous liquids. (Courtesy of the Hydraulic Institute, Cleveland, Ohio.)

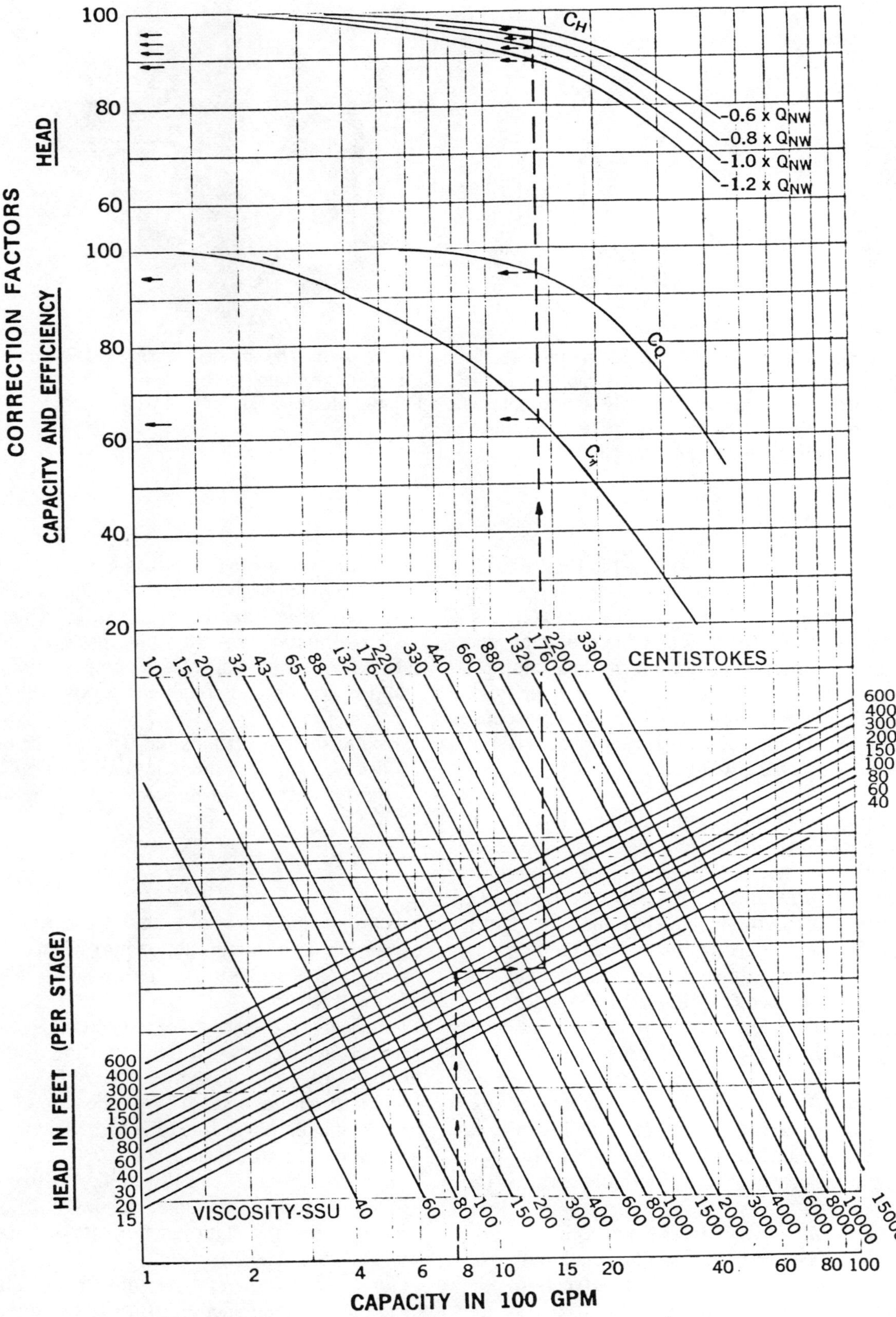

Figure 6-8. Performance correction chart for viscous liquids. (Courtesy of the Hydraulic Institute, Cleveland, Ohio.)

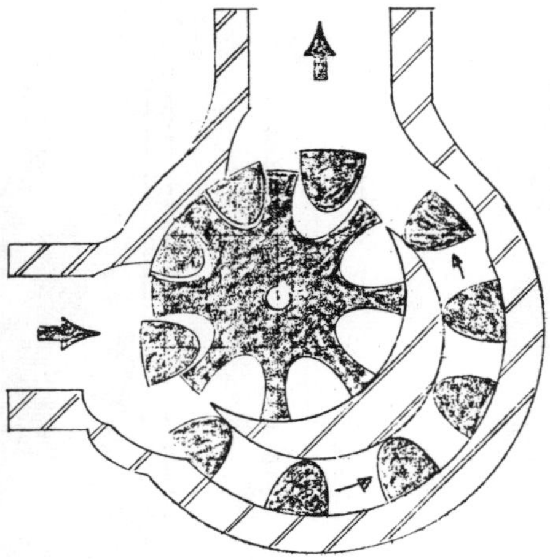

Figure 6-9. This drawing of a rotary gear pump illustrates the positive-displacement principle. The fluid is captured in the gear teeth and displaced to the suction port. The crescent acts as a seal between the suction and discharge ports. An application of this type of pump is illustrated in Example 6-2.

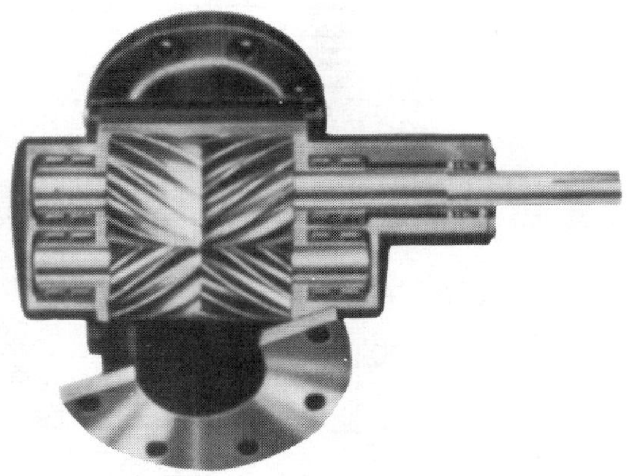

Figure 6-10. The internal bearing gear pump is a variant of the rotary gear pump in Figure 6-9. (Courtesy of Worthington Pumps, McGraw Edison Company.)

3. *Screw* pumps—these pumps, depicted in Figure 6-11, are used where large flow capacities, 4,000 gpm and 3,000 psi, are required. Screw pumps can handle viscosities up to 10 × 10^7 SSU and have bearing and timing gear requirements similar to gear pumps. Screw pumps come in various designs, and one type, shown in Figure 6-12, can handle highly viscous, non-Newtonian fluids such as glues, molasses, tar, asphalt, and wastewater with ease.

Positive displacement (PD) pumps come in a vast variety and you should refer to the manufacturers' literature to best determine the selection of the particular pump to be used. However, PD pumps are sized very much like centrifugal pumps, and the calculation sheet in Figure 6-5 can safely be used for sizing either type. Pump sizing is focused upon here to illustrate the various ways in which a pump may be specified. Figure 6-13 shows various installations for a pump. Some properties and characteristics illustrated in Figure 6-13 are

Static suction lift—the vertical distance in feet (expressed in psi) between the liquid level of the liquid to be pumped and the centerline of the pump suction port when the pump is located *above* the liquid level of the liquid to be pumped.

Static suction head—the vertical distance in feet (expressed in psi) between the liquid level of the liquid to be pumped and the centerline of the pump suction port when the pump is located *below* the liquid level of the liquid to be pumped.

Friction head—the pressure (psi) required to overcome frictional resistance of a piping system.

Velocity head—expressed in psi, see Chapter 1.

Total suction lift—the total pressure below atmospheric (in Hg or psi) at the pump suction port during pump operation and equals the following:

1. Static suction lift plus the frictional head, or
2. Frictional head minus the static suction head (only if the frictional head is greater than the static suction head).

Total suction head—the total pressure (psi) above atmospheric at the pump suction port when the pump is operating and is equal to the static suction head minus the frictional head.

Static discharge head—expressed in psi, is the vertical distance in feet between the centerline of the pump and the point of liquid discharge.

Total discharge head (TOH)—the sum of the frictional head in the discharge line (discharge frictional head) and the static discharge head.

Total static head—the difference between the static discharge head and the static suction head or the difference between the static suction lift and the static discharge head.

Total dynamic head—the sum of the total discharge head and the total suction lift or the difference between the total discharge head and the total suction head.

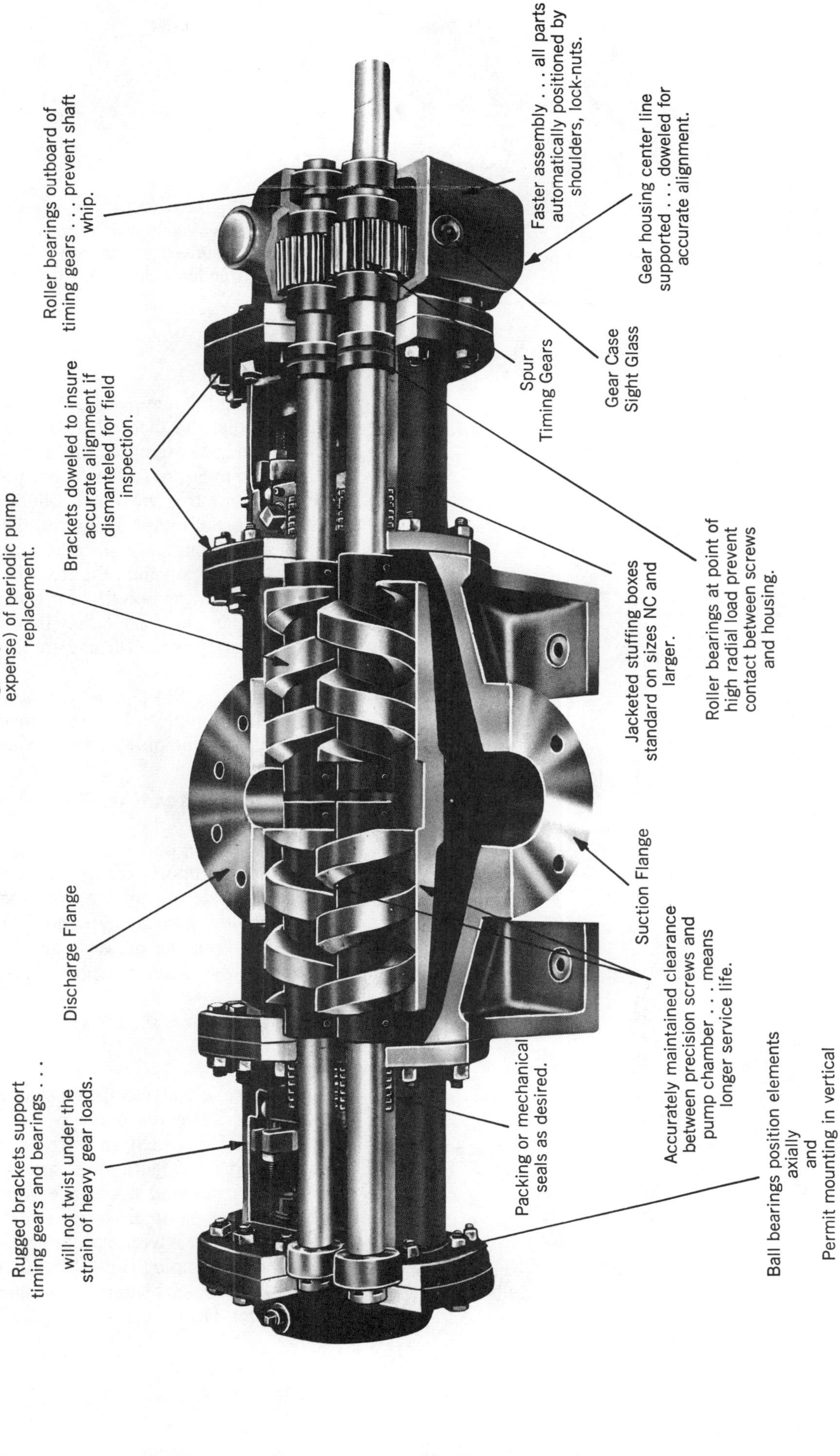

Figure 6-11. Cross-sectional view of hot oil screw pump. (Courtesy of Worthington Pumps, McGraw Edison Company.)

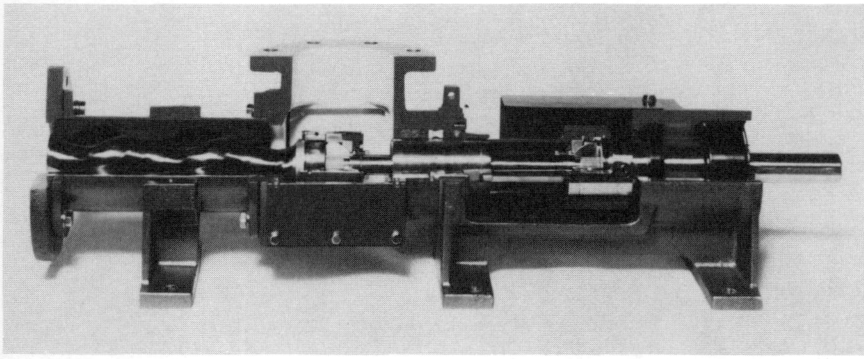

Figure 6-12. A cavity screw pump is ideal for handling highly viscous non-Newtonian liquids. (Courtesy of Moyno® Industrial Products, Fluids Handling Division, Robbins and Meyers, Inc.)

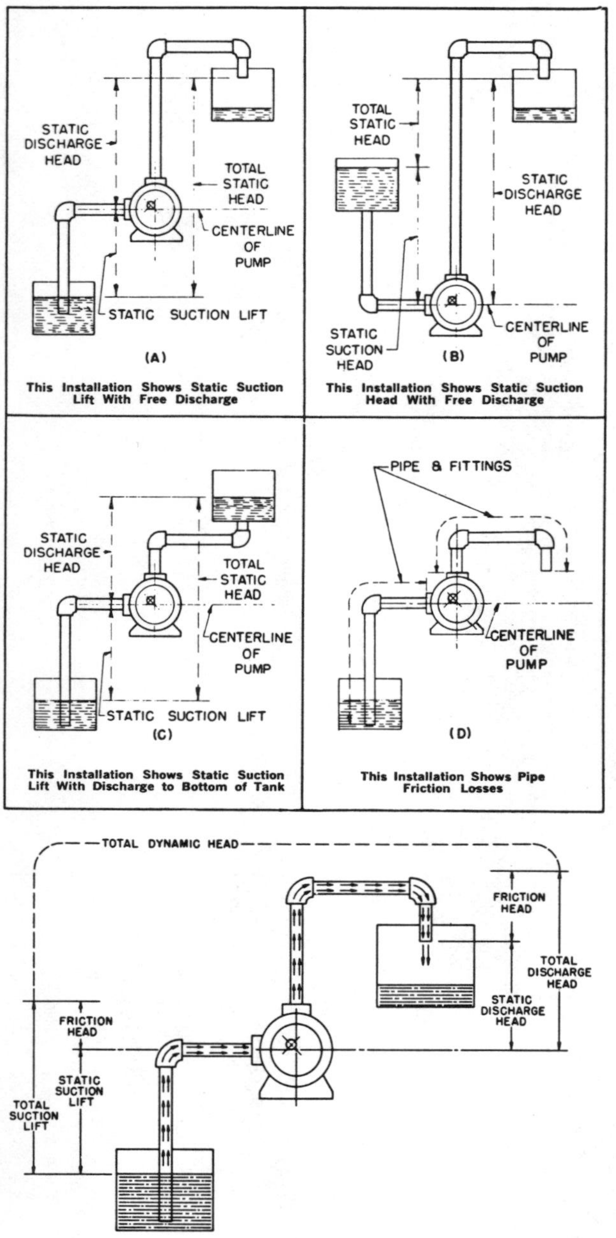

Figure 6-13. The principal parameters of pump selection. (Courtesy of Viking Pump Division, Houdaille Industries, Inc.)

When using PD pumps where a suction lift is required, remember that the theoretical height to which a liquid can be lifted at any temperature is the difference between atmospheric pressure and the vapor pressure of the liquid at that temperature, when both values of pressure are expressed in feet of liquid. However, the suction lift practical for actual pumping installations is somewhat less than the theoretical value. Figure 6-14 shows the theoretical and practical suction lifts for water. Also, remember that the higher the installation is above sea level, the lower the vapor pressure, and the lower the maximum suction lift.

Application of PD pumps to practical installations is given in the examples. The unit conversions included in Appendix D are helpful in pump calculations.

Pressure Protection For PD Pumps

By definition, a positive-displacement pump transfers fluid at a rate proportional to the speed of displacing action and this rate of transfer is independent of the pressure differential across the pump. Thus, means must be provided to limit the pressure and the pump discharge side should the discharge piping become restricted or blocked.

There are various methods used to prevent overpressure:

1. Install a relief valve at the discharge of the pump with the relief valve discharge being piped back to the pump inlet in which an intercooler is placed in the line. Such a configuration is shown in Figure 6-15. In such an arrangement a temperature sensor device is placed at the pump discharge to detect excessive temperatures. The intercooler, or heat exchanger, is used to cool the pumping fluid. Normally, temperature becomes a problem when the instantaneous discharge and inlet flows are equal. Gear and multiplex

(plunger, diaphragm, and piston) pumps are examples of such pumps in which this situation occasionally develops.

2. Place a pressure switch in the discharge side of the pump piping, interlocked to shut off the pump driver. Since pressure switch set points are not as reliable as relief valves, a relief valve must be added to the discharge piping and set at a pressure slightly greater than the pressure switch to ensure adequate protection. The relief valve would be piped-up similarly to that shown in Figure 6-15.
3. Install a torque-limiting device in the pump driver when a relief is not practical, such as slurry service. A torque-limiting device can come in the forms of a shear-pin or torque limiting coupling. These devices have advantages other than protecting the system against overpressure; they protect the pump against foreign material or whenever the pumped fluid might tend to solidify.

Overpressure protection is essential in positive-displacement pumps. Relief valves applied should be added to the discharge piping itself, because built-in relief valves on the pump that are not removable for testing are undependable.

COMPRESSORS

The three types of compressors used in the process industries are centrifugal, reciprocating, and axial flow compressors. Like pumps, depending on the application, the type of compressor is roughly a function of the gas capacity, action, and discharge pressure. Figure 6-16 shows the operating ranges of the three basic types of compressors. As clearly shown, one type of compressor, despite its disadvantages or advantages compared to other types, is usually the obvious choice.

Reciprocating compressors are normally used when a relatively low flow rate is required, but high discharge pressures are expected. This situation is common in the gas processing industry where high discharge pressures are needed for process conditions. The need and use of reciprocating compressors is unavoidable in many process system applications.

Centrifugal compressors are the most common type in hydrocarbon processing plants and are to some extent the workhorse of chemical process compression needs. There are four basic advantages a centrifugal compressor has over a reciprocating compressor:

1. Lower initial capital investment. The cost advantage is increased as the power demand is increased.

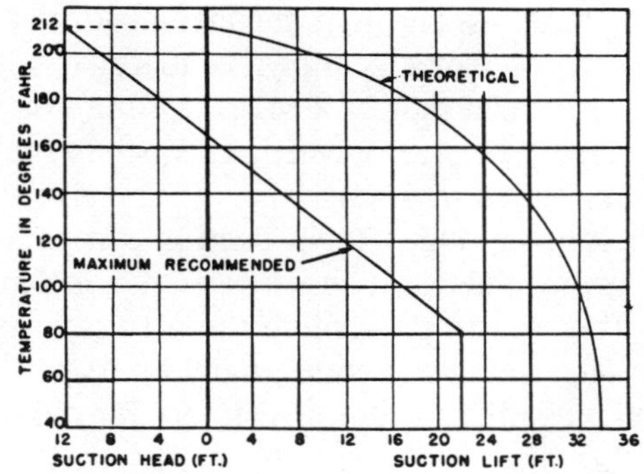

Figure 6-14. The theoretical and maximum recommended suction lift for water at various temperatures, °F. (Courtesy of Viking Pump Division, Houdaille Industries, Inc.)

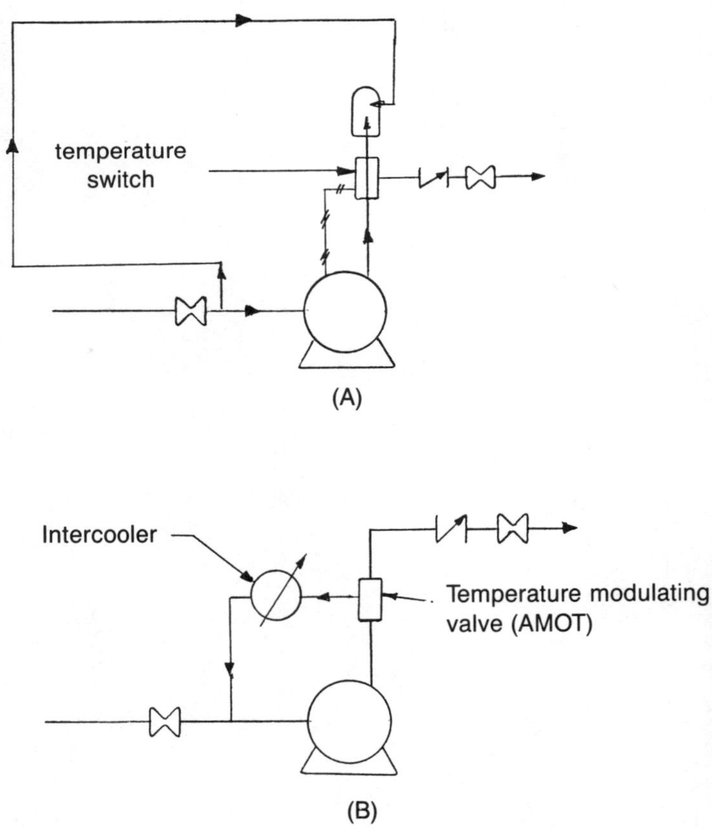

Figure 6-15. A temperature switch can be used in lieu of an intercooler (heat exchanger) in which the switch can shut off the pump driver when liquid temperatures become excessive as in (A) or can be used with an intercooler in (B) to divert flow through the exchanger. In either case, a pressure safety valve should be used on discharge. (B) assumes the suction temperature is constant. To prevent overheating on low flow rate conditions, a flow switch is often used.

2. Lower operating and maintenance cost. The operating and maintenance cost of a centrifugal is approximately one-third that of a reciprocating compressor.
3. Compactness of size. Centrifugals occupy less space and make much less noise.
4. Simplicity of piping. Reciprocating compressors can cause severe pulsation shock response in piping systems. The cost in preventing the effects of pulsation in piping systems can entail many hours of engineering and a healthy capital investment for either analog or digital simulation tests. Centrifugals do not have this problem.

Axial-flow compressors operate at greater capacities and are often used in series with centrifugal units. Axial-flow compressors are governed by the same formulas that apply to centrifugals. The axial units are more efficient than the centrifugals, but the latter have a much wider operating range. Axials are used primarily for clean gases such as air, because they are much more susceptible to corrosion, erosion, and deposits than centrifugals.

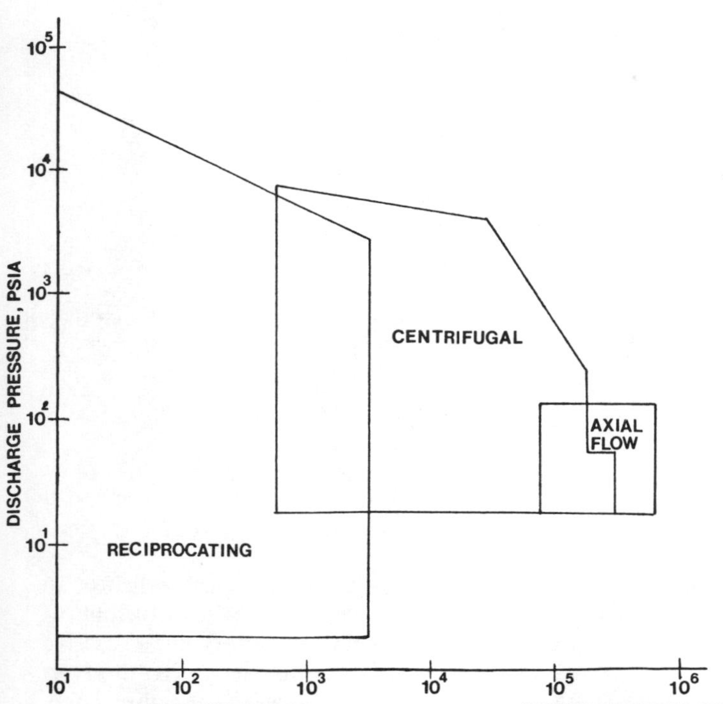

Figure 6-16. Approximate ranges of application for reciprocating, centrifugal, and axial-flow compressors [2].

Principles of Compression

The general gas law that applies to all gases can be written in several forms:

$$PV = zmRt \tag{6-6}$$

$$= \frac{zm\bar{R}t}{mw} \tag{6-7}$$

$$= zM_o\bar{R}t \tag{6-8}$$

$$Pv = zRt \tag{6-9}$$

where
- P = absolute pressure, psia
- V = volume of gas, ft^3
- z = compressibility factor for real gases ($z = 1$ for a perfect gas)
- R = $\bar{R}/mw$ = gas constant of the particular gas
- $\bar{R}$ = universal gas constant = 1,545 ft-lb_f/lb_m mole − °R
- t = absolute temperature, °R = °F + 459.7
- m = mass of gas, lb_m
- mw = molecular weight of gas
- M_o = number of moles of gas = m/mw
- v = specific volume of gas, ft^3/lb_m

A very important gas property is the specific heat ratio, k. This property is determined from the following:

$$k = \frac{C_p}{C_v} = \frac{C_p}{C_p - 1.986} \tag{6-10}$$

where
- C_v = specific heat at constant volume, Btu/lb_m-mole-°F = 4.97 Btu/lb_m-mole-°F for ideal monatomic gases
- C_p = specific heat at constant pressure, Btu/lb_m-mole-°F = 7.00 Btu/lb_m-mole-°F for most diatomic gases

Reversible Adiabatic (Isentropic) Compression

The reversible adiabatic (isentropic) compression of an ideal gas is obtained when no heat is added to, or removed from, the gas during compression. The process is reversible when no friction exists. The formulations differ for a perfect gas versus a real gas.

Perfect Gas (z = 1)

$$P_1V_1^k = P_2V_2^k \tag{6-11}$$

$$\frac{t_2}{t_1} = \left(\frac{P_2}{P_1}\right)^{\frac{k-1}{k}} \tag{6-12}$$

Real Gas ($z \neq 1$)

$$P_1V_1\gamma = P_2V_2\gamma \tag{6-13}$$

$$\frac{t_2}{t_1} = \left(\frac{P_2}{P_1}\right)^{\frac{\gamma-1}{\gamma}} \tag{6-14}$$

where, for any system of units
P = absolute pressure
V = volume or specific volume, v
k = specific heat ratio
γ = isentropic exponent for real gases, C_p/C_v
t = absolute temperature

subscripts 1 and 2 denote inlet and discharge conditions, respectively

To determine the exponent, γ, real gas properties must be used. These properties can be obtained from gas property charts and used in the following formulation:

$$\frac{\gamma - 1}{\gamma} = \frac{R}{JC_p}\left[z + t\left(\frac{\partial z}{\partial t}\right)_p\right] \tag{6-15}$$

where J = mechanical equivalent of heat = 778 ft-lb_f/Btu

$\left(\frac{\partial z}{\partial t}\right)_P$ = rate of change of compressibility factor, z, with respect to the required temperature, t, along a constant pressure, P, path

To determine a mean value of the isentropic exponent for a real gas, γ, over a compression range, Equation 6-15 must be solved by iteration.

In Equation 6-15 if we have a perfect gas in which

$$\left(\frac{\partial z}{\partial t}\right)_p = 0 \text{ and } z = 1.0$$

then Equation 6-15 becomes

$$\frac{\gamma - 1}{\gamma} = \frac{R}{JC_p} = \frac{k - 1}{k}$$

$$\gamma = k$$

For a compression ratio $P_2/P_1 < 2.0$, γ is approximately equal to k for most real gases.

For isentropic compression of an ideal gas the theoretical horsepower requirement is as follows:

$$hp_T = \frac{0.642Q}{\left(\frac{k-1}{k}\right)}\left[\left(\frac{P_d}{P_s}\right)^{\frac{k-1}{k}} - 1\right]\left(\frac{t_s}{520}\right)\bar{z} \tag{6-16}$$

where Q = gas flow rate in standard cubic feet per minute of gas (60°F, 14.7 psia)
P_s = absolute pressure at suction, psia
P_d = absolute pressure at discharge, psia
t_s = absolute temperature at suction, °R

$$\bar{z} = \frac{z_s + z_d}{z} = \text{mean compressibility factor}$$

z_s = compressibility factor at suction
z_d = compressibility factor at discharge

For a gas capacity of Q = 100 scfm, Equation 6-16 becomes

$$hp_T = \frac{6.42}{\frac{k-1}{k}}\left[\left(\frac{P_d}{P_s}\right)^{\frac{k-1}{k}} - 1\right]\left(\frac{t_s}{520}\right)\bar{z}$$

In applying these formulations that deal with the isentropic compression of an ideal gas, efficiency factors must be defined in order to apply the equations to real world compressors. These efficiencies are as follows:

η_a = adiabatic efficiency
= the isentropic horsepower, hp_T, delivered by the actual horsepower delivered to the gas, or

$$\eta_a = \frac{hp_T}{ghp} \tag{6-17}$$

η_m = mechanical efficiency
= the ratio of the actual horsepower delivered to the gas to the brake horsepower, or

$$\eta_m = \frac{ghp}{bhp} \tag{6-18}$$

η_{ao} = overall adiabatic efficiency
= the ratio of the isentropic horsepower, hp_T, for a stage of compression to the brake horsepower, or

$$\eta_{ao} = \frac{hp_T}{bhp} \tag{6-19}$$

In defining the horsepower input for a single stage of compression, utilize the overall efficiency as follows:

$$bhp = \frac{hp_T}{\eta_{ao}} = \frac{0.642Q}{\left(\frac{k-1}{k}\right)}\left[\left(\frac{P_d}{P_s}\right)^{\frac{k-1}{k}} - 1\right]\left(\frac{t_s}{520}\right)\bar{z}\left(\frac{1}{\eta_{ao}}\right) \tag{6-20}$$

For bhp at 100 scfm, Equation 6-20 becomes

$$\text{bhp} = \frac{64.2Q}{\left(\frac{k-1}{k}\right)}\left[\left(\frac{P_d}{P_s}\right)^{\frac{k-1}{k}} - 1\right]\bar{z}\left(\frac{1}{\eta_{ao}}\right)$$

The isentropic energy transmitted to the compressed gas in ft-lb_f/lb_m of gas represents the adiabatic head, or

$$H_a = \left(\frac{Rts}{mw}\right)\left(\frac{k}{k-1}\right)\left[\left(\frac{P_d}{P_s}\right)^{\frac{k-1}{k}} - 1\right](\bar{z}) \quad (6\text{-}21)$$

The compressor driver horsepower (bhp or ghp) is related to the adiabatic head by the following:

$$\text{ghp} = \frac{\dot{m}H_a}{33{,}000\eta_a} \quad (6\text{-}22)$$

$$\text{bhp} = \frac{\dot{m}H_a}{33{,}000\eta_{ao}} \quad (6\text{-}23)$$

where $\dot{m}$ = mass flow rate of the gas, lb_m/min

The adiabatic efficiency can be defined in terms of the polytropic efficiency by the following:

$$\eta_a = \frac{\left(\frac{P_d}{P_s}\right)^{\left(\frac{k-1}{k}\right)} - 1}{\left(\frac{P_d}{P_s}\right)^{\frac{1}{\eta_p}\left(\frac{k-1}{k}\right)} - 1} \quad (6\text{-}24)$$

Equation 6-24 is discussed in more detail below.

For a single stage of compression, neglecting any changes in potential and kinetic energy, the temperature change from the inlet and discharge is given by

$$\Delta t = t_d - t_s = \frac{6.33(2{,}547\text{bhp} - q)}{QC_p} \quad (6\text{-}25)$$

where q = total heat energy lost to the surroundings or to any available cooling water or cooling jackets. This value does not include thermal energy for intercoolers or aftercoolers.

For a multistage compressor, Equations 6-20 through 6-25 must be applied separately for each stage.

Polytropic Compression

This type of compression occurs when a gas is reversibly compressed along a path that is defined by the following relation:

$$PV^n = \text{constant} \quad (6\text{-}26)$$

When Equation 6-26 is expressed between the initial and final conditions we have

$$P_1V_1^n = P_2V_2^n \quad (6\text{-}27)$$

where n = the polytropic exponent, $n \neq 1$ or $n \neq k$

Expressing Equation 6-27 in terms of temperature and pressure we have

$$\frac{t_2}{t_1} = \left(\frac{P_2}{P_1}\right)^{\frac{n-1}{n}} \quad (6\text{-}28)$$

The value of n depends on whether the gas is a perfect gas (z = 1) or a real gas ($z \neq 1$) as previously discussed.

For a perfect gas the relationship between adiabatic and polytropic efficiencies is given by Equation 6-24. Similarly, the polytropic exponent, n, for a perfect gas is related to the polytropic efficiency and adiabatic exponent, k, as follows:

$$\frac{n-1}{n} = \frac{k-1}{k}\left(\frac{1}{\eta_p}\right) \quad (6\text{-}29)$$

$$\frac{k-1}{k} = \frac{R}{JC_p} \quad (6\text{-}30)$$

since $k = C_p/C_v$

The relationship between the polytropic efficiency and adiabatic (isentropic) efficiency of a perfect gas is shown in Figure 6-17. The polytropic efficiency, η_p, is usually determined by the compressor manufacturer using either an old design or testing a new design.

The polytropic exponent, n, for a real gas is determined from real gas properties or with using real gas data and using the following expression:

$$\frac{n-1}{n} = \frac{R}{JC_p}\left[\frac{z}{\eta_p} + t\left(\frac{\partial z}{\partial t}\right)\right] \quad (6\text{-}31)$$

Equation 6-31 is identical to Equation 6-15 except that the isentropic exponent for a real gas, γ, is replaced by the polytropic exponent, n, and the compressibility factor for real gases, z, is divided by the polytropic efficiency, η_p.

Similarly to Equation 6-15, Equation 6-31 must be solved by iteration for a mean value of the polytropic exponent, n, over a compression range.

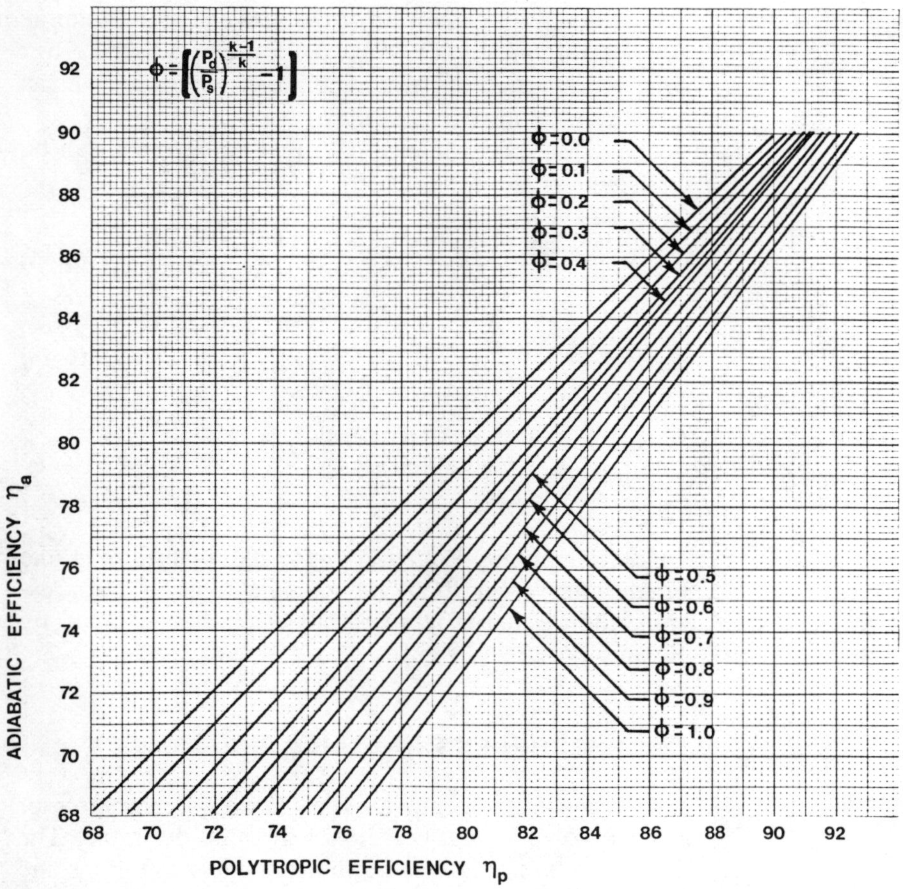

Figure 6-17. The relationship between the polytropic efficiency and the adiabatic efficiency for a perfect gas (Z = 1).

In Equation 6-31, if we have

$$\left(\frac{z}{t}\right)_p = 0 \text{ and } z = 1 \text{ for a perfect gas}$$

then,

$$\frac{n-1}{n} = \frac{R}{JC_p\eta_p} = \frac{k-1}{k\left(\frac{1}{\eta_p}\right)} \tag{6-32}$$

For most real gases below a compression ratio of approximately 2, then

$$\frac{n-1}{n} \approx \frac{k-1}{k\left(\frac{1}{\eta}\right)}$$

The basic horsepower and head expressions for polytropic compression are similar to those for isothermal compression, Equation 6-20. Thus, we have

$$ghp = \frac{0.0642Q}{\left(\frac{1}{\eta_p}\right)\left(\frac{k-1}{k}\right)}\left[\left(\frac{P_d}{P_s}\right)^{\left(\frac{1}{\eta_p}\right)\left(\frac{k-1}{k}\right)} - 1\right]\left(\frac{t_s}{520}\right)(\bar{z})\left(\frac{1}{\eta_p}\right) \tag{6-33}$$

For ghp at 100 scfm,

$$ghp = \frac{64.2Q}{\left(\frac{1}{\eta_p}\right)\left(\frac{k-1}{k}\right)}\left[\left(\frac{P_d}{P_s}\right)^{\left(\frac{1}{\eta_p}\right)\left(\frac{k-1}{k}\right)} - 1\right]\left(\frac{t_s}{520}\right)(\bar{z})\left(\frac{1}{\eta_p}\right) \tag{6-34}$$

The equations for polytropic head are similar to those for adiabatic head, Equation 6-21. Thus,

$$H = \left(\frac{Rts}{mw}\right)\left(\frac{\eta_p K}{k-1}\right)\left[\left(\frac{P_d}{P_s}\right)^{\left(\frac{1}{\eta_p}\right)\left(\frac{k-1}{k}\right)} - 1\right](\bar{z}) \tag{6-35}$$

If the polytropic head is known, the compressor horsepower (ghp or bhp) can be obtained from the following:

$$bhp = \frac{\dot{m}H}{33{,}000\eta_{po}} \tag{6-36}$$

$$ghp = \frac{\dot{m}H}{33{,}000\eta_{p}} \tag{6-37}$$

where η_{po} = overall polytropic efficiency
$= \eta_p\eta_m$

The outlet and inlet temperatures for polytropic compression are related by the following expression:

$$\frac{t_d}{t_s} = \left(\frac{P_d}{P_s}\right)^{\left(\frac{1}{\eta_p}\right)\left(\frac{k-1}{k}\right)} \tag{6-38}$$

Equations 6-35 through 6-38 are used separately for each stage of a multistage compressor. Equations 6-38 and 6-39 can be used to calculate the polytropic efficiency directly (provided t_s, t_d, P_s, P_d and k are known values):

$$y' = \frac{1}{\eta_p}\left(\frac{k-1}{k}\right) = \frac{y}{\eta_p} \tag{6-39}$$

$$\text{where } y = \frac{k-1}{k} \tag{6-40}$$

Normally, the value of η_p is estimated from data supplied by the manufacturer. For initial or preliminary values of the polytropic efficiency, η_p, Figure 6-17 may be used.

Isothermal Compression

This compression occurs when the temperature of the gas being compressed remains constant during compression. For a perfect gas in which z = 1.0 and $(\partial z/\partial t)p = 0$ we have

$$P_1V_1 = P_2V_2, \tag{6-41}$$

or

$$PV = \text{constant} \tag{6-42}$$

The theoretical horsepower developed during a reversible isothermal compression process is

$$hp_t = \frac{Qt\bar{z}}{8{,}110}\ln\left(\frac{P_d}{P_s}\right) \tag{6-43}$$

Equation 6-43 assumes that the heat of compression is fully removed by cooling. In practice this is not achieved, because the heat of compression causes the gas to exceed the inlet temperature.

The actual performance of a real compressor can be evaluated by the following:

$$bhp = \frac{hp_T}{\eta_t\eta_m}\frac{hp_T}{\eta_{to}} \tag{6-44}$$

$$ghp = \frac{hp_T}{\eta_t} \tag{6-45}$$

where η_t = isothermal efficiency
η_{to} = overall efficiency
$= \eta_t\eta_m$

After applying Equation 6-44 and determining the brake horsepower (bhp) for a single stage of compression, the discharge temperature can be determined by Equation 6-25.

Dimensionless Reference Numbers

In sizing and selecting the type of pump or compressor to be used, a logical correlation is often desirable. The following dimensionless parameters apply to pumps and compressors and are the specific speed and specific diameter, as defined as follows:

$$N_s = \frac{N(Q)^{0.5}}{H^{0.75}} \tag{6-46}$$

where N_s = specific speed, dimensionless
N = speed, rpm
Q = capacity of flow rate, ft^3/sec
H = head, $ft\text{-}lb_f/lb_m$

$$D_s = \frac{D(H)^{0.25}}{Q^{0.5}} \tag{6-47}$$

where D_s = specific diameter, dimensionless
D = diameter of impeller or rotor, ft
H = head, $ft\text{-}lb_f/lb_m$

Figure 6-18 shows the dimensionless parameters as originally presented by Balje [3]. This figure is the graphical combination of Equations 6-46 and 6-47. Past experience often dictates what type of pump or compressor is to be used and in cases of uncertainty or new applications, this figure will be useful in equipment selection.

Figure 6-18 must be applied to each stage separately, as each impeller or stage must be chosen with each separate inlet capacity or head for that stage.

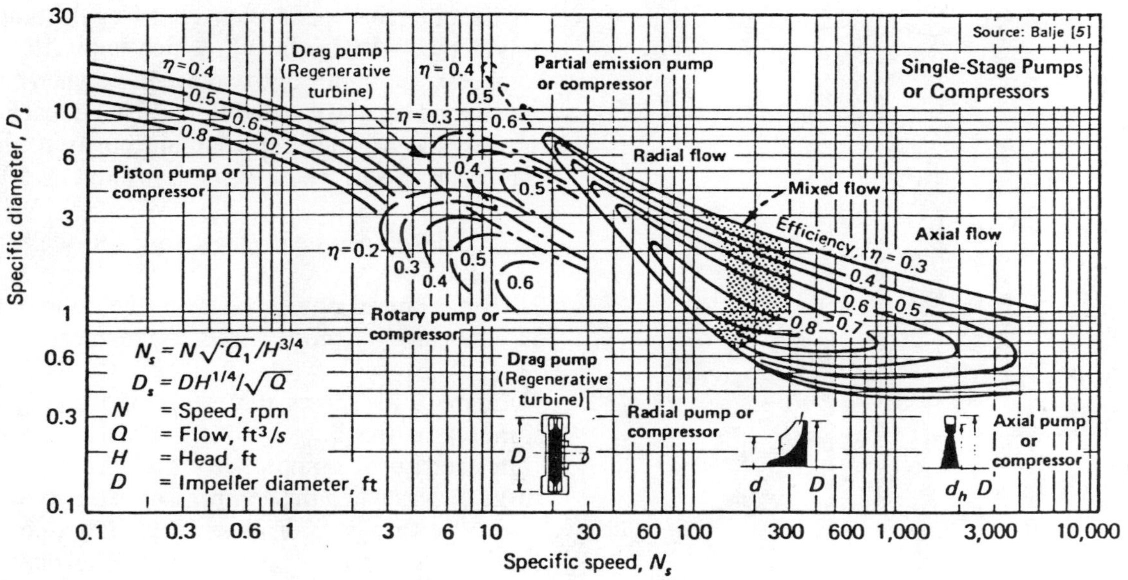

Figure 6-18. The initial selection of a single-stage compressor is made using the specific speed and specific diameter parameters [3].

Centrifugal Compressors

The centrifugal compressor powered the first turbojet-powered aircraft and is still used today in jet engines as a supercharger. The main advantage of the centrifugal compressor is that it produces a large pressure ratio for a single stage of compression, and is easily manufactured. Its advantages over the reciprocating design were cited previously.

Most centrifugal compressors are designed so that the gas enters the impeller axially—parallel to the rotating shaft—as shown in Figure 6-19. The gas flow is then changed to the radial direction and is accelerated in a peripheral direction as it moves along the impeller. As the gas exits the impeller, it enters a stationary diffuser where the gas velocity is reduced. This process is repeated at each stage on multistage compressors. Most of the pressure increase in the gas occurs in the impeller and the greatest pressure drop occurs in the diffuser. In multistage compressors, cooling the gas between stages is quite common and many such compressors have water-cooled separators or diaphragms.

The polytropic relations, Equations 6-26 through 6-40, are usually preferred for centrifugal compressor calculations. Figure 6-20 shows why with a schematic plot of the centrifugal compression process on a temperature-entropy graph. Using the adiabatic (isentropic) process, the actual discharge temperature is underestimated

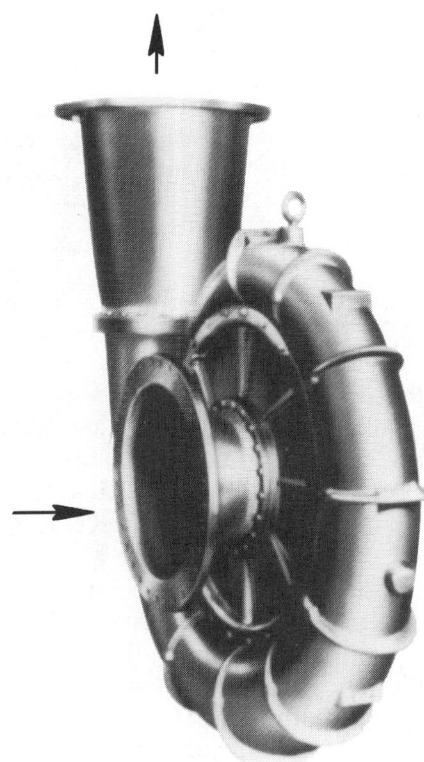

Figure 6-19A. Centrifugal compressor—single-stage. (Courtesy of Dresser Industries, Inc., Roots Blower Operation.)

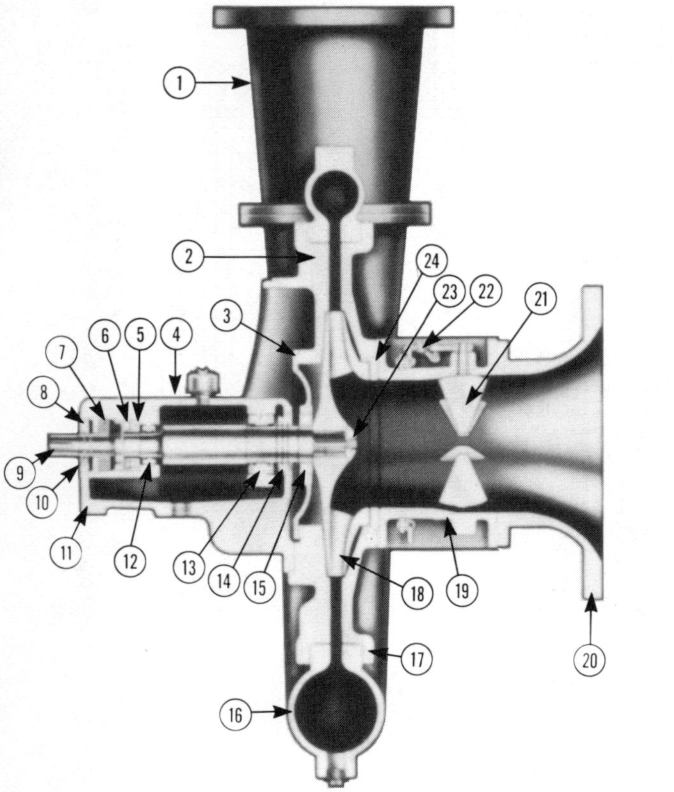

1. Discharge Nozzle	9. Shaft	17. Inlet Section
2. Casing Cover	10. Oil Retainer	18. Impeller
3. Sub-Cover Section	11. Bearing Stand	19. Guide Vane Housing
4. Bearing Stand Cap	12. Coupling End Bearing	20. Inlet Nozzle
5. Steel Shim	13. Impeller End Bearing	21. Guide Vane
6. Thrust Bearing	14. Oil Retainer	22. Guide Vane Linkage
7. Thrust Housing Seal	15. Seal	23. Impeller End Plate
8. Spacing Ring	16. Volute	24. Inlet Wearing Ring

Figure 6-19B. Cross-section of a single-stage centrifugal compressor. (Courtesy of Dresser Industries, Inc., Roots Blower Operation.)

(ideal). Since the polytropic compression process, by definition, is the path connecting the inlet and actual discharge conditions, the polytropic formulations are preferred by compressor manufacturers. This factor becomes extremely important in sizing intercoolers, since using the adiabatic discharge temperature would result in undersizing the cooler. The larger the compression ratio of the machine, the more severe the mistake of undersizing the cooler becomes.

Gas inlet conditions can change and when they do they affect a centrifugal compressor differently from a positive-displacement compressor, such as a reciprocating machine. Table 6-1 lists the effects of changing inlet parameters on a centifugal compressor operating at a constant volumetric flow rate and a constant speed.

Changing the speed of a centrifugal compressor involves the "affinity laws," which apply to single-stage compressors, multistage compressors when each stage is considered separately, and to multistage machines over a narrow speed range representing no more than a 15% change in speed. These laws are stated as follows:

1. The developed head (feet) varies to the square of the speed.
2. The required power varies to the cube of the speed.
3. The capacity (cfm) varies to the speed.

Figure 6-21 shows the effect of varying centrifugal compressor speed.

In centrifugal compressors a phenomenon known as surge occurs when the compressor capacity is lower than a specific flow rate. This specific flow rate is shown in Figure 6-22 as the "surge limit." The phenomenon of surging is manifested by cyclic vibration of gas flow, which can even result in reversal of flow direction, power requirement, and discharge pressure. The phenomenon normally is associated with excess noise and

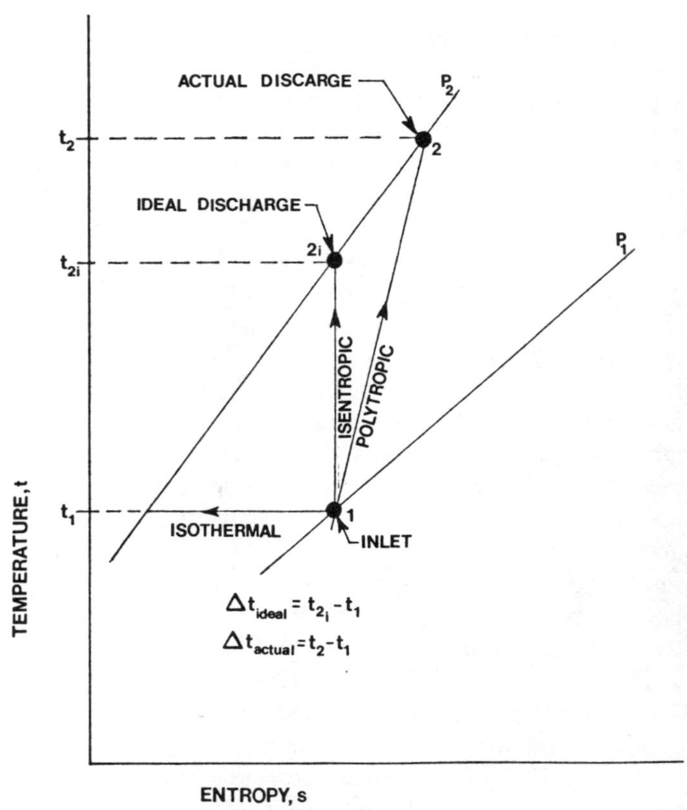

Figure 6-20. Centrifugal compression process.

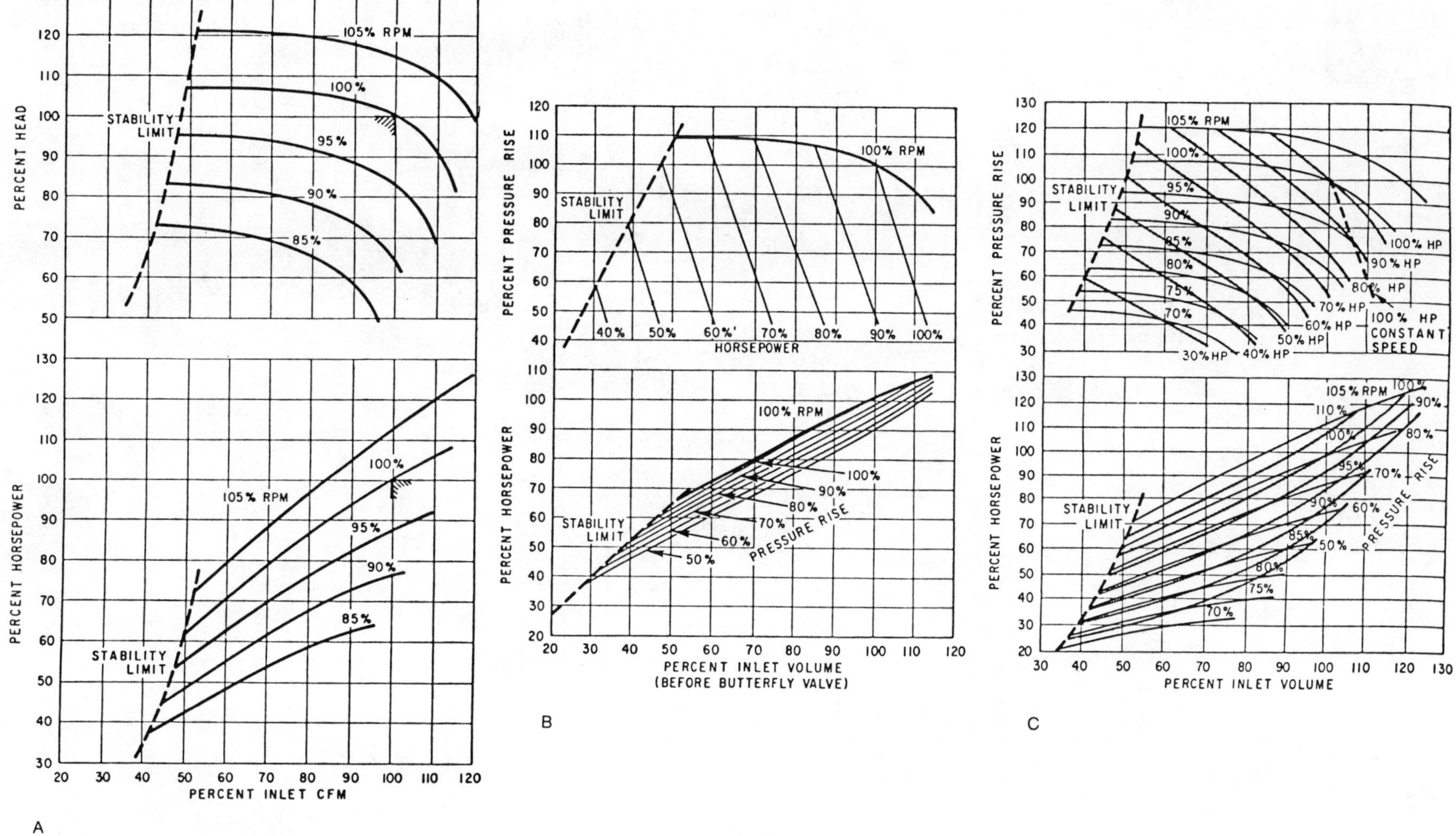

Figure 6-21. (A) Typical centrifugal compressor performance curves; (B) typical centrifugal compressor constant-speed performance curves; (C) typical centrifugal compressor variable-speed performance curves [4].

Table 6-1
Effects of Varying Various Inlet Parameters on a Centrifugal Compressor

	Increasing Inlet Temperature	Increasing Inlet Pressure	Increasing Molecular Weight of Gas	Increasing value of Polytropic n or Adiabatic k Exponent
Pressure Differential	Decreases	Increases	Increases	Decreases
Compression Ratio	Decreases	Constant	Increases	Decreases
Inlet Density	Decreases	Increases	Increases	Constant
Discharge Pressure	Decreases	Increases	Increases	Decreases
Discharge Temperature	Decreases	Constant	Increases	Increases
Power Required	Decreases	Increases	Increases	Constant
Head Developed	Constant	Constant	Constant	Constant
Mass Flow Rate	Decreases	Increases	Increases	Constant

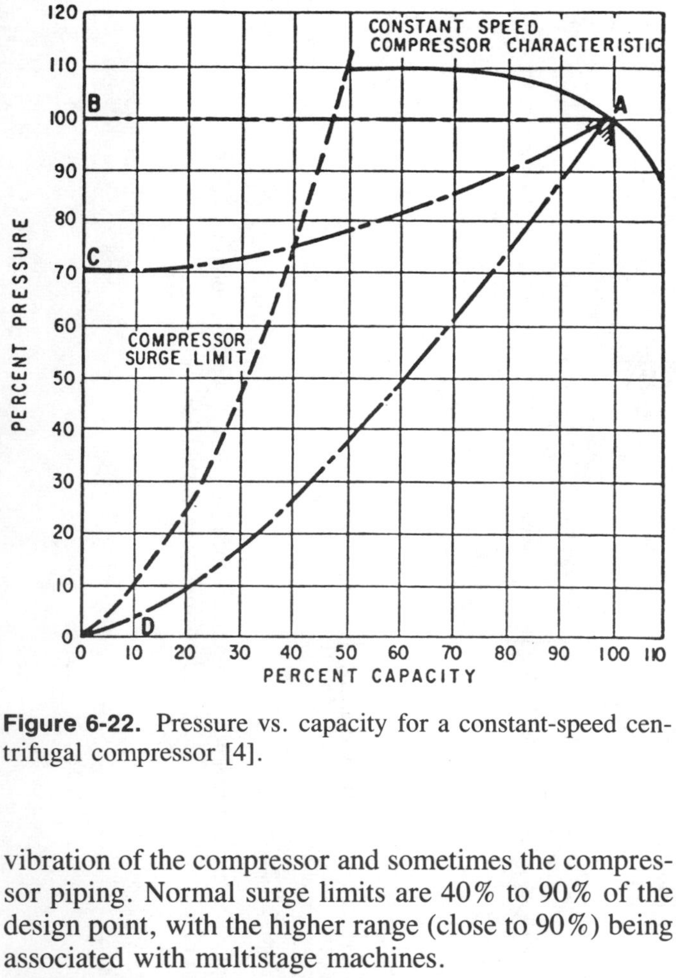

Figure 6-22. Pressure vs. capacity for a constant-speed centrifugal compressor [4].

vibration of the compressor and sometimes the compressor piping. Normal surge limits are 40% to 90% of the design point, with the higher range (close to 90%) being associated with multistage machines.

Controlling surge in centrifugal compressors is more difficult than in centrifugal pumps, but the following factors ease the problem considerably:

1. Throttling at the discharge flange.
2. Throttling at the inlet flange, which is usually more efficient than throttling at the discharge flange.
3. Using a variable speed driver, usually accomplished by the turbine driver.
4. Bypassing or blowing off excess gas to avoid surge.

These steps will help in alleviating surge problems, but if a variable rate operation is required, the compressor manufacturer should be consulted.

Anti-surge devices can be incorporated into compressor systems. For nontoxic or inexpensive gases the compressor discharge can be vented to the atmosphere as shown in Figure 6-23. For expensive or toxic gases an automatic anti-surge system can be installed as shown in Figure 6-24. In this type of arrangement a heat exchanger is placed in the system to remove the heat of compression from the vented discharge gas to prevent a loss of compressor performance caused by the temperature rise above the design value at the inlet.

Compressor manufacturers use standard cubic (scfm) feet to specify compressor performance, just as pump manufacturers use water to determine pump performance. The manner in which scfm and altitude correction is handled is discussed later.

Impellers are critical in the selection of centrifugal compressors. The three basic types of impellers for centrifugal compressors are shown in Figure 6-25. The conventional closed impeller shown in Figure 6-25 is used for adiabatic heads up to approximately 12,000 ft-lb_f/lb_m. The open, radial-bladed impeller shown in Figure 6-25 develops more head with the same impeller diameter and shaft speed. The open inducer impeller can produce heads up to 20,000 ft-lb_f/lb_m. Whenever the head requirement becomes too great for a single impeller, then one must think in terms of multistage compressors. Each stage of compression of a multistage compressor is treated as a single stage compressor and the same formulations hold.

Reciprocating Compressors

These compressors normally are sized according to the adiabatic expressions of Equations 6-11 through 6-25. Normal practice in calculations for reciprocating compressors is to use the adiabatic exponent, $k = C_p/C_v$, then adjust the results according to the specific compressor design and configuration. The parameters that affect the compressor horsepower, cylinder capacity, and discharge temperature are length of stroke, shaft rotation speed, cooling efficiency, and fixed clearance of cylinders. All of these parameters vary for each given application, but have the same basic cylinder design and cycle. Figure 6-26 shows the reciprocating compressor cycle. This cycle involves this displacement of gas, hence the classification of a reciprocating compressor as a positive displacement type of unit. The compressor is unable to exhaust all gas from the cylinders and the residual gas remaining in the compressor at discharge conditions expands to inlet conditions. This phenomenon is shown in Figure 6-27.

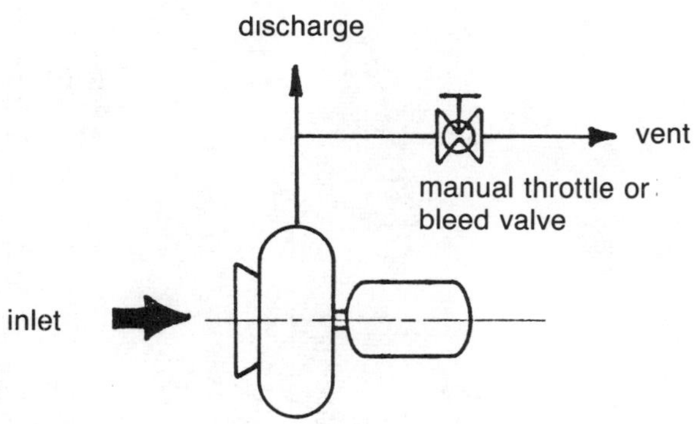

Figure 6-23. Manual surge control system for centrifugal compressor.

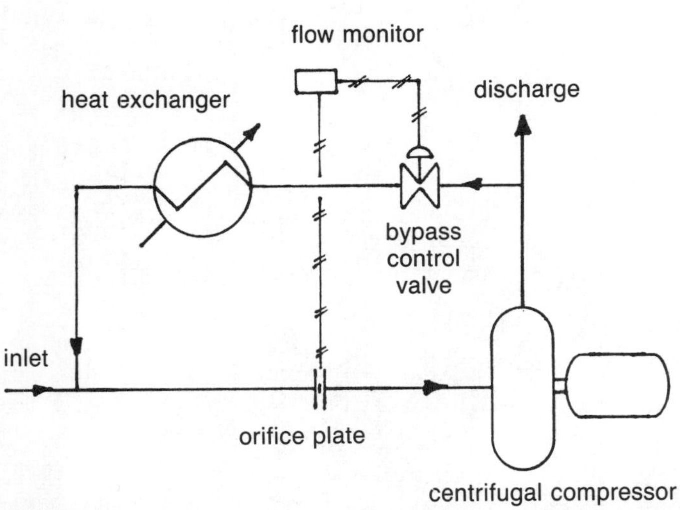

Figure 6-24. Automatic surge control with recirculating bypass.

The clearance volume is usually set by the compressor manufacturer and is specified to match the specified capacity with the standard size compressor unit. Power consumption is not affected by the clearance volume or the volumetric efficiency.

The use of "clearance pockets" is used in some compressors to vary the volumetric efficiency. These clearance pockets can be sized to affect the capacity of the compressor, as in Figure 6-28. Power consumption at reduced flow rates is minimized by use of capacity control. The use of a clearance pocket (additional clearance volume) reduces the volumetric efficiency of the compressor, because the re-expanding gas fills most of the cylinder, and the suction valve opens further in the stroke. This mechanism is economical, because the energy expended in gas compression is retrieved in expansion. The clearance pocket is separated from the cylinder by a stop valve. Figure 6-28 shows how varying the cylinder clearance affects the numeric value of the volumetric efficiency at constant compression ratio. The volumetric efficiency for a reciprocating compressor is given by:

$$\eta_v = \frac{\text{actual capacity}}{\text{piston displacement}} \quad (6\text{-}48)$$

The parameters that affect the volumetric efficiency are as follows:

1. The ratio of a relative clearance volume, $\bar{C}$, which is the ratio of clearance to theoretical displacement expressed as percent.
2. The compression ratio, C_r, of discharge to inlet pressure.
3. The various exponents of the polytropic curve of re-expansion. Such a curve is shown in Figure 6-29. Here the cylinder is normally cooled by a water jacket or surrounding air. The small volume of gas that remains in the clearance volume expands and contracts with a cooling surface. Consequently, the re-expansion curve (curve 3-4) is initially steeper than the adiabatic curve (curve 1-2). With continuing expansion of the gas, the gas temperature falls below that of the piston and walls, and heat is transferred from these surfaces to the gas. Thus, the exponent of the re-expansion curve (curve 3-4) is variable. For re-expansion of lower compression ratios, Chlumsky [5]

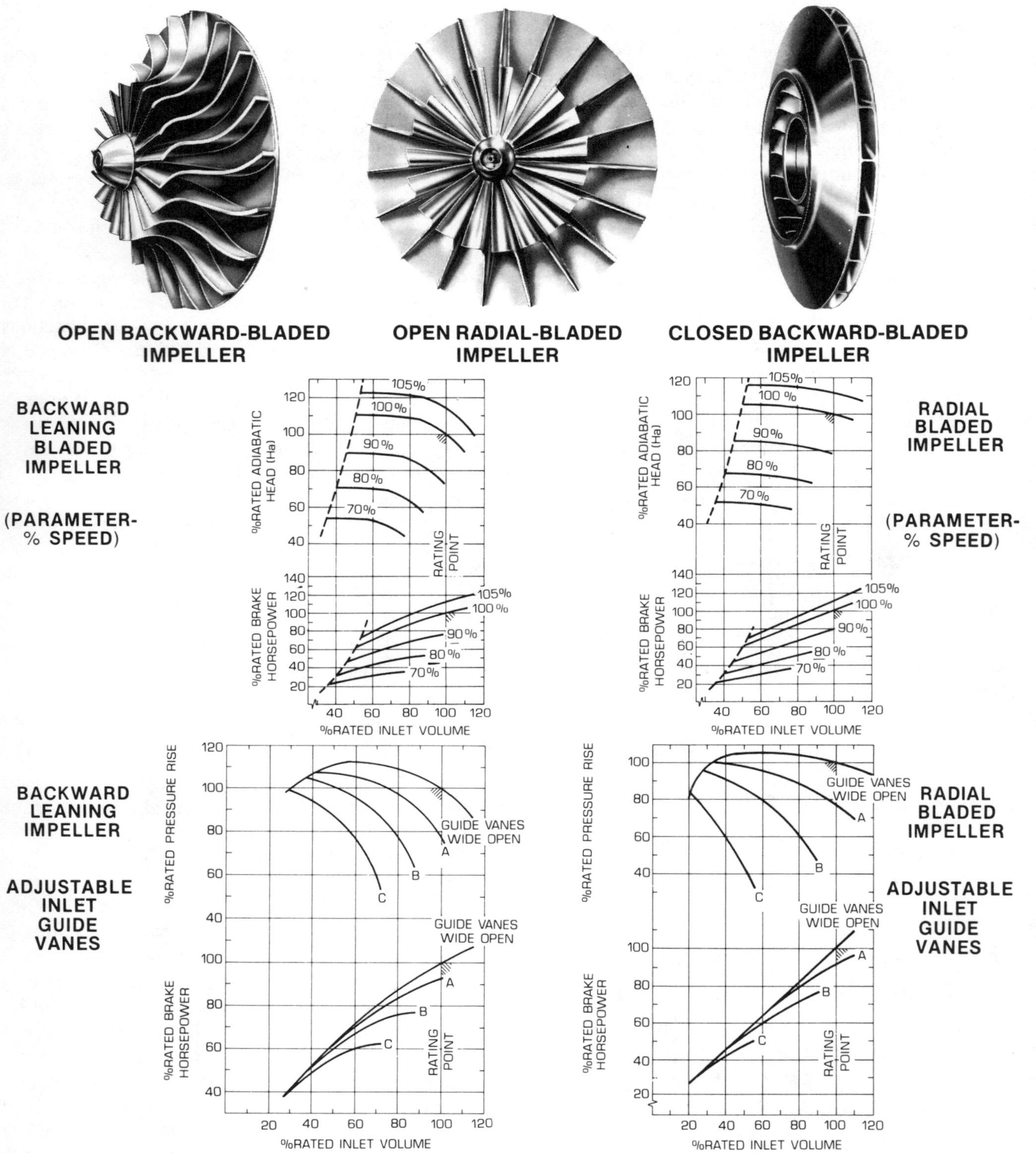

Figure 6-25. Basic types of impellers for centrifugal compressors. (Courtesy of Dresser Industries, Inc., Roots Blower Operation.)

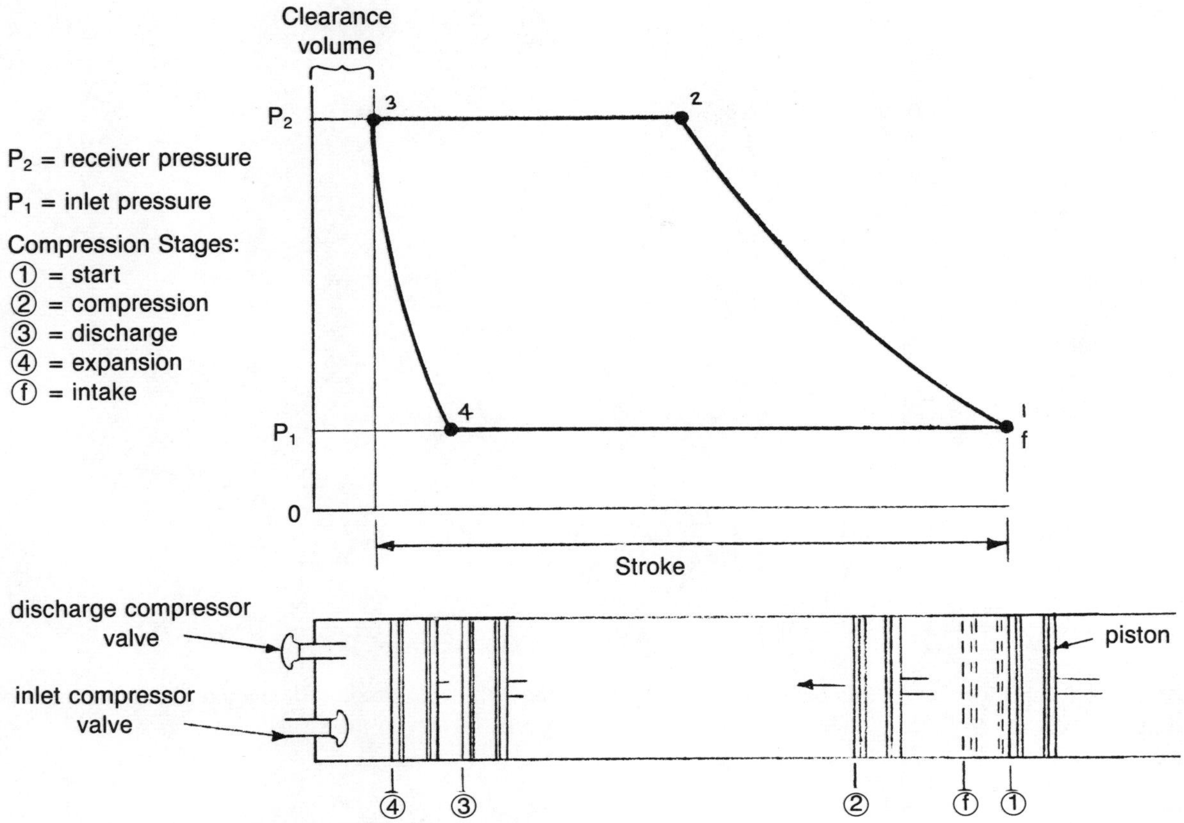

Figure 6-26. Reciprocating compressor cycle.

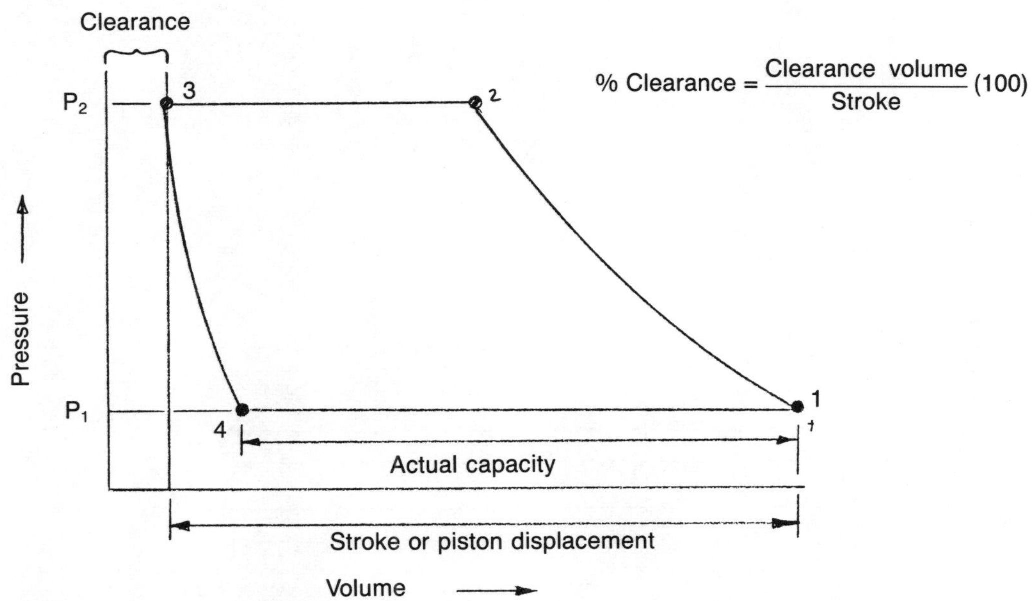

Figure 6-27. The effect of clearance capacity.

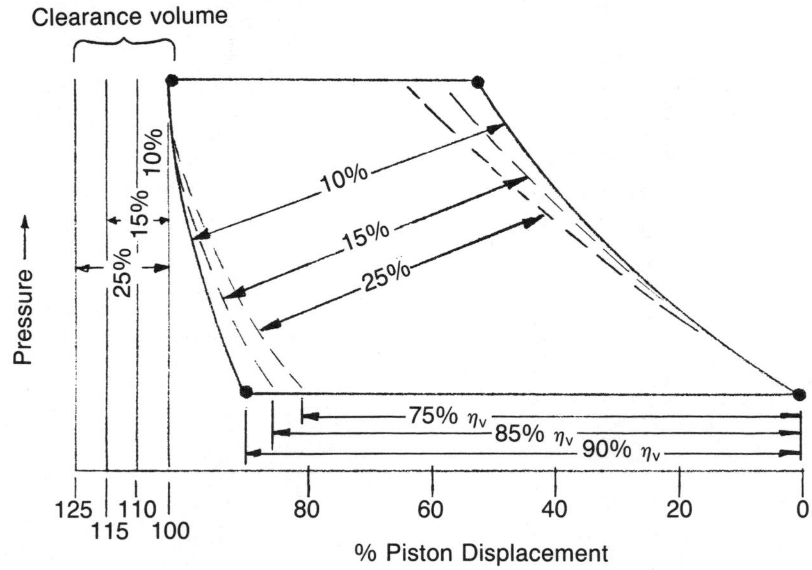

Figure 6-28. A clearance pocket (additional clearance volume) reduces the volumetric efficiency of the compressor because the re-expanding gas fills most of the cylinder, and the suction valve opens further in the stroke.

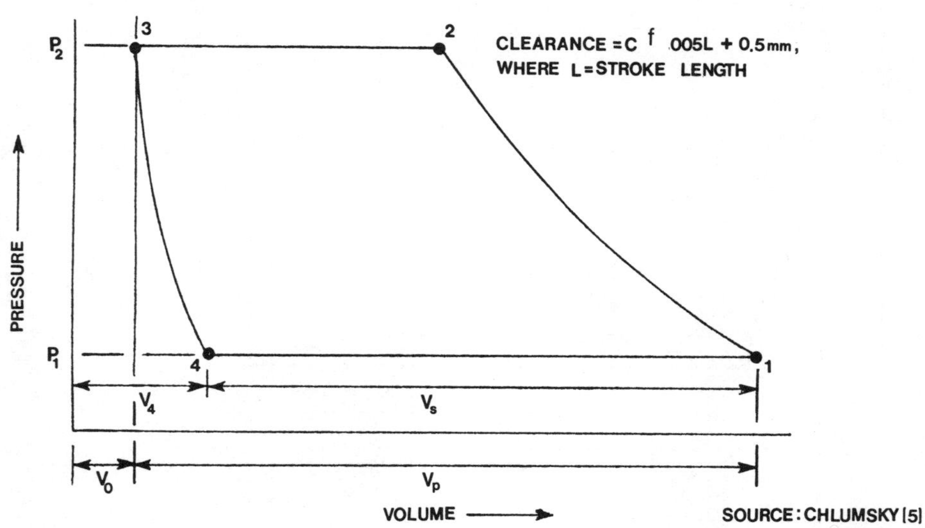

Figure 6-29. A pressure-volume diagram of a compresor with clearance (zero flow resistances) [5].

recommends for compression ratios of approximately 2; the re-expansion may be approximated as an adiabatic process. For the volume, V_4—the volume to which the gas expands during the pressure drop from P_2 to P_1—we have the expression

$$V_4 = V_o \left(\frac{P_2}{P_1}\right)^{\frac{1}{k}} \qquad (6\text{-}49)$$

Substituting Equation 6-49 into the expression for volumetric efficiency, we have

$$\eta_v = \frac{V_p + V_o - V_4}{V_p} = \frac{V_p + \epsilon V_p - \epsilon V_p \left(\frac{P_2}{P_1}\right)^{\frac{1}{k}}}{V_p}$$

or

$$\eta_v = 1 - \epsilon \left[\left(\frac{P_2}{P_1}\right)^{\frac{1}{k}} - 1\right] \qquad (6\text{-}50)$$

where $\epsilon = \frac{V_o}{V_p}$ = ratio of the clearance volume, V_o, to the volume swept by the piston stroke, V_p

$\eta_v = \frac{V_s}{V_p}$ = expression for volumetric efficiency, Equation 6-48, the ratio of gas volume pumped to the volume swept by the piston (compressor displacement)

Figure 6-30 shows the graphical solutions of Equation 6-50 for various compression ratios and exponents of the polytropic curve of re-expansion and clearance values.

For compression ratios of 4 and higher, the re-expansion cannot be considered as an adiabatic process. For these compression ratios the polytropic exponent m (where m denotes the difference between the re-expansion PV^m (constant) and the compression PV^n (constant). For diatomic gases, m = 1.25.

The value of the polytropic curve exponent, m, varies with pressure. Chlumsky [5] recommends for a compression ratio of 3:4 the following values of m be used:

First stage	m = 1.20
Second stage	m = 1.25
Third stage	m = 1.30
Fourth stage	m = 1.35
Fifth and further stages	m = k

These values are given at different pressure levels, as exist in multistage compressors with the suction of the first stage at atmospheric pressure.

The volumetric efficiency for a perfect gas (z = 1), not realistic, is given by

$$\eta_{v_T} = 100 - \bar{C}(C_R^{1/k} - 1) \qquad (6\text{-}51)$$

where η_{v_T} = theoretical volumetric efficiency

The volumetric efficiency for a perfect gas (z = 1) with realistic effects,

$$\eta_v = 100 - C_R - \bar{C}(C_R^{1/k} - 1) \qquad (6\text{-}52)$$

C_R = compression ratio = P_2/P_1

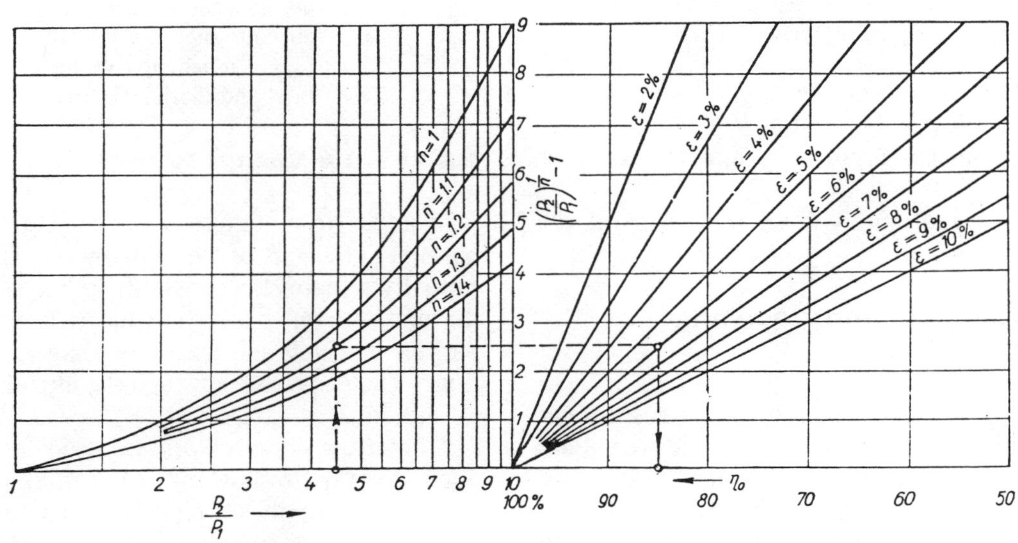

Figure 6-30. Curves for determining volumetric efficiency [5].

The difference between Equation 6-52 and Equation 6-51 is that the theoretical volumetric efficiency should be reduced by a value equal to the compression ratio to obtain an actual value for a perfect gas. This is a factor that has been determined from field experience.

For a real gas ($z \neq 1$) with realistic effects,

$$\eta_V = 100 - C_R - \bar{C}\frac{z_1}{z_2}(C_R)^{1/k} - 1 \tag{6-53}$$

where

z_1, z_2 = inlet and discharge compressibility factors, respectively

As stated previously, reciprocating compressors follow the expressions for an adiabatic process. The work required for the adiabatic compression of a perfect gas ($z = 1$) is found by the following expression:

$$W = P_1V_1\left(\frac{k}{k-1}\right)\left[\left(\frac{P_2}{P_1}\right)^{\frac{k-1}{k}} - 1\right] \tag{6-54}$$

The theoretical horsepower may be found by Equation 6-16 or by the following:

$$hp_t = \frac{(P_1V_1)144}{33{,}000}\,\frac{k}{k-1}\left[\left(\frac{P_2}{P_1}\right)^{\frac{k-1}{k}} - 1\right]\left(\frac{z_1 + z_2}{2z_1}\right) \tag{6-55}$$

For an ideal gas, $z_1 = z_2$

where P_1, P_2 = inlet and discharge pressures, respectively, psia

V_1, V_2 = inlet and discharge gas flow rates, respectively, acfm

In Equation 6-55, the theoretical horsepower may be varied by the following parameters:

1. Increasing the compression ratio, C_R
2. Increasing the specific heat ratio, k
3. Increasing the inlet pressure at a constant compression rate.
4. Increasing the actual inlet volume (*not* standard volume).

Multiple Staging of Reciprocating Compressors

Multiple staging is the compression of a gas from one pressure to another involving more than one step. Each step acts in series with the others and entails a basic machine element. In multiple staging of reciprocating compressors, increasing the cylinder size is less expensive than increasing the number of cylinders, thus the tendency has been to increase the cylinder size using a smaller number of cylinders. Multistage reciprocating compressors have the following advantages:

1. Operating at high speeds, they can be coupled directly at high shaft speeds thus utilizing cheap electric motors.
2. Better balance of inertia forces.
3. The mass of the flywheel, which rotates at high speeds, can be made smaller, resulting in a smaller fluctuation of torque. The more cylinders, the less the fluctuation of torque.
4. Starting multistage compressors is easier because they have small moving masses and thus can be driven by electric motors with less inertia torque and lighter construction.
5. Variations of pressure and flow velocity in the intercooler or oil separator are less, thus making these parts smaller.
6. Machines of various capacities can be manufactured using identical parts, making interchangeability efficient.
7. Multistage compressors are better suited to automatic operation.

Gas Temperature for Reciprocating Compressors

The discharge temperature of a positive displacement compressor, a class of which the reciprocating is included, can be predicted by the following expression:

$$\frac{t_2}{t_1} = \left(\frac{P_2}{P_1}\right)^{\frac{k-1}{k}} \tag{6-56}$$

where t = absolute temperature for any system

P = absolute pressure for any system

k = C_p/C_v, adiabatic exponent

1, 2 = inlet and discharge conditions, respectively

Axial Flow Compressors

In axial flow compressors, the flow enters the unit parallel to the axis of the shaft and the flow direction essentially remains unchanged from the inlet to the exit of the unit. Airfoil blades are located on the rotor shaft, varying in pitch and size according to the flow conditions. The gas passes through the airfoil blades in an axial direction.

Axial flow compressors are used for applications of about 25,000 cfm upward. The formulas for centrifugal compressors apply to axial flow machines. Axial flow compressors can handle greater capacities, which is the primary reason why they have replaced centrifugal compressors in aircraft gas turbine units. The characteristic curve (head versus flow) for an axial flow compressor is much steeper than for a centrifugal compressor and the surge limit is a function of design capacity. Contrary to a centrifugal compressor, the required horsepower for an axial flow compressor at constant speed and pressure decreases with increased flow. Axial flow compressors are not as common in the process industries as centrifugal or reciprocating types of machines.

Fans and Blowers

Fans and blowers are basically compressors. They fall under two types of compressors—centrifugal and axial flow. If one understands the basics of centrifugal or axial flow compressors, fans and blowers come easy, for they are less complicated than compressors.

Specifying Compressor Flow Conditions

Specifying compressor flow conditions is a major source of confusion in applying compressors to process systems. There are three basic ways to specify compressor flow conditions:

1. Mass flow—define the mass flow rate of the gas, $lb_m/$ in the English system and kg/hr-m in the SI/metric.
2. Actual, or inlet, volume flow—volumetric flow rate of the gas at the inlet conditions, expressed as acfm or icfm in the English system and m^3/hr in the SI and MKGFS systems.
3. Standard volumetric flow—the volumetric flow rate of the gas at the inlet conditions expressed in terms of standard cubic feet of gas per minute (scfm) or millions of standard cubic feet of gas per day (MMscfd) in the English system and m^3/hr in the SI and MKGFS systems.

Mass Flow

The method of defining the mass flow rate of the gas in terms of the inlet conditions of the compressor is favored by many and is mandatory in calculating gas properties between stages. Mass flow rate *must* be specified as either dry gas or wet gas. If the gas, for example, contains water vapor, this could drastically change the compressor design. One of the problems of using mass flow is not specifying the flow conditions as a dry gas, which in reality is a two-phase or multiphase flow.

Another disadvantage to using mass flow is that it does not allow one to appreciate the physical size of the system. An intuitive feel for any system is essential to its successful design.

Actual or Inlet Volumetric Flow

Actual flow rate conditions at the inlet to the compressor is denoted as acfm or icfm—acfm meaning actual cubic feet per minute and icfm meaning inlet cubic feet per minute.

The disadvantage to specifying acfm is in the internal components of the compressor, e.g., a side-load refrigeration compressor, or in a multistage compressor. In a multistage compressor the previous stage's discharge temperature is a function of the previous stage's compression efficiency, and mass flow rates are better for such conditions.

Acfm is best for plotting compressor performance curves, because the impeller is sensitive only to the actual volumetric flow and is insensitive to the gas state conditions.

Mass flow and acfm volumetric flow should be used because mass flow is invaluable in communicating with the compressor manufacturer and in dealing with internal machine flow conditions, and acfm is essential in getting a feel for the physical size of the system. The use of mass flow and acfm should counter the disadvantages of both approaches.

In computing pressure drop through connecting piping systems to compressors, it is imperative that acfm be used to avoid any confusion in designing the piping systems.

Standard Volumetric Flow

Specifying gas conditions in terms of standard volumetric flow is done extensively throughout industry. The gas flow conditions are based on standard inlet conditions—pressure, molecular weight, temperature, and compressibility—all based on "standard" conditions. Thus, the standard specific volume is constant being that

$$V_{std} = \frac{z_{std} R\, t_{std}}{P_{std}} = \text{constant} \tag{6-57}$$

where z_{std} = compressibility factor at standard conditions
R = universal gas constant, which is a function of the molecular weight of the gas
t_{std} = temperature at standard conditions
P_{std} = pressure at standard conditions

Volume flow is expressed as

$$Q_{std} = mV_{std} \tag{6-58}$$

where the standard volumetric flow is directly proportional to the mass flow rate.

As with using mass flow, when using standard flow conditions one cannot appreciate the physical size of the system. And worse still, using scfm does not provide any of the advantages of using either mass flow or acfm. To specify something as "standard" one thing is essential, that all parties agree on what is "standard." Unfortunately, this is not the case with using scfm, as the following "standards" cited by Lapina [6] indicate:

English system	Metric system
1. $P_{std} = 14.7$ psia $t_{std} = 60°F$	1. $P_{std} = 101.3$ kPa $t_{std} = 0°C$
2. $P_{std} = 14.7$ psia $t_{std} = 70°F$	2. $P_{std} = 101.3$ kPa $t_{std} = 15°C$
3. $P_{std} = 14.7$ psia $t_{std} = 32°F$	

Thus, what is considered "standard," as Lapina [6] writes, varies from industry to industry and engineer to engineer. In the net result what is often gained is confusion.

Properly Specifying Compressor Flow Conditions

To properly size or select a compressor, the capacity—no matter how it is given—must be converted to the inlet conditions. To do this the following expressions are used:

$$\frac{P_1V_1}{t_1z_1} = \frac{P_2V_2}{t_2z_2} \tag{6-59}$$

where V = volume
P = absolute pressure
t = absolute temperature
z = compressibility factor

In Equation 6-59, if z = 1.0 for a perfect gas, and P and t are at standard conditions, then

$$\text{acfm} = Q_s = \dot{m}V = \frac{\dot{m}}{\rho} \tag{6-60}$$

where $\dot{m}$ = mass flow rate, lb_m/min
V = specific volume, ft^3/lb_m
ρ = density, lb_m/ft^3

The specific volume, V, may be determined by

$$V = z\left(\frac{1545}{mw}\right)\left(\frac{t}{144P}\right) \tag{6-61}$$

where, as before, mw = molecular weight

$$\text{scfm} = \frac{(379.46)mh}{60} \tag{6-62}$$

where mh = moles/hour

and

$$\dot{m} = (mh)(mw) \tag{6-63}$$

and finally,

$$\text{acfm} = Q_s = \left[\frac{(\text{MMscfd})(10^6)}{(60)(924)}\right]\left(\frac{14.96}{P_s}\right)\left(\frac{460 + t_{std}}{460 + t_{std}}\right)\left(\frac{z_{std}}{z_{std}}\right) \tag{6-64}$$

where the subscript, s, denotes properties at the inlet (or suction) conditions.

Equation 6-64 may be expressed as follows:

$$Q_s = \text{acfm} = \text{scfm}\left(\frac{z_1}{z_{std}}\right)\left(\frac{t_1}{t_{std}}\right)\left(\frac{P_{std}}{P_s}\right)\left(\frac{MW_{std}}{MW_i}\right) \tag{6-65}$$

where the scfm is based on a dry gas.

To convert the standard volumetric flow to mass flow the following relations are used:

English system:

$$\dot{m} = \text{scfm}\left(\frac{144\ P_{std}}{z_{std}\ R_{std}\ t_{std}}\right) \tag{6-66}$$

SI/metric system:

$$\dot{m} = \text{scfm}\left(\frac{P_{std} \times 10^3}{z_{std}\ R_{std}\ t_{std}}\right) \tag{6-67}$$

PIPING SYSTEMS FOR ROTATING EQUIPMENT

For rotary equipment to be functional and contribute to the process system, it must be connected to the system with piping. The science of connecting piping systems to rotary equipment is a relatively new field and has drawn the stalwarts of academe to join with industry in solving problems of piping and equipment.

The two problems focused upon here are nozzle loadings and pulsation response spectra distributed to the attached piping system by reciprocating machines.

Nozzle Loadings

In earlier years various rotating equipment manufacturers would define allowable nozzle loadings as "zero force and zero moments." Such statements were not only ludicrous, but showed how little confidence some rotary equipment manufacturers had in their products. Ultimately, the pipe stress engineer was left to use his (or her) sole judgment to determine if the piping loads were substantial enough to damage the attached equipment.

There are several standards for handling nozzle loadings on rotating equipment, and probably the best known are those of NEMA (National Electrical Manufacturers Association). NEMA provides guidelines for nozzle loadings for steam turbines for mechanical drive service. Unfortunately, its guidelines are applied to *every* piece of rotating equipment by eager customers and engineering contractors. For example, what is valid for steam turbines is not valid for inline pumps. Because steam turbines are more fragile than most types of rotary equipment, using the NEMA standard produces over-conservative designs for most types of rotary equipment.

The American Petroleum Institute (API) also has standards for rotating equipment: API 611—General-Purpose Steam Turbines For Refinery Service; API 612—Special-Purpose Steam Turbines For Refinery Service; API 617—Centrifugal Compressors For General Refinery Services; and API 618—Reciprocating Compressors for General Refinery Service.

Applying API standards to nozzle loadings on rotating equipment leads to the argument in which rotating equipment specialists claim that the API standards are only intended for procurement purposes, and the pipe stress engineers, having no other guidelines to follow, assert that the API standards are what is to be used in practice.

The best criterion for judging nozzle loadings is experience with a given piece of equipment. For example, my several years of practical experience with turbo expanders dictate they can withstand three times the nozzle loadings allowed by NEMA (remember—only for *steam* turbines!)

Allowables for inline pumps, as above, did not exist a few years ago. Such pumps were regarded as piping components, e.g., valves, and allowables were considered unnecessary. But "thinning-up" casings to reduce material and costs makes such allowables possible, although controversial at times.

Table 6-2
Typical Manufacturer Allowables for Nozzle Loadings for Inline Pumps

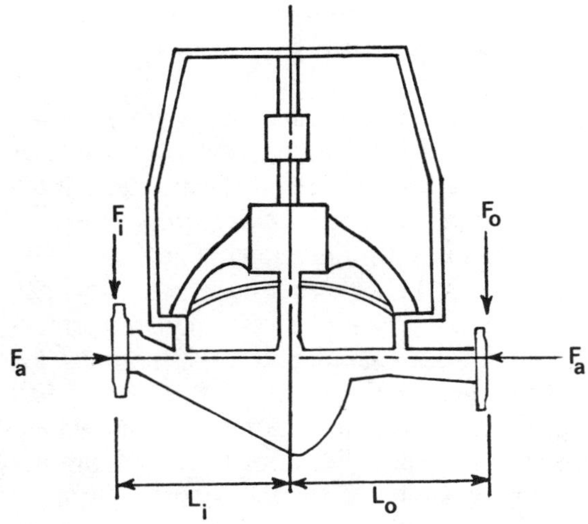

$M_i = F_i L_i$

$M_o = F_o L_o$

PUMP SIZE (in)	F_a lb	$M_{i_{max}}$ ft-lb	$M_{o_{max}}$ ft-lb
2x3x6	4000	5000	4000
3x4x6	6000	6000	5000
2x3x8	4000	5000	4000
3x4x8	5000	6000	5000
4x6x8	6000	7000	6000
4x6x10	6000	7000	6000
6x8x10	8000	9000	8000
6x6x20	5000	6000	5000
10x10x20	8000	9000	6000
12x12x20	12000	13000	10000

$$\frac{F}{F_a} + \frac{M_{i_{act}}}{M_{i_{max}}} + \frac{M_{o_{act}}}{M_{o_{max}}} < 2.0$$

Where,

F = resultant of actual force applied, lb

$M_{i_{act}}$ = actual bending moment on suction nozzle, ft-lb

$M_{o_{act}}$ = actual bending moment on discharge nozzle, ft-lb

There are three basic options to solving nozzle loadings on rotating equipment.

1. A detailed finite element study of the equipment.
2. Destructive testing of the equipment.
3. Close interface between the rotating equipment manufacturer and the piping stress engineer.

The problem with finite element analyses is who is going to pay for it—the client, the engineering contractor, or the rotating equipment manufacturer? Next, can the rotating equipment manufacturer disclose proprietary information often required in finite element analyses? Destructive testing poses the same question, who will pay for it? The third option—the pipe stress engineer conferring with the equipment manufacturer—is perhaps the most viable of the three, because if the NEMA and API criteria cannot be met, then the rotating equipment manufacturer can at least expect extra loadings and can design for it, if time permits. Thus, the rotary equipment vendor working as a team with the piping stress engineer(s) can help to alleviate most nozzle loading problems.

NEMA and API standards are very safe and a piece of equipment that meets their requirements should not have any nozzle loading problems, such as leaks. The problem comes in modular skid construction, where the values provided by the standards are very conservative.

Manufacturers often give allowable values for their equipment, and Table 6-2 presents some typical ones. A generalized standard taken from several pump manufacturers' allowable standards is shown in Figure 6-31. Reasonable nozzle loadings for turbo expanders worked out by the author and several turbo expander manufacturers are listed in Table 6-3.

Neither Table 6-2 nor Table 6-3 should be substituted for the manufacturer's allowables, if the vendor has his own. However, the information can be a valuable tool. Rules of thumb often are not only invalid but are often based on special situations that may not be true for every case.

One must be extra careful in piping steam turbines, because these units are usually fragile. Example 2-2 in Chapter 2 illustrates a piping arrangement connected to a steam turbine. If expansion joints are allowed, the configuration shown in Figure 6-32 is ideal.

PULSATION RESPONSE SPECTRA INDUCED BY RECIPROCATING EQUIPMENT

Reciprocating machinery often induces pulsation response spectra in attached piping systems. This subject alone is comprehensive to fill several volumes, so we will just outline the problem here.

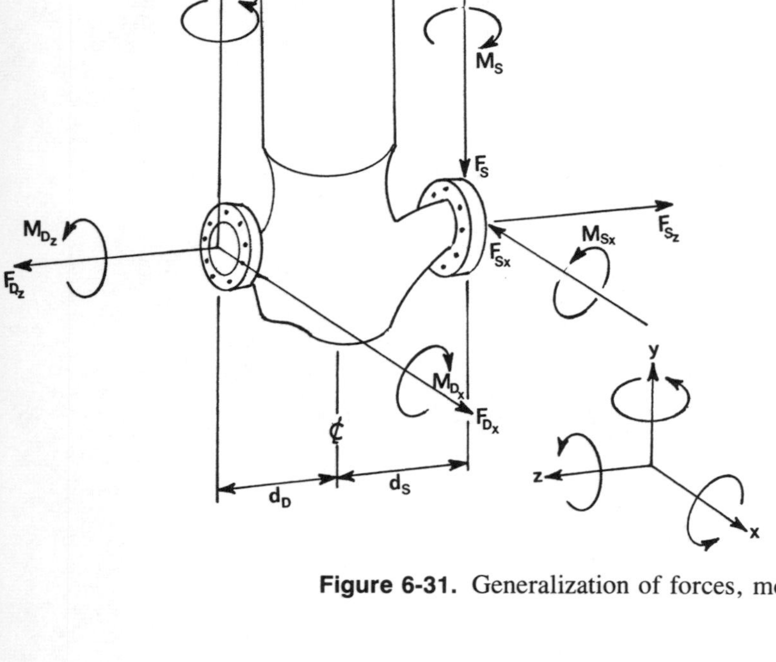

$$M_{R_D} = \sqrt{M_{D_x}^2 + M_{D_y}^2 + M_{D_z}^2} \qquad M_{R_s} = \sqrt{M_{s_x}^2 + M_{s_y}^2 + M_{s_z}^2}$$

M_{R_N} = greater of M_{R_D} & M_{R_s}, where M_{R_D} & M_{R_s} are resultant moments applied at nozzles

$M_{R\text{℄}}$ = resultant bending moment about

$$\sum M_y = F_{s_x}(d_s) + F_{D_x}(d_D) + M_{s_y} + M_{D_y}$$

$$\sum M_{x\text{℄}} = F_{s_y}(d_s) + F_{D_y}(d_D) + M_{s_y} + M_{D_y}$$

$$M_{R\text{℄}} = \left[\left[\sum M_{x\text{℄}}\right]^2 + \left[\sum M_{y\text{℄}}\right]\right]^{0.5};$$

$$F_{R_s} = [F_{s_x}^2 + F_{s_y}^2 + F_{s_z}^2]^{0.5};$$

$$F_{R_D} = [F_{D_x}^2 + F_{D_y}^2 + F_{D_z}^2]^{0.5}$$

F_R = greater of F_{R_s} or F_{R_D}

$$\frac{F_R}{F_{all}} + \frac{M_{R_n}}{M_{n_{all}}} + \frac{M_{R\text{℄}}}{M_{\text{℄}\,all}} < 2.0$$

Figure 6-31. Generalization of forces, moments, and allowable nozzle loadings.

Table 6-3
Reasonable Turbo Expander Nozzle Loadings

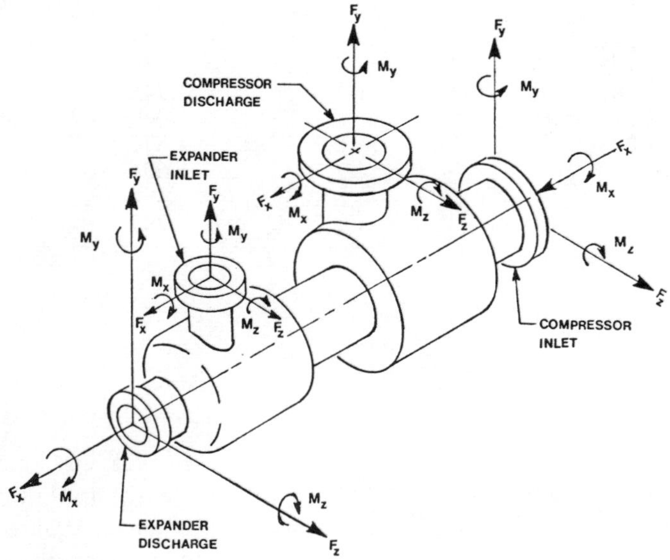

Expander Inlet

Nozzle Size (in.)	F_x	F_y	F_z	F_R	M_x	M_y	M_z	M_R
4	649	1,299	1,299	1,948	1,624	2,436	3,383	4,474
6	974	1,948	1,948	2,922	2,436	3,654	5,074	6,710
8	1,299	2,597	2,597	3,896	3,247	4,870	6,764	8,947
10	1,623	3,246	3,246	4,869	4,059	6,088	8,455	11,184
12	1,948	3,895	3,895	5,843	4,871	7,306	10,146	13,421
14	2,272	4,545	4,545	6,817	5,683	8,524	11,838	15,658
16	2,595	5,189	5,189	7,784	6,486	9,730	13,513	17,870

Expander Discharge

Nozzle Size (in.)	F_x	F_y	F_z	F_R	M_x	M_y	M_z	M_R
4	649	1,299	1,299	1,948	1,624	3,383	2,436	4,474
6	974	1,948	1,948	2,922	2,436	5,074	3,654	6,710
8	1,299	2,597	2,597	3,896	3,247	6,764	4,870	8,947
10	1,623	3,246	3,246	4,869	4,059	8,455	6,088	11,184
12	1,948	3,895	3,895	5,843	4,871	10,146	7,306	13,421
14	2,272	4,545	4,545	6,817	5,683	11,838	8,524	15,658
16	2,595	5,189	5,189	7,784	6,486	13,513	9,730	17,870

Compressor Inlet

Nozzle Size (in.)	F_x	F_y	F_z	F_R	M_x	M_y	M_z	M_R
4	648	1,080	1,080	1,659	1,620	2,699	2,699	4,147
6	972	1,620	1,620	2,488	2,429	4,049	4,049	6,220
8	1,296	2,160	2,160	3,318	3,239	5,399	5,399	8,294
10	1,620	2,699	2,699	4,147	4,049	6,748	6,748	10,367
12	1,944	3,239	3,239	4,976	4,859	8,098	8,098	12,441
14	2,268	3,779	3,779	5,806	5,669	9,448	9,448	14,514
16	2,592	4,319	4,319	6,635	6,479	10,798	10,798	16,588
18	2,915	4,859	4,859	7,464	7,289	12,147	12,147	18,661
20	3,240	5,399	5,399	8,294	8,099	13,497	13,497	20,735
24	3,892	6,486	6,486	9,964	9,730	16,216	16,216	24,912

Table 6-3 (continued)

Compressor Discharge								
Nozzle Size (in.)	F_x	F_y	F_z	F_R	M_x	M_y	M_z	M_R
4	650	1,444	1,299	2,048	1,624	2,165	3,007	4,046
6	974	2,165	1,949	3,072	2,436	3,249	4,512	6,070
8	1,300	2,888	2,599	4,097	3,248	4,331	6,016	8,093
10	1,624	3,610	3,249	5,121	4,060	5,414	7,519	10,116
12	1,949	4,331	3,899	6,145	4,872	6,496	9,023	12,139
14	2,274	5,053	4,548	7,169	5,684	7,579	10,527	14,162
16	2,599	5,775	5,198	8,193	6,496	8,662	12,030	16,185
18	2,919	6,486	5,838	9,202	7,297	9,730	13,514	18,181

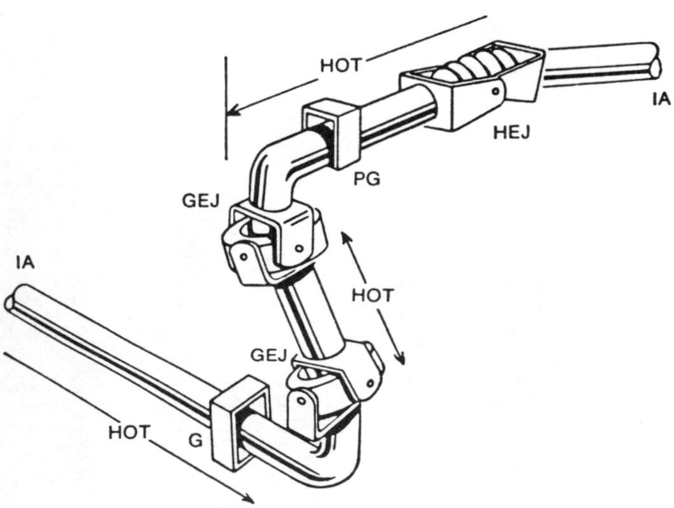

PG: Planar Guide
IA: Intermediate Anchor
G: Guide
HEJ: Hinge Expansion Joint
GEJ: Gimbal Expansion Joint

Figure 6-32. An expansion joint arrangement ideal for steam turbines where nozzle loadings must be kept low (almost always the case with steam turbines) and the use of expansion joints is practical. (Courtesy of Pathway Bellows, Inc.)

Currently, two methods are used to predict pulsation problems: (a) modeling the system on an analog computer and (b) simulating it on a digital computer. Basically, the piping system is modeled with support and soil stiffness values input at every pipe support as discussed in Chapter 2. Then the system is excited with various forcing functions that represent the reciprocating machine or machines. The piping supports are moved around, deleted, or added to decrease the amplitudes generated by the forcing functions. This analysis can be done on either an analog or digital computer.

There are two methods available on existing computer software that can help head off pulsation problems. These methods are *modal extraction analysis* and *time spectra (time history) analysis.* Modal extraction is computing the natural frequency of the piping system, after modeling the pipe support and soil stiffness values, and comparing this frequency to that of the shaft speed of the equipment. Time spectra analysis is a transient analysis that basically does exactly what modal extraction does except on a transient basis for every time interval over a specified period of time. In other words, we compute the system's natural frequency for every second over a period of one hour. Over the period of one hour we excite the system with a forcing function that accurately defines the rotating equipment.

Figure 6-33 shows a piping system excited by pulsations from a reciprocating machine. A complete investigation of the pulsation frequencies and surge capacity is normally required, which involves the compressor bottles (surge drums), compressor suction header, and suction compressor bottle, the discharge header, and discharge compressor bottle. Two companies are engaged separately in investigating these problems—Southern Gas Association's compressor analog computer at Southwest Research Institute and the Structural Dynamics Research Corporation (SDRC). The compressor bottle (or surge drum) acts as a pulsation dampener. A typical bottle is shown in Figure 6-34. The compressor bottle acts as an acoustic filter designed for all frequencies induced as the reciprocating engine speed varies. The compressor bottle cannot damp out all frequencies, but should store energy generated from the various frequencies and reduce them to produce a relatively smooth and continu-

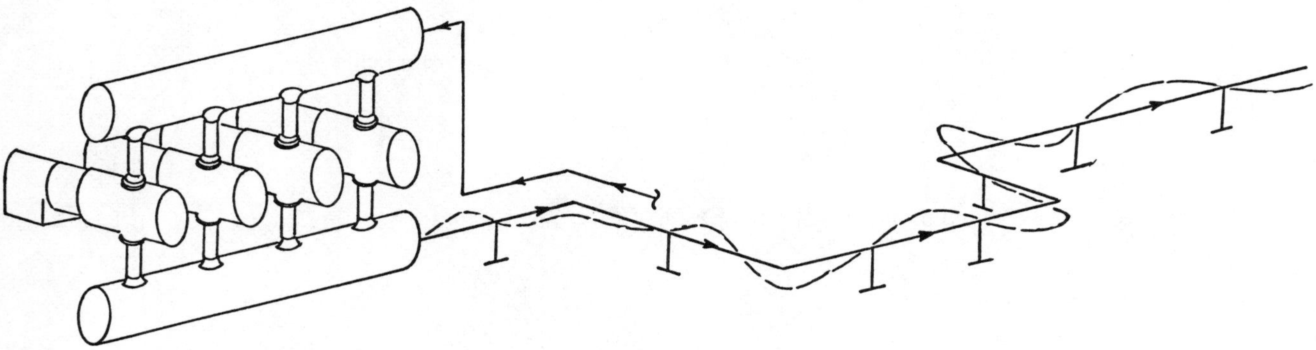

Figure 6-33. Piping system excited by pulsations from a reciprocating machine.

ous operation. Sizing the compressor bottles should be done by a specialist who has worked in this field for several years.

In the days before analog and digital simulations, pulsation problems were solved (and still are) with orifice plates. These plates were placed in the piping system and the orifice diameter was approximately 0.53 times the internal diameter of the pipe. These plates, distributed throughout the piping system, acted as pulsation dampeners. Although orifice plates produce huge pressure drops, they are effective in many installations.

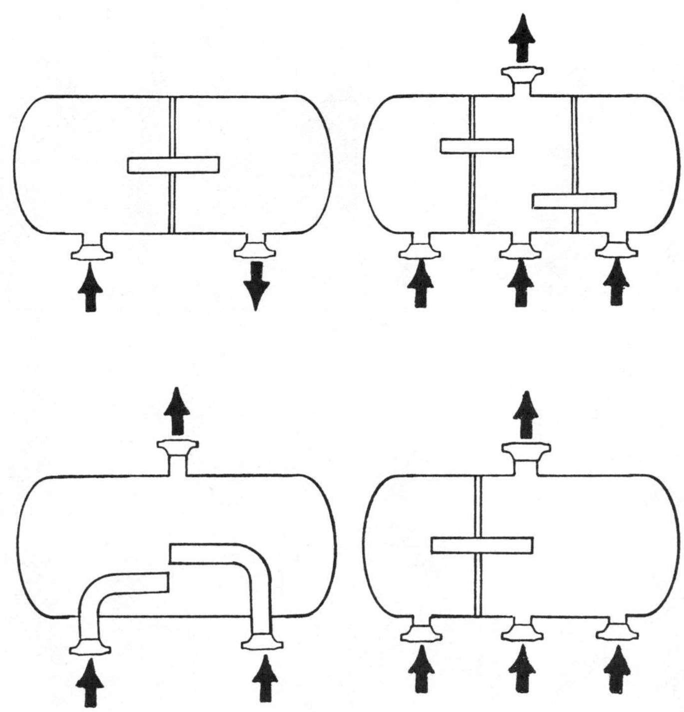

Figure 6-34. Typical pulsation bottle (or drum) configurations that act as pulsation dampeners.

EXAMPLE 6-1: HORIZONTAL, CENTRIFUGAL PUMP SYSTEM DESIGN

A food processing plant is having a cooking kettle installed to process molasses into refined syrup for breakfast foods. A horizontal centrifugal pump is to be installed next to a fuel tank to supply fuel oil to a burner in the cooking kettle. The fuel oil tank is to have a 50 psig nitrogen pad because the tank cannot be raised for higher head at the pump. The cooking kettle is 200 ft downstream and 15 ft above the discharge flange of the pump. It is desired to select and size the burner feed pump shown in Figure 6-35. The discharge pressure at the burner end is to be 40 psig.

Suction Line Pressure Drop

Fluid = fuel oil
Temperature = 90°F
Pressure = 50 psig
$\rho = 54.725\ lb_m/ft^3$
$\mu = 139.53\ cp = (139.53)(6.72 \times 10^{-4}) = 0.094\ lb_m/ft\text{-}sec$
$\epsilon = 0.0018$
L = 1.0 ft
Suction line = 3″ϕSch 40, D_i = 3.068 in.
Q = 150 gpm

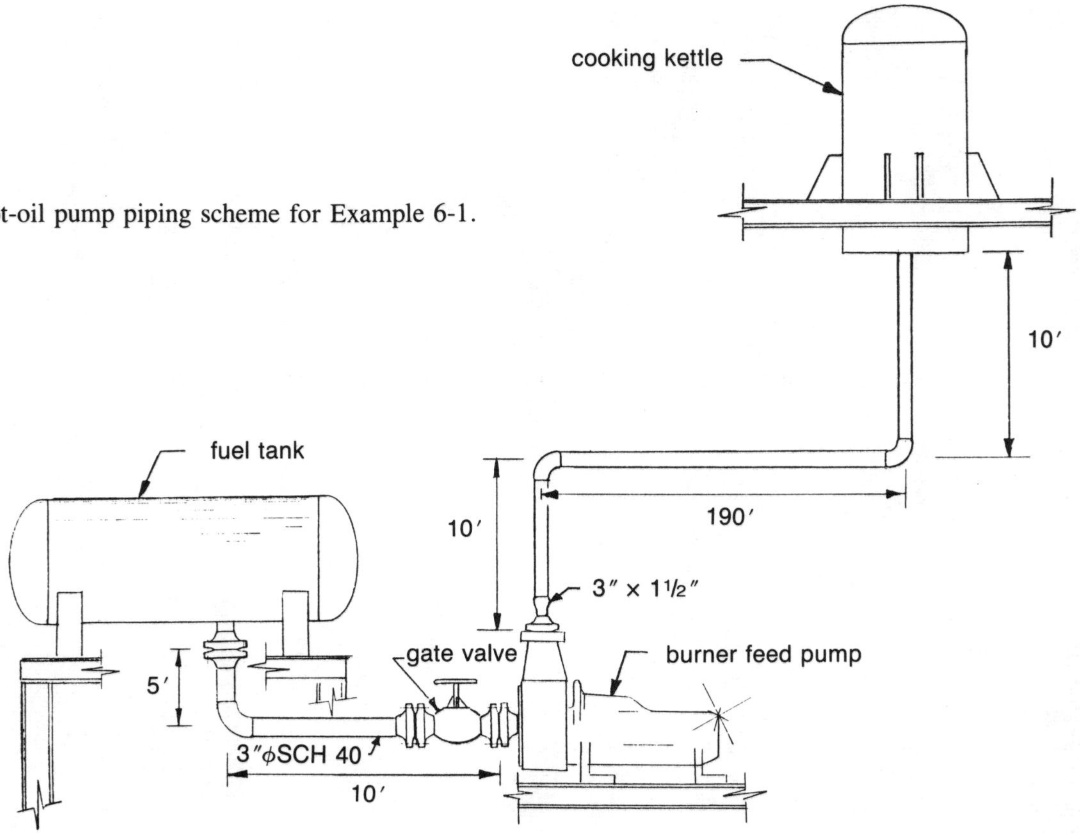

Figure 6-35. Hot-oil pump piping scheme for Example 6-1.

$$V = \frac{(150)\,\frac{gal}{min}\left(\frac{1\ ft^3}{7.479\ gal}\right)\left(\frac{1\ min}{60\ sec}\right)}{(7.393)\ in.^2\left(\frac{1\ ft^2}{144\ in.^2}\right)} = 6.511\ ft/sec$$

$$N_{Re} = \frac{DV\rho}{\mu} = \frac{\left(\frac{3.068}{12}\right)ft(6.511)\frac{ft}{sec}(54.725)\frac{lb_m}{ft^3}}{(0.094)\frac{lb_m}{ft\text{-}sec}} = 969.1$$

With $N_{Re} = 969.1$, the flow is laminar.

From Equation 1-6b we compute the friction factor as follows:

$$f = \frac{64}{N_{Re}} = \frac{64}{969.1} = 0.066$$

K-Values (Velocity Heads)

Referring to Figures 1-7 and 1-11 we have the following:

$$\begin{aligned} \text{Entrance and exit} &= K = 1.78 \\ \text{1–3-in.}\phi\ 90^\circ\ \text{std ell} &= K = 0.30 \\ \text{1–3-in.}\phi\ \text{gate valve} &= K = \underline{0.14} \\ \textstyle\sum K &= 2.22 \end{aligned}$$

From Equation 1-4 we compute the frictional pressure drop as follows:

$$\Delta\rho_f = \left(\frac{fL}{D} + \sum K_i\right)\frac{\rho V^2}{2g_c}$$

$$\Delta\rho_f = \left[\frac{(0.066)(15.00)(12)}{(3.068)} + 2.22\right] \times \frac{(54.725)\,\frac{lb_m}{ft^3}\,(6.511)^2\,\frac{ft^2}{sec^2}\left(\frac{1\ ft^2}{144\ in.^2}\right)}{2(32.2)\,\frac{ft\text{-}lb_m}{sec^2\text{-}lb_f}}$$

$$\Delta\rho_f = 1.524\ psi$$

Discharge Line Pressure Drop

The conditions are the same as the suction line except for the following:

Line size = 2-in. Schedule 40 for which D_i = 2.067
For $1^1/2$-in. ϕ pump discharge,

$$V = \frac{(150)\ \frac{gal}{min}\left(\frac{1\ ft^3}{7.479\ gal}\right)\left(\frac{1\ min}{60\ sec}\right)}{(2.036)\ in.^2\left(\frac{1\ ft^2}{144\ in.^2}\right)} = 23.642\ \frac{ft}{sec}$$

For 2-in. S/40 discharge line,

$$V = \frac{(150)\ \frac{gal}{min}\left(\frac{1\ ft^3}{7.479\ gal}\right)\left(\frac{1\ min}{60\ sec}\right)}{(3.356)\ in.^2\left(\frac{1\ ft^2}{144\ in.^2}\right)} = 14.343\frac{ft}{sec}$$

For 2-in. ϕ S/40 pipe,

$$N_{Re} = \frac{\left(\frac{2.067}{12}\right) ft\ (14.343)\frac{ft}{sec}(54.725)\frac{lb_m}{ft^3}}{(0.094)\ \frac{lb_m}{ft\text{-}sec}} = 1{,}438.3$$

$$f = \frac{64}{N_{Re}} = \frac{64}{1{,}438.3} = 0.044$$

K-Values for $1^1/2$-in. Portion

Entrance = K = 0.78
From Table 1-7, for a 2-in. × $1^1/2$-in. diffuser, K = 0.055

$\sum K = 0.835$

L = 3.0 in., d = 1.610 in.

$$N_{Re} = \left[\frac{\left(\frac{1.500}{12}\right) ft\ (23.642)\ \frac{ft}{sec}\ (54.725)\ \frac{lb_m}{ft^3}}{(0.094)\ \frac{lb_m}{ft\text{-}sec}}\right] = 1{,}720.5$$

$$f = \frac{64}{N_{Re}} = 0.037$$

$$\Delta\rho_f = \left[\frac{(0.037)(3.0)}{(1.610)} + 0.835\right] \times \frac{(54.725)\ \frac{lb_m}{ft^3}\ (23.642)^2\ \frac{ft^2}{sec^2}\left(\frac{1\ ft^2}{144\ in.^2}\right)}{2(32.2)\ \frac{ft\text{-}lb_m}{sec^2\text{-}lb_f}}$$

$\Delta\rho_f$ = 2.982 psi

K-Values for 2-in. Portion

2–2-in.-std 90° elbows = K = 0.40
exit = K = 1.0

$\sum K = 1.40$

L = 200 ft

$$\Delta\rho_f = \left[\frac{(0.044)(200.0)(12)}{(2.067)} + 1.40\right] \times \frac{(54.725)\ \frac{lb_m}{ft^3}\ (14.343)^2\ \frac{ft^2}{sec^2}\left(\frac{1\ ft^2}{144\ in.^2}\right)}{2(32.2)\ \frac{ft\text{-}lb_m}{sec^2\text{-}lb_f}}$$

$\Delta\rho_f$ = 63.72 psi = too high—choose a $1^1/2$-in. × 3-in. diffuser
With 3-in.ϕ Sch 40 pipe,

$$V = \frac{(150)\ \frac{gal}{min}\left(\frac{1\ ft^3}{7.479\ gal}\right)\left(\frac{1\ min}{60\ sec}\right)}{(7.393)\ in.^2\left(\frac{1}{144\ in.^2}\right)} = 6.511\ \frac{ft}{sec}$$

K-Values for 3-in. ϕ Pipe

2–2-in.-std 90° elbow = K = 0.54
exit = K = 1.00

$\sum K = 1.54$

$$N_{Re} = \frac{\left(\frac{3.068}{12}\right) ft\ (6.511)\ \frac{ft}{sec}\ (54.725)\ \frac{lb_m}{ft^3}}{(0.094)\ \frac{lb_m}{ft\text{-}sec}} = 969.125$$

$$f = \frac{64}{N_{Re}} = 0.066$$

$$\Delta\rho_f = \left[\frac{(0.066)(200.0)(12)}{(3.068)} + 1.54\right] \times \frac{(54.725)\frac{lb_m}{ft^3}(6.511)^2\frac{ft^2}{sec^2}\left(\frac{1\ ft^2}{144\ in.^2}\right)}{2(32.2)\frac{ft\text{-}lb_m}{sec^2\text{-}lb_f}}$$

$\Delta\rho_f = 13.309$ psi ⇒ use 3-in. ϕ S/40 pipe

New K-Values for 1½-in. Pipe

Entrance = K = 0.78
From Table 1-7, for a 3-in. × 1½-in. diffuser,
K = 0.337
$\sum$K = 1.117

L = 3.0 in.; d = 1.610 in.
N_{Re} = 1,720.5; f = 0.037

$$\Delta\rho_f = \left[\frac{(0.037)(3.0)}{(1.610)} + 1.117\right] \times \frac{(54.725)\frac{lb_m}{ft^3}(23.642)^2\frac{ft^2}{sec^2}\left(\frac{1\ ft^2}{144\ in.^2}\right)}{2(32.2)\frac{ft\text{-}lb_m}{sec^2\text{-}lb_f}}$$

$\Delta\rho_f = 3.912$ psi
Total pressure loss in discharge line
= 13.309 + 3.912 = 17.221 psi

Using the pump manufacturer's curve in Figure 6-36, we can enter data on the Hydraulic Design Calculation Sheet in Figure 6-37 to size the pump.

The Effects of Liquid Viscosity on Centrifugal Pumps

From the previous analysis and Figure 6-36 we know the hydraulic performance required of the pump. Before the actual horsepower requirement for the motor and the impeller size can be determined, the viscosity effects of the liquid being handled must be considered. One requirement of a centrifugal pump is that the handled liquid be relatively clean of suspended particles. Obviously, for the same size pump and motor a highly viscous liquid will tax the unit more than would a low viscous liquid. Thus, the viscosity is an important property that affects the horsepower of the pump motor. To account for this, the Hydraulic Institute has prepared charts shown in Figures 6-38 and 6-39 for determining viscosity effects. To use the charts, the fluid being handled should be Newtonian. Gels, slurries, asphalt, and other non-Newtonian fluids should not be considered with these charts. In handling such fluids a positive-displacement pump is usually required. (Example 6-2 is an illustration of how to handle such a liquid.)

To use Figure 6-39 we must convert the absolute viscosity to kinematic viscosity. This is done as follows:

$\mu = 139.53$ cp at 90°F

w = 54.725 lb/ft³

$$\nu = \frac{(139.53)cp\left(\frac{0.0000209}{1\ cp}\right)\frac{lb_f\text{-}sec}{ft^2}(32.17)\frac{ft\text{-}lb_m}{lb_f\text{-}sec^2}}{(54.725)\frac{lb_m}{ft^3}}$$

$$\nu = 0.0017\frac{ft^2}{sec}$$

or

$$\nu = \frac{0.0017\ \frac{ft^2}{sec}}{0.0000107639\ \frac{ft^2}{sec}}\ \text{centistokes}$$

$\nu = 159.261$ centistokes

Using Table 1-8 we make the viscosity conversion from centistoke to SSU as follows:

$$0.226t - \frac{195}{t} = \nu$$

$$t^2 - 704.695t - 862.832 = 0$$

t = 706 SSU

Now, looking at Figure 6-39 we see that for 150 gpm, TDH = 82 feet, and 706 SSU we obtain the following coefficients:

$C_\eta = 0.56$

$C_Q = 0.90$

$C_H = 0.90$ for $1.0 \times Q_{NW}$, where Q_{NW} is the water capacity at which maximum efficiency is obtained

The corrected flow rate becomes

$$Q_C = \frac{gpm}{C_Q} = \frac{150}{0.90} = 166.67 \approx 167\ gpm$$

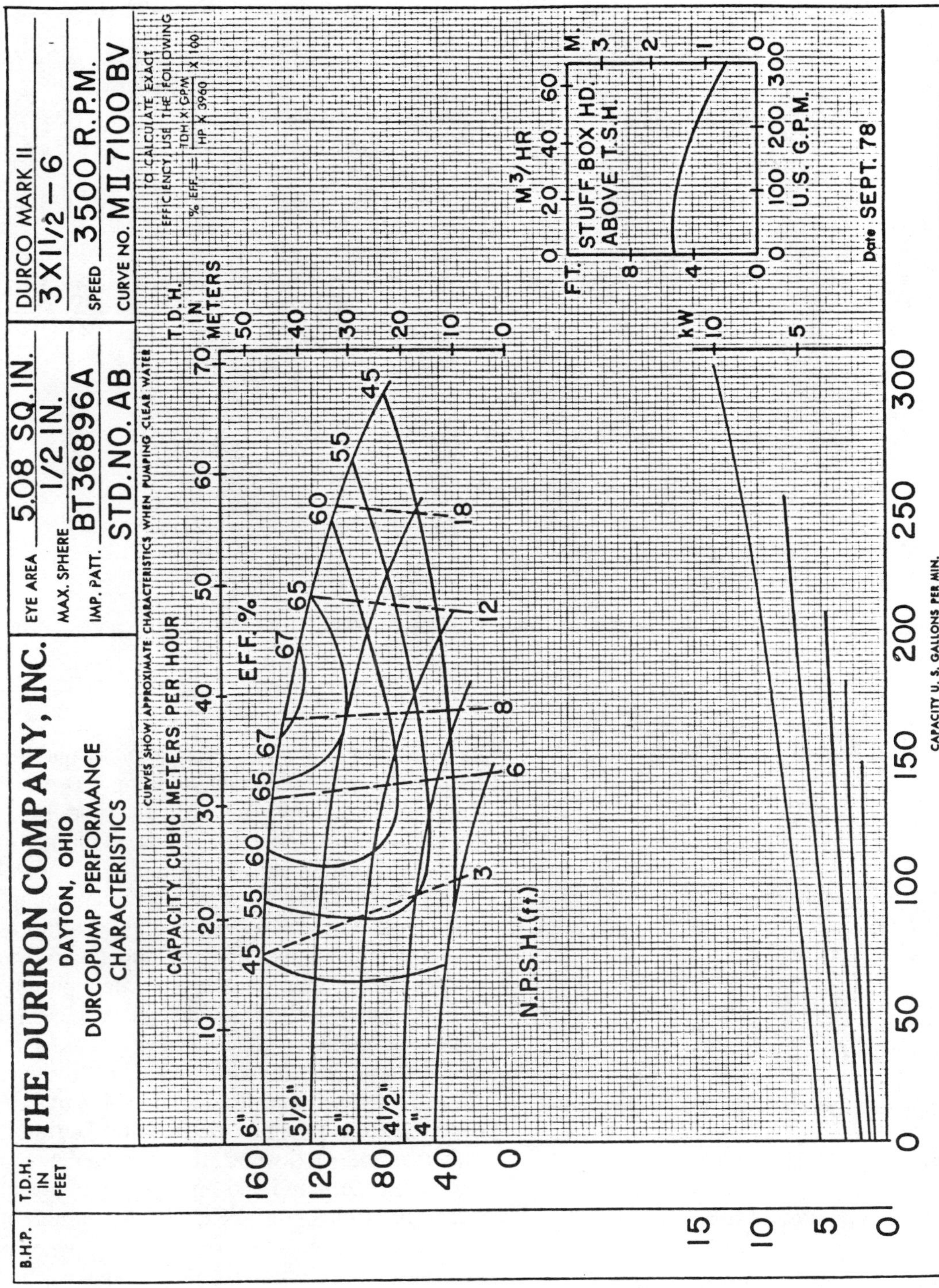

Figure 6-36. Pump performance curve for fuel oil pump of Example 6-1. (Courtesy of the Duriron Company, Inc.)

Pump Hydraulic Design Calculation Sheet

Liquid	fuel oil	
Viscosity at P.T. (Pumping Temp.)	139.53	cp
Vapor pressure at P.T.	0.010	psia
Sp. gr. (γ) at P.T.	0.877	
Flow at ambient temp.	150	gpm
Operating flow at P.T.	150	gpm
Design flow at P.T.	150	gpm

Suction			Discharge		
Source pressure	64.7	psia	Terminal pressure	71.38	psia
Static head	1.9	psi	Static head	1.9	psi
$-\Delta P_f$, line loss	1.52	psi	$-\Delta P_f$ discharge		
Suction pressure	65.08	psia	Piping system	17.221	psi
− Vapor pressure	−0.01	psia	Other		psi
NPSH avail	65.07	psia	Discharge press.	96.201	psia
NPSH avail	171.2	ft	− Suction press.	65.08	psia
NPSH req'd	3.2	ft	TDH	31.12	psia
			TDH	82.017	feet

bhp at Duty Condition

$$bhp_D = \frac{(gpm)(TDH)(\gamma)}{(3{,}960)(\eta)} = \frac{(150)(84)(0.877)}{(3{,}960)(0.542)} = 5.15 \text{ hp} \approx 5^1/_4 \text{ hp}$$

bhp at Back-Pressure Condition

$$bhp_{BP} = \frac{(gpm)(TDH)(\gamma)}{(3{,}960)(\eta)} = \frac{(150)(45)(1.0)}{(3{,}960)(0.46)} = 3.706 \text{ hp} \approx 4 \text{ hp with water}$$

Figure 6-37. Pump hydraulic design calculation sheet for Example 6-1.

The total dynamic head becomes

$$H_C = \frac{TDH}{C_H} = \frac{82}{0.90} = 91.11 \approx 91 \text{ ft}$$

Now, referring to the manufacturer's curve in Figure 6-40, for $Q_C = 167$ gpm and TDH = 91 ft, we determine the pump efficiency as

$$\eta = 63\%$$

The NPSH required = 8 ft

To correct the efficiency for viscosity we have

$$\eta_c = \eta C_\eta = (63\%)(0.56) = 35.28\% \text{ efficiency}$$

The brake horsepower for pumping the liquid is

$$bhp_{VIS} = \frac{QH\gamma}{3{,}960\ \eta_c} = \frac{(167)(91)(0.877)}{(3{,}960)(0.353)} = 9.53 \text{ hp}$$

Referring to Table 6-4, we see that the next larger motor size is a 10 hp motor, thus we select a 3 × 1½-in. centrifugal pump with a 10-hp motor and a 5-in. impeller. In selecting a centrifugal pump it is desirable for the required flow rate to fall in the middle of the pump curve. Avoid extreme sides of the manufacturer's performance curves. Select an impeller that is at least two sizes below the largest size available for the pump, because if greater head is later required, e.g., if additional piping is added to the system, changing impellers is much cheaper and expedient than purchasing a new pump.

In the final analysis the design engineer must not forget the potential problem of back pressure that the pump could be exposed to under varying conditions. For example, if the discharge line contained a bypass valve that diverted flow to either the cooking kettle or to a reservoir that collected water, the reservoir would be used if and when the pump and piping system are cleaned with water or a cleaning agent. In this situation the pump would have to be sized for handling water or whatever cleaning is to be used. When the bypass valve is shut off, closing the discharge piping connecting the pump to the cooking kettle, the flow conditions are changed, resulting in a lower TDH. With the same size impeller, as the TDH lowers, the flow rate increases as the curve shifts

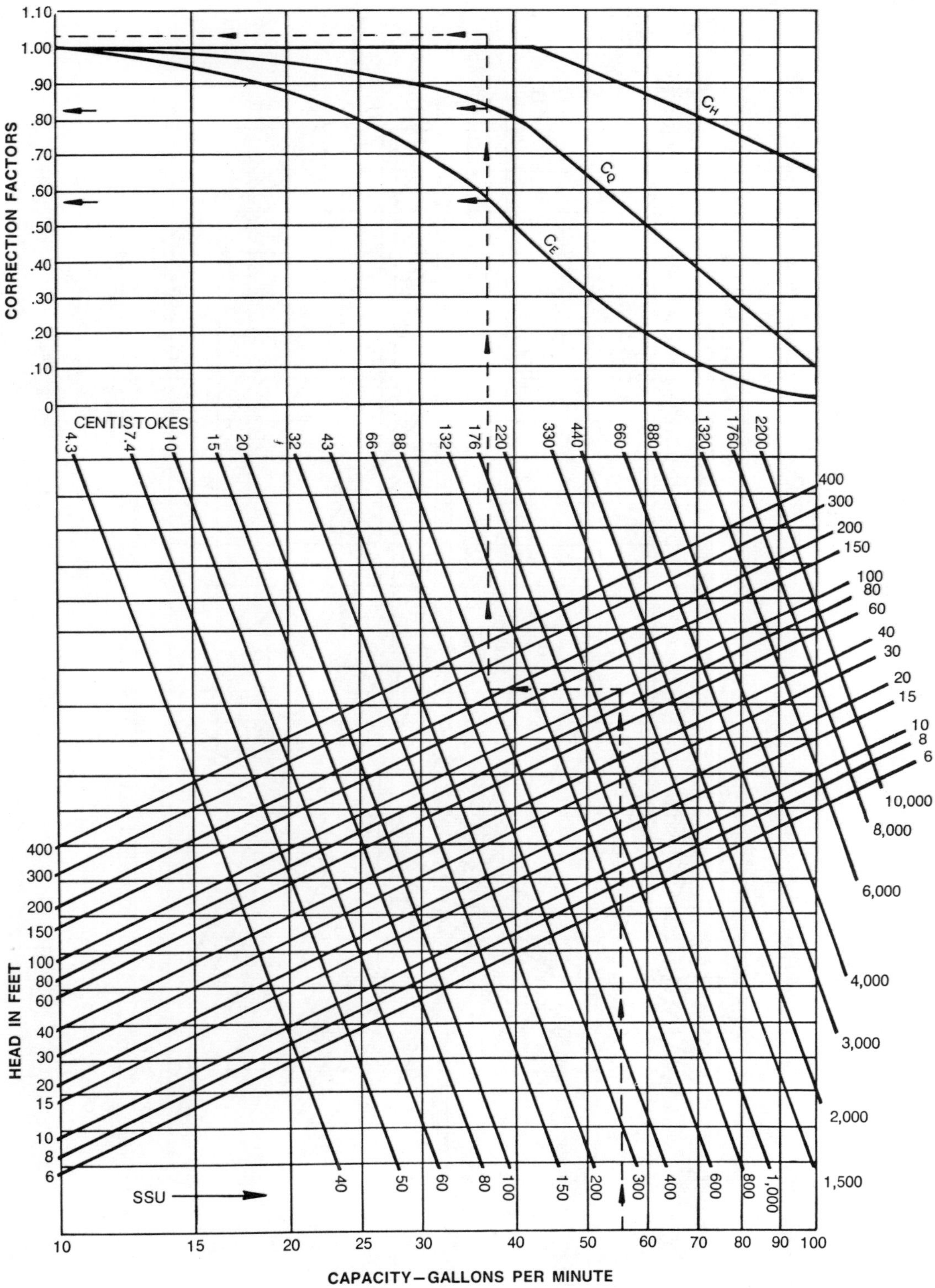

Figure 6-38. Viscosity corrections for capacities of 100 gpm or less (Courtesy of the Hydraulic Institute, Cleveland Ohio.)

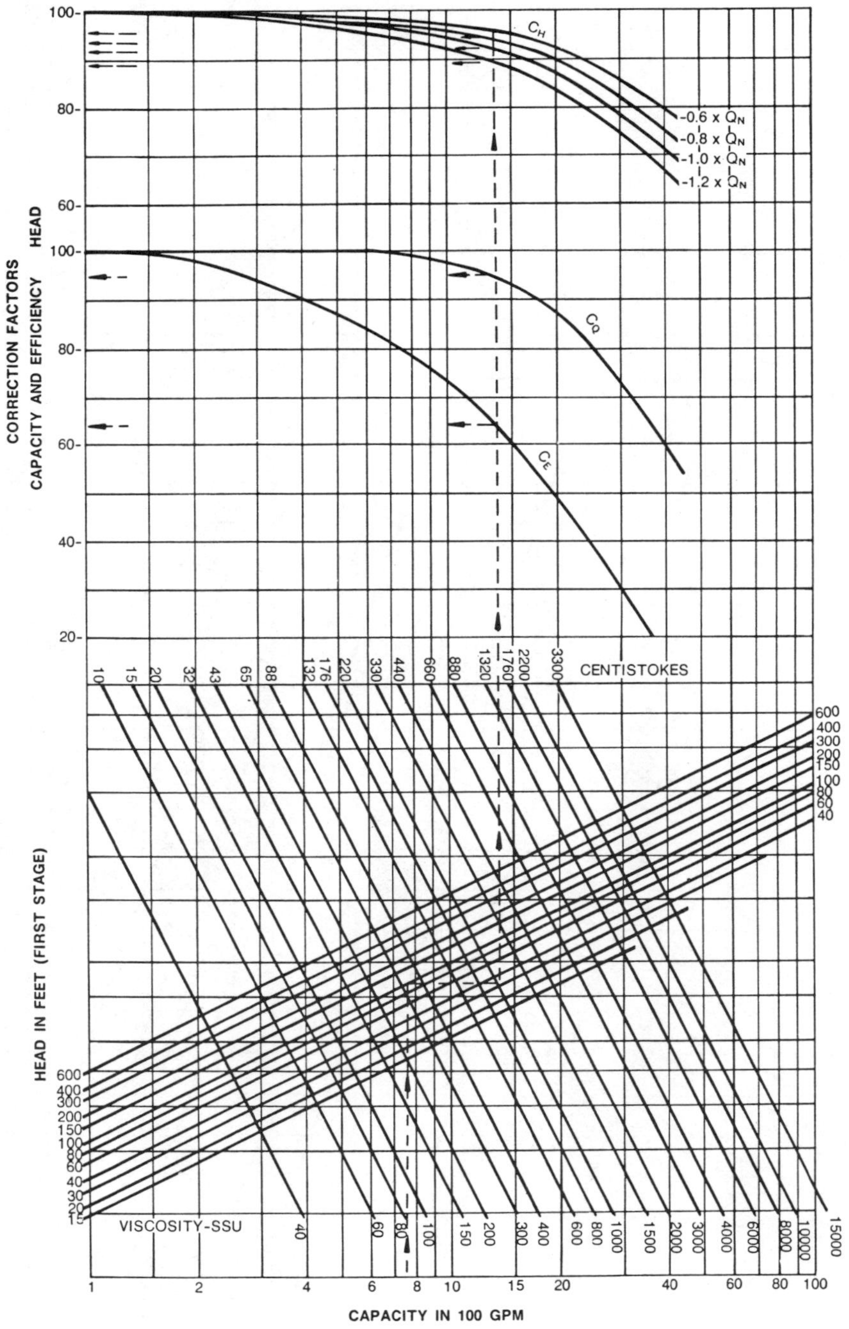

Figure 6-39. Performance correction chart for viscous liquids. (Courtesy of the Hydraulic Institute, Cleveland, Ohio.)

Table 6-4
NEMA Frame Dimensions

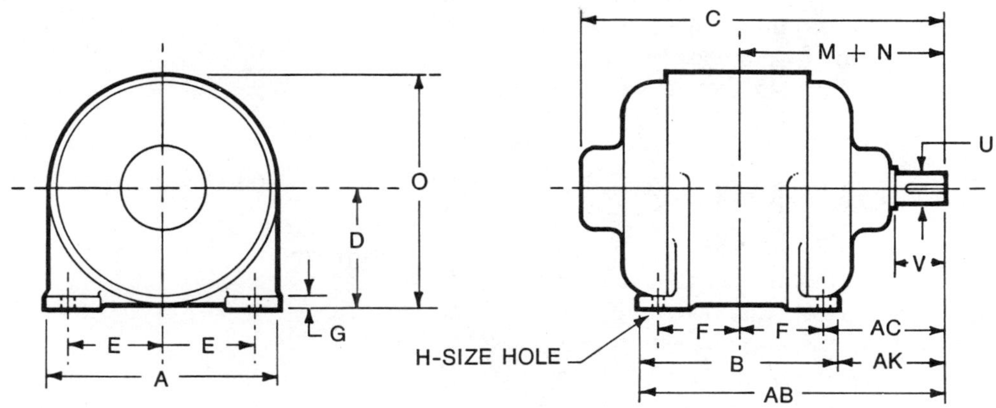

Motor Frame	H.P. (Open)				H.P. (Enclosed)				A Max.	B Max.	C (Approx.)		D	E	F	G Appr.	H	M+N	O (Approx.)		U	Keyway	V Min.	AB Max.	AC	AK Min.	Bolts		Wt. (Approx.)	
	900	1200	1800	3600	900	1200	1800	3600			Open	Encl.							Open	Encl.							Dia.	Lg.	Open	Encl.
143T	½	¾	1	1½	½	¾	1	1½	7	6	12	12½	3½	2¾	2		11/32	6½	6⅞	7	⅞	3/16 × 3/32	2	9½	4½	3½	¼		40	45
145T	¾	1	1½–2	2–3	¾	1	1½–2	2	7	6	12½	13½	3½	2¾	2½		11/32	7	6⅞	7	⅞	3/16 × 3/32	2	10	4½	4	¼		45	50
182T	1	1½	3	5	1	1½	3	3	9	6½	13	14½	4½	3¾	2¼	⅝	13/32	7¾	9⅛	9¼	1⅛	¼ × ⅛	2½	11	5½	4½	5/16	1	65	79
184T	1½	2	5	7½	1½	2	5	5	9	7½	14	15½	4½	3¾	2¾	⅝	13/32	8¼	9⅛	9¼	1⅛	¼ × ⅛	2½	12	5½	4½	5/16	1	80	95
213T	2	3	7½	10	2	3	7½	7½	10½	7½	16	18	5¼	4¼	2¾	¾	13/32	9⅝	10¾	10⅞	1⅜	5/16 × 5/32	3⅛	13⅜	6⅞	5⅞	5/16	1	120	140
215T	3	5	10	15	3	5	10	10	10½	9	17½	19½	5¼	4¼	3½	¾	13/32	10⅜	10¾	10⅞	1⅜	5/16 × 5/32	3⅛	14⅞	6⅞	5⅞	5/16	1	140	160
254T	5	7½	15	20	5	7½	15	15	12½	10¾	20½	22½	6¼	5	4⅛	⅞	17/32	12⅜	12⅝	12¾	1⅝	⅜ × 3/16	3¾	17¾	8¼	7	⅜	1¼	200	235
256T	7½	10	20	25	7½	10	20	20	12½	12½	22½	24	6¼	5	5	⅞	17/32	13¼	12⅝	12¾	1⅝	⅜ × 3/16	3¾	19½	8¼	7	⅜	1¼	235	270
284T	10	15	25		10	15	25		14	12½	23½	25½	7	5½	4¾	1	17/32	14⅛	14	14⅜	1⅞	½ × ¼	4⅜	20⅜	9⅜	7⅞	⅜	1½	295	370
284TS				30				25	14	12½	22	24½	7	5½	4¾	1	17/32	12¾	14	14⅜	1⅝	⅜ × 3/16	3	19	8	6½	⅜	1½	255	340
286T	15	20	30		15	20	30		14	14	25	27	7	5½	5½	1	17/32	14⅞	14	14⅜	1⅞	½ × ¼	4⅜	21⅞	9⅜	7⅞	⅜	1½	340	405
286TS				40				30	14	14	23½	26	7	5½	5½	1	17/32	13½	14	14⅜	1⅝	⅜ × 3/16	3	20½	8	6½	⅜	1½	295	395
324T	20	25	40		20	25	40		16	14	26	28½	8	6¼	5¼	1⅛	21/32	15¾	16	16⅝	2⅛	½ × ¼	5	22¾	10½	8¾	½	1¾	440	520
324TS				50				40	16	14	24½	27	8	6¼	5¼	1⅛	21/32	14¼	16	16⅝	1⅞	½ × ¼	3½	21¼	9	7¼	½	1¾	445	500
326T	25	30	50		25	30	50		16	15½	27½	30	8	6¼	6	1⅛	21/32	16½	16	16⅝	2⅛	½ × ¼	5	24¼	10½	8¾	½	1¾	435	580
326TS				60				50	16	15½	26	28½	8	6¼	6	1⅛	21/32	15	16	16⅝	1⅞	½ × ¼	3½	22¾	9	7¼	½	1¾	480	560
364T	30	40			30	40			18	15¼	29	33	9	7	5⅝	1¼	21/32	17⅜	18	18½	2⅜	⅝ × 5/16	5⅝	25	11¾	9¾	½	1¾	605	755
364TS			60	75			60	60	18	15¼	27	31	9	7	5⅝	1¼	21/32	15¼	18	18½	1⅞	½ × ¼	3½	22⅞	9⅝	7⅝	½	1¾	670	740
365T	40	50			40	50			18	16¼	30	34	9	7	6⅛	1¼	21/32	17⅞	18	18½	2⅜	⅝ × 5/16	5⅝	26	11¾	9¾	½	1¾	665	835
365TS			75	100			75	75	18	16¼	28	32	9	7	6⅛	1¼	21/32	15¾	18	18½	1⅞	½ × ¼	3½	23⅞	9⅝	7⅝	½	1¾	730	820
404T	50	60			50	60			20	16¼	32½	37	10	8	6⅛	1½	13/16	20	20	20⅝	2⅞	¾ × ⅜	7	28⅛	13⅞	11⅞	⅝	2¼	830	1050
404TS			100	125					20	16¼	29½	34	10	8	6⅛	1½	13/16	17	20	20⅝	2⅛	½ × ¼	4	25⅛	10⅞	8⅞	⅝	2¼	870	1050
405T	60	75			60	75			20	17¾	34	38½	10	8	6⅞	1½	13/16	20¾	20	20⅝	2⅞	¾ × ⅜	7	29⅝	13⅞	11⅞	⅝	2¼	930	1160
405TS			125	150			100	100	20	17¾	31	35½	10	8	6⅞	1½	13/16	17¾	20	20⅝	2⅛	½ × ¼	4	26⅝	10⅞	8⅞	⅝	2¼	950	1150
444T	75	100			75	100			22	18½	38	42½	11	9	7¼	1½	13/16	23¼	22⅜	23⅛	3⅜	⅞ × 7/16	8¼	32½	16	14	⅝	2¼	1165	1440
444TS			150	200			125	125	22	18½	34	38½	11	9	7¼	1½	13/16	19½	22⅜	23⅛	2⅜	⅝ × 5/16	4½	28¾	12¼	10¼	⅝	2¼	1050	1440
445T	100	125			100	125			22	20½	40	44½	11	9	8¼	1½	13/16	24½	22⅜	23⅛	3⅜	⅞ × 7/16	8¼	34½	16	14	⅝	2¼	1370	1650
445TS			200	250			150	150	22	20½	36	41	11	9	8¼	1½	13/16	20½	22⅜	23⅛	2⅜	⅝ × 5/16	4½	30¾	12¼	10¼	⅝	2¼	1250	1615
447T									22	23¼	43½	48	11	9	10	1½	13/16	26	22⅜	23⅛	3⅜	⅞ × 7/16	8¼	37⅝	16	14⅜	⅝	2¼	1800	2260
447TS									22	23¼	40½	46½	11	9	10	1½	13/16	22¼	22⅜	23⅛	2⅜	⅝ × 5/16	4½	33⅞	12¼	10⅝	⅝	2¼	1800	2260
56									6½	3⅞	10½		3½	2 7/16	1½	⅛	11/32	6 3/16	6⅞		⅝	3/16 × 3/32	1⅞	8⅛	4⅝	4¼	¼	1		
182	½	¾	1	1½	½	¾	1	1½	9	6½	12½	14½	4½	3¾	2¼	⅝	13/32	7¼	9	9	⅞	3/16 × 3/32	2	10½	5	4	5/16	1	60	70
184	¾	1–1½	1½–2	2–3	¾	1–1½	1½–2	2–3	9	7½	13½	15½	4½	3¾	2¾	⅝	13/32	7¾	9	9	⅞	3/16 × 3/32	2	11½	5	4	5/16	1	70	80
213	1–1½	2	3	5	1–1½	2	3	5	10½	7½	15½	17½	5¼	4¼	2¾	¾	13/32	9¼	10½	10⅝	1⅛	¼ × ⅛	2¾	13	6½	5½	5/16	1	105	125
215	2	3	5	7½	2	3	5	7½	10½	9	17	19	5¼	4¼	3½	¾	13/32	10	10½	10⅝	1⅛	¼ × ⅛	2¾	14½	6½	5½	5/16	1	115	140
254U	3	5	7½	10	3	5	7½	10	12½	10¾	20½	22	6¼	5	4⅛	⅞	17/32	12⅛	12⅝	13⅛	1⅜	5/16 × 5/32	3½	17½	8	6¾	⅜	1¼	180	210
256U	5	7½	10	15	5	7½	10	15	12½	12½	22½	24	6¼	5	5	⅞	17/32	13	12⅝	13⅛	1⅜	5/16 × 5/32	3½	19¼	8	6¾	⅜	1¼	210	245
284U	7½	10	15	20	7½	10	15		14	12½	24	25	7	5½	4¾	1	17/32	14⅜	14	14⅝	1⅝	⅜ × 3/16	4⅝	20⅝	9⅝	8⅛	⅜	1½	280	330
286U	10		20	25	10		20	20	14	14	25½	26½	7	5½	5½	1	17/32	15⅛	14	14⅝	1⅝	⅜ × 3/16	4⅝	22⅛	9⅝	8⅛	⅜	1½	325	365
324U		15	25			15	25	25	16	14	26½	28	8	6¼	5¼	1⅛	21/32	16⅛	16	16¾	1⅞	½ × ¼	5⅜	23⅛	10⅞	9⅛	½	1¾	380	480
324S				30					16	14	24½	25½	8	6¼	5¼	1⅛	21/32	13¾	16	16¾	1⅝	⅜ × 3/16	3	20¾	8½	6¾	½	1¾	380	480
326U	15	20	30		15	20	30		16	15½	28	29½	8	6¼	6	1⅛	21/32	16⅞	16	16¾	1⅞	½ × ¼	5⅜	24⅝	10⅞	9⅛	½	1¾	430	560
326S				40				30	16	15½	26	27	8	6¼	6	1⅛	21/32	14½	16	16¾	1⅝	⅜ × 3/16	3	22¼	8½	6¾	½	1¾	430	560
364U	20	25	40		20	25	40		18	15¼	29½	34	9	7	5⅝	1¼	21/32	17⅞	18¼	18¾	2⅛	½ × ¼	6⅛	25½	12¼	10¼	½	1¾	525	720
364US				50				40	18	15¼	27	31	9	7	5⅝	1¼	21/32	15¼	18¼	18¾	1⅞	½ × ¼	3½	22⅞	9⅝	7⅝	½	1¾	670	710
365U	25	30			25	30			18	16¼	30½	35	9	7	6⅛	1¼	21/32	18⅜	18¼	18¾	2⅛	½ × ¼	6⅛	26½	12¼	10¼	½	1¾	580	785
365US			50	60			50	50	18	16¼	28	32	9	7	6⅛	1¼	21/32	15¾	18¼	18¾	1⅞	½ × ¼	3½	23⅞	9⅝	7⅝	½	1¾	730	780
404U	30	40			30	40			20	16¼	32½	37½	10	8	6⅛	1½	13/16	19⅞	20¼	20⅞	2⅜	⅝ × 5/16	6⅞	28	13¾	11¾	⅝	2¼	725	965
404US			60	75					20	16¼	30	34½	10	8	6⅛	1½	13/16	17	20¼	20⅞	2⅛	½ × ¼	4	25⅛	10⅞	8⅞	⅝	2¼	860	1075
405U	40	50			40	50			20	17¾	34	39	10	8	6⅞	1½	13/16	20⅝	20¼	20⅞	2⅜	⅝ × 5/16	6⅞	29½	13¾	11¾	⅝	2¼	810	1110
405US			75	100			60	60	20	17¾	31½	36	10	8	6⅞	1½	13/16	17¾	20¼	20⅞	2⅛	½ × ¼	4	26⅝	10⅞	8⅞	⅝	2¼	970	1165
444U	50	60			50	60			22	18½	38	43	11	9	7¼	1½	13/16	23⅜	22¼	23⅛	2⅞	¾ × ⅜	8⅜	32⅝	16⅛	14⅛	⅝	2¼	985	1315
444US			100	125			75	75	22	18½	34	38½	11	9	7¼	1½	13/16	19	22¼	23⅛	2⅛	½ × ¼	4	28¼	11¾	9¾	⅝	2¼	1175	1355
445U	60	75			60	75			22	20½	40	45	11	9	8¼	1½	13/16	24⅜	22¼	23⅛	2⅞	¾ × ⅜	8⅜	34⅝	16⅛	14⅛	⅝	2¼	1135	1550
445US			125	150			100	100	22	20½	36	40½	11	9	8¼	1½	13/16	20	22¼	23⅛	2⅛	½ × ¼	4	30¼	11¾	9¾	⅝	2¼	1340	1620

Source: Goulds Pumps, Inc.

to the right in Figure 6-40. Since the impeller does not change, more horsepower is required for the lower TDH. This condition is known as the break horsepower (bhp) required at the end of the pump curve, or maximum flow capacity condition. In our case we have a minimum TDH of approximately 45 feet in which the bhp becomes

$$\text{bhp} = \frac{(150)(45)(1.0)}{3{,}960(0.46)} = 3.706 \text{ or 4 hp with water}$$

Thus, we see that our 10-hp motor is sufficient against back pressure. Often, the water condition requires more horsepower, and thus a larger motor, than the process liquid condition. The design engineer must be always cognizant of any other fluid that the specified pump may have to handle.

EXAMPLE 6-2: POSITIVE DISPLACEMENT PUMP DESIGN

A positive-displacement pump is required to transfer a adhesive coating mix from a storage tank to a bin in which the mix is dropped onto a nylon sheet (see Example 3-6). The adhesive coating mix adheres the particles together to form roofing shingles.

First, we must perform a fluid analysis of the system shown in Figure 6-41.

Suction Line Pressure Drop

Fluid = coating mix $\rho = 95.909 \text{ lb}_m/\text{ft}^3$
Temperature = 400°F L = 11.0 ft
Pressure = 20 psig Q = 150 gpm
Suction line = 4 in. Schedule 40 $\Rightarrow D_i = 4.026$ in.
$\epsilon = 0.0018$

$$\mu = 938.08 \text{ cp} = (938.08)(6.72 \times 10^{-4}) = 0.630 \frac{\text{lb}_m}{\text{ft-sec}}$$

$$V = \frac{(150)\dfrac{\text{gal}}{\text{min}}\left(\dfrac{1 \text{ ft}^3}{7.479 \text{ gal}}\right)\left(\dfrac{1 \text{ min}}{60 \text{ sec}}\right)}{(12.73) \text{ in.}^2\left(\dfrac{1 \text{ ft}^2}{144 \text{ in.}^2}\right)} = 3.781 \text{ ft/sec}$$

$$N_{Re} = \frac{DV\rho}{\mu} = \frac{\left(\dfrac{4.026}{12}\right) \text{ft } (3.781)\dfrac{\text{ft}}{\text{sec}}(95.909)\dfrac{\text{lb}_m}{\text{ft}^3}}{(0.630)\dfrac{\text{lb}_m}{\text{ft-sec}}}$$

$$= 193.116$$

From Equation 1-6b we compute the friction factor as

$$f = \frac{64}{N_{Re}} = 0.332$$

K-Values (Velocity Heads) for Suction Line

Referring to Figures 1-7 and 1-11 we have the following:

Entrance and exit = K = 1.0 + 0.78 = 1.78
2–4-in. plug valves = K = 2(18)(0.017) = 0.612
1–4-in.-90° standard elbow = K
= 30(0.017) = 0.510

$\sum K = 2.902$ velocity heads

From Equation 1-4 we compute the frictional pressure drop as follows:

$$\Delta\rho_f = \left[\frac{(0.253)(90.0)(12)}{(3.068)} + 4.048\right] \times \frac{(95{,}909)\dfrac{\text{lb}_m}{\text{ft}^3}(6.511)^2\dfrac{\text{ft}^2}{\text{sec}^2}\left(\dfrac{1 \text{ ft}^2}{144 \text{ in.}^2}\right)}{2(32.2)\dfrac{\text{ft-lb}_m}{\text{sec}^2\text{-lb}_f}}$$

$$\Delta\rho_f = 40.822 \text{ psi}$$

Referring to the pump hydraulic calculation sheet, Figure 6-42, we summarize our results. From this we compute a total dynamic head (TDH) of 93.76 feet. Past experience indicates that a rotary gear pump of the type shown in Figure 6-43 is excellent for handling high viscosity liquids. The pump manufacturer has the performance curves rated in terms of kinematic viscosity in SSU. Now converting our viscosity to SSU's we have

$$\text{SSU} = \frac{\mu(\text{cp})(4.635)}{W/g} \frac{(938.08)(4.635)}{\left(\dfrac{95.909}{32.2}\right)} = 1{,}459.78 \text{ SSU}$$

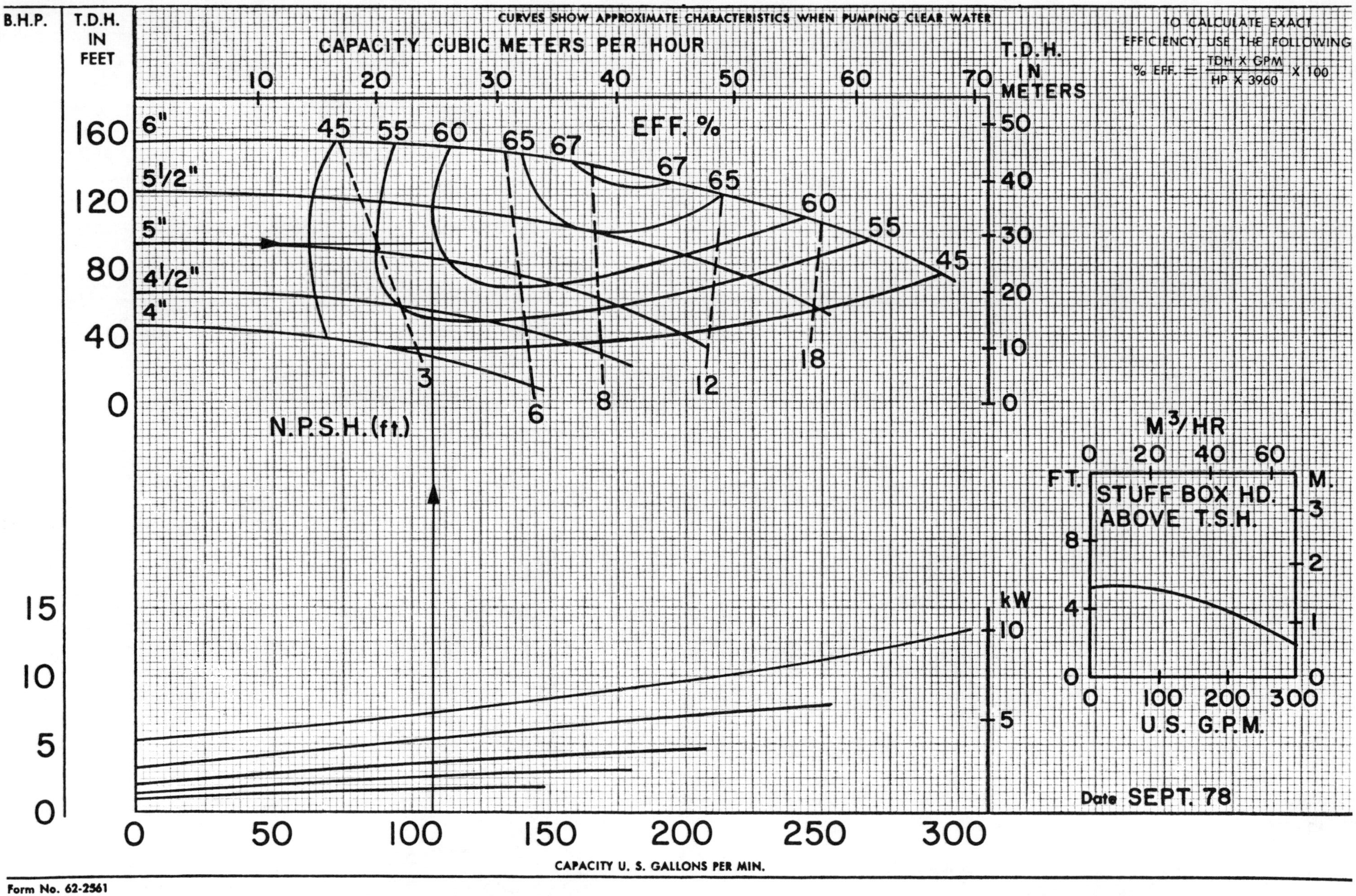

Figure 6-40. Manufacturer's pump performance curve correction for pump in Example 6-1. (Courtesy of the Duriron Company, Inc.)

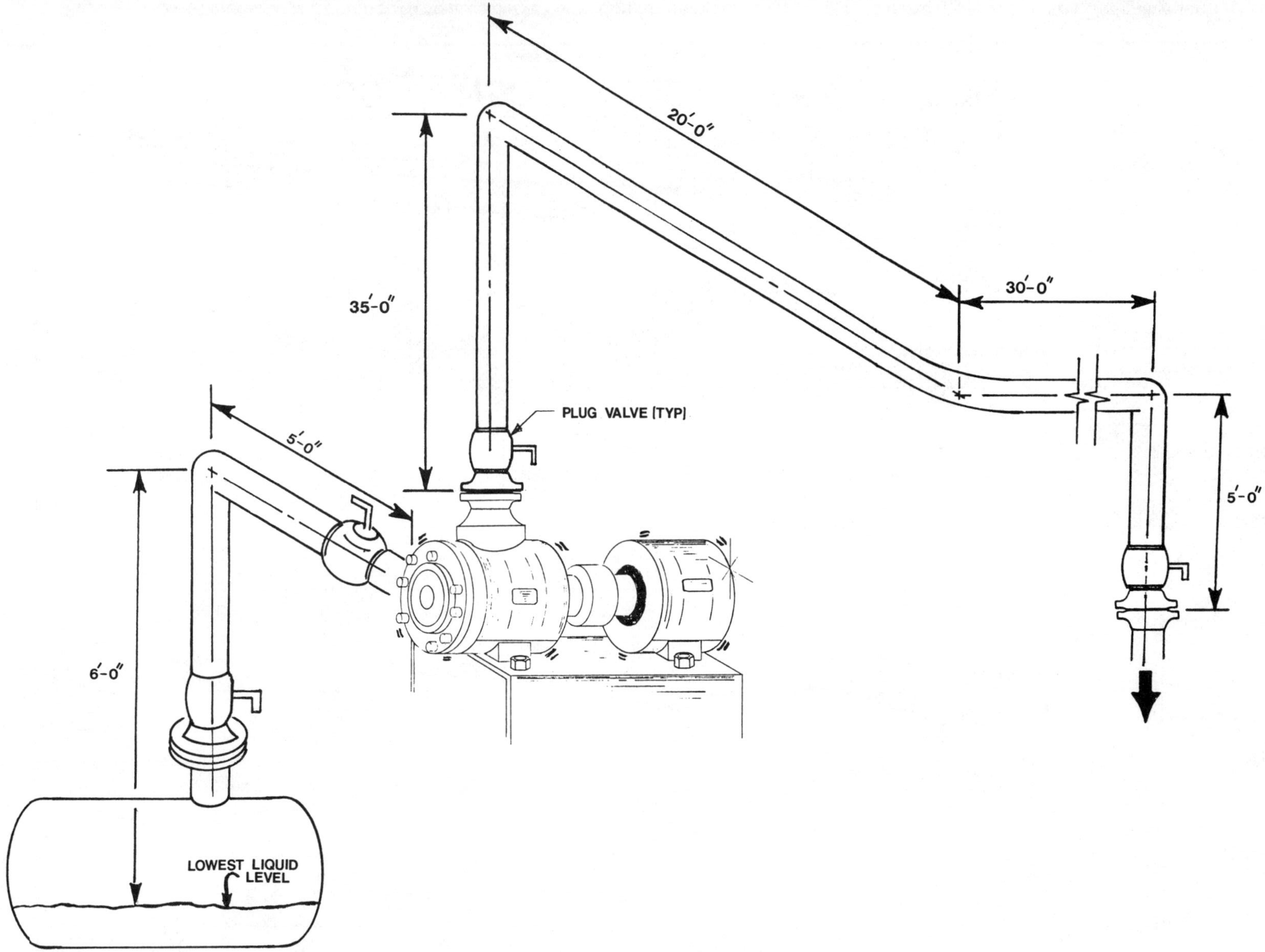

Figure 6-41. Adhesive coating mix rotary gear pump system of Example 6-2.

Pump Hydraulic Design Calculation Sheet

Liquid	adhesive coating mix	
Viscosity at P.T. (Pumping Temp.)	938.08	cp
Vapor pressure at P.T.	—	psia
Sp. gr. (γ) at P.T.	1.537	
Flow at ambient temp.	—	gpm
Operating flow at P.T.	150	gpm
Design flow at P.T.	150	gpm

Suction				**Discharge**			
Source* pressure	=	14.7	psia	Terminal pressure	=	16.70	psia
Static + (head-lift)	=	−4.0	psi	Static lift	=	23.31	psi
$-\ \Delta P_f$ line loss	=	−2.0	psi	$-\ \Delta P_f$ discharge			
Suction pressure	=	8.70	psia	Piping system	=	13.74	psi
− Vapor pressure	=	—	psia	Other	=	—	psi
NPSH avail	=	8.70	psia	Discharge press.	=	53.75	psia
NPSH avail	=	13.05	ft	− Suction press.	=	−8.70	psia
NPSH req'd	=	6.90	ft	TDH	=	45.05	psia
				TDH	=	67.58	feet

2 + 2 = 4 ft required

Min NPSH avail ⩾ NPSH req'd + 2 ft

*Initial press., e.g., ATM press. or press. pad

bhp at Duty Condition

$$bhp_D = \frac{(gpm)(TDH)(\gamma)}{(3{,}960)(\eta)} \Rightarrow \frac{(150)(67.58)(1.537)}{(3{,}960)(10)} = \eta = 39.34\%$$

bhp at Maximum Capacity Condition

$$bhp_{MC} = \frac{(gpm)(TDH)(\gamma)}{(3{,}960)(\eta)}$$

TDH = total dynamic head
TDH = discharge press. − suction press.
η = pump efficiency, %

Figure 6-42. Pump hydraulic design calculation sheet for Example 6-2.

We now refer to the manufacturer's performance curves which, in this case, are rated to the viscosity of the service fluid. The closest curve is that shown in Figure 6-41. As a starting point, it is always desirable to start at the middle of the curve. Extreme ends of any pump performance curve should be avoided, as the pump's performance varies significantly at either end of the curve. Thus, we select a very common speed for this type of pump—155 rpm. Now for 150 gpm and 62.45 psi TDH, we find that we need approximately an 11-hp motor. Solving for the pump efficiency we have

$$bhp = \frac{Q(TDH)\gamma}{(3{,}960)\eta} \qquad (6\text{-}2)$$

Thus, we have

$$\eta = \frac{(150)(93.76)(1.537)}{(3{,}960)(10)} = 0.496 \text{ or } 49.6\%$$

This efficiency rating is quite common with a rotary gear pump handling a highly viscous liquid. Now, referring to Table 6-4 one can observe the classifications of electric motors. From Figure 6-44 we see that the viscosity of our fluid, 1,460 SSU, is about mid-way between the two curves shown. Thus, the required horsepower is between 8 hp and 10 hp. Looking at Table 6-4 we see that electric motors are 7½ hp and 10 hp. To meet our requirements, we select a 10-hp motor, because 7½ hp is too small. Notice that the pump has built-in jacketed enclosures to match the piping, which is hot-oil traced, to keep the fluid in the piping and pump liquid. These jacketed systems are discussed in Chapter 3.

In this problem we have a suction lift on the suction side of the pump. It is important to remember that the theoretical height to which a liquid can be lifted at any specified temperature is the atmospheric pressure at the installation site minus the vapor pressure of the liquid at the specified temperature minus the friction loss in the piping. The theoretical and maximum suction lift for water is shown for various temperatures in Figure 6-14. For non-volatile liquids, the maximum allowable suction lift should never exceed 15 in. Hg (7.4 psia) under ideal conditions. For volatile liquids, the maximum allowable

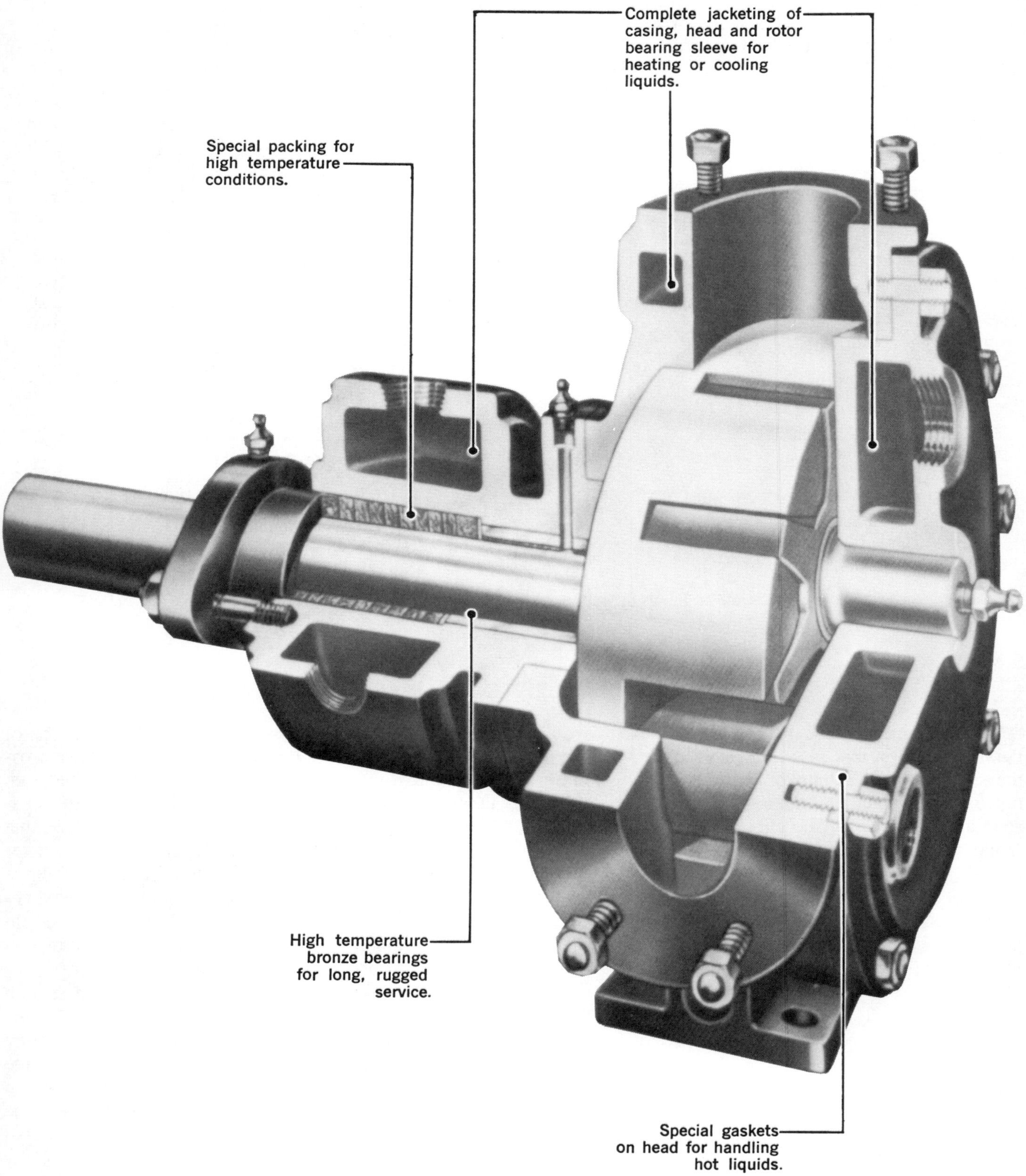

Figure 6-43. The type of gear rotary pump selected in Example 6-2. (Courtesy of Viking Pump Division, Houdaille Industries, Inc.)

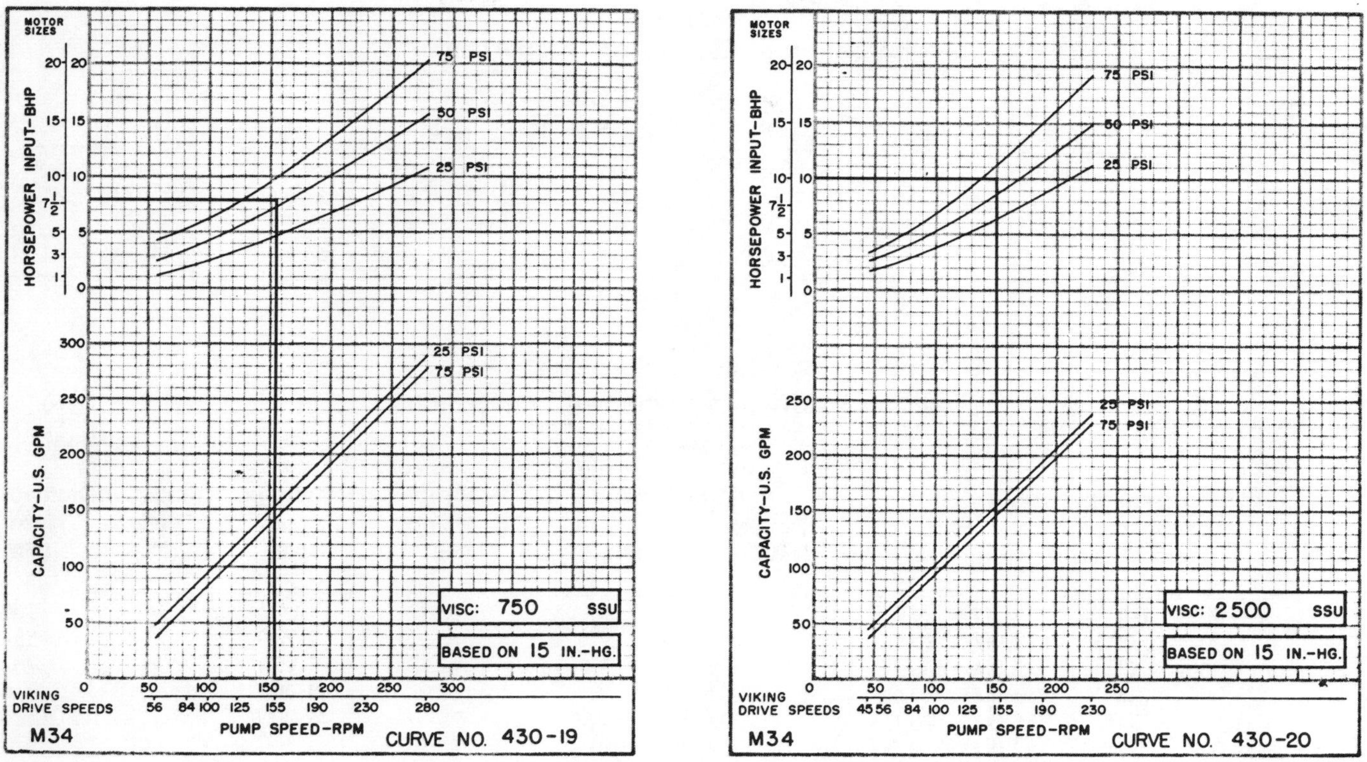

Figure 6-44. Rotary gear pump performance curve. (Courtesy of Viking Pump Division, Houdaille Industries, Inc.)

suction lift should never exceed 10 in. Hg. If these values are exceeded, then the suction source should be pressurized with a neutral gas (inert nitrogen) to offset any pressure that may fall below the vapor pressure of the liquid. At the liquid vapor pressure, vaporization occurs, resulting in possible cavitation and pump damage.

A Word About Priming

A positive-displacement pump, like the rotary gear pump in this example, must be primed when pumping low viscosity liquids. This is done by a vacuum device or by using a foot valve. Also, with a low viscous liquid, the fluid drains back to the suction when the pump is idle. For a viscous liquid, like the one in this example, the liquid is retained in the rotary gear clearances and thus acts as a seal when the pump is restarted. However, before restarting the pump, the liquid being pumped should be introduced through the discharge side of the pump to lubricate the rotating components.

Since the coating mix is not a clean service, a centrifugal pump is impractical because it cannot handle a non-Newtonian fluid containing suspended particles.

EXAMPLE 6-3: CENTRIFUGAL COMPRESSOR SELECTION

A centrifugal compressor is to be specified for a gas plant, which is at sea level. The unit is to compress 3,000 lb_m/min of gas mixture at 50 psia at 60°F to 150 psia. The gas mixture is composed of 40% propane, 30% ethane, and 30% methane.

The reduced pressure, P_c, the reduced temperature, t_c, the molecular weight, and the specific heat of the mixture is determined as shown in Table 6-5. Using the data in the table we calculate the ratio of specific heats for the mixture as follows:

$$k_{mix} = \frac{C_{P_{mix}}}{C_{P_{mix}} - 1.986} \qquad (6\text{-}10)$$

$$k_{mix} = \frac{13.08}{13.08 - 1.986} = 1.18$$

The compressibility factor for the mixture is determined from the reduced pressure and reduced temperature. Thus,

Table 6-5
Tabulation of Gas Mixture Properties

Gas	Mol %	mw	P_c (psia)	t_c (°R)	$C_p \left(\frac{Btu}{lb_m\text{-mol-}°R}\right)$	mw	P_c	t_c	C_p
Propane	40	44.10	616	666	17.14	17.64	246.40	266.40	6.86
Ethane	30	30.07	708	550	12.27	9.02	212.40	165.00	3.68
Methane	30	16.07	668	343	8.45	4.81	200.40	102.90	2.54
					Gas Mixture	31.47	659.20	534.30	13.08

Table 6-6 [6]
Typical Centrifugal Compressor Frame Data*

Frame	Nominal Inlet Volume Flow English (icfm)	Nominal Inlet Volume Flow Metric (m³/h)	Nominal Polytropic Head English (ft-lbf/lbm)	Nominal Polytropic Head Metric (k·Nm/kg)	Nominal Polytropic Efficiency (%)	Rotational Speed (rpm)	Nominal Impeller Diameter English (in.)	Nominal Impeller Diameter Metric (mm)
A	1,000–7,000	1,700–12,000	10,000	30	76	11,000	16	406
B	6,000–18,000	10,000–31,000	10,000	30	76	7,700	23	584
C	13,000–31,000	22,000–53,000	10,000	30	77	5,900	30	762
D	23,000–44,000	39,000–75,000	10,000	30	77	4,900	36	914
E	33,000–65,000	56,000–110,000	10,000	30	78	4,000	44	1,120
F	48,000–100,000	82,000–170,000	10,000	30	78	3,300	54	1,370

**While this table is based on a survey of currently available equipment, the instance of any machinery duplicating this table would be purely coincidental.*

$$P_r = \frac{P}{P_c} = \frac{50}{659.20} = 0.076$$

$$t_r = \frac{t}{t_c} = \frac{60 + 460}{534.30} = 0.973$$

Now from Figure 6-45, we have

$z_1 = 0.972$ = inlet compressibility factor

Using Equation 6-6 the inlet volumetric flow is

$$V = \frac{zm\bar{R}t_i}{(mw)P_i} \tag{6-6}$$

$$V = \frac{(0.972)(3{,}000)(1{,}545)(60 + 460)}{(144)(31.47)(50)}$$

$V = 10{,}339.276$ icfm (or acfm at the inlet)

Using Table 6-6 from Lapina [6], we find our unit to be a Frame B with nominal values to be as follows:

$H_{p_n} = 10{,}000$ ft-lb$_f$/lb$_m$
$N_n = 7{,}700$ rpm
$\eta_{p_n} = 76\%$

Computing the compression ratio we have

$$C_R = \frac{P_2}{P_1} = \frac{150}{50} = 3.0$$

Assuming that we have a perfect gas (z = 1), we can use Equation 6-14 to find the average discharge temperature. Thus, we have

$$\frac{t_2}{t_1} = (C_R)^{\frac{n-1}{n}} \tag{6-14}$$

Now from Equation 6-32 we have

$$\frac{n-1}{n} = \left(\frac{k-1}{k}\right)\eta_p$$

From above,

$k_1 = 1.18$
$\eta_{p_n} = 0.76$

Thus,

$$\frac{n-1}{n} = \left(\frac{0.18}{1.18}\right)(0.76) = 0.116$$

from which

$$t_2 = t_1(C_R)^{\frac{n-1}{n}} = (60 + 460)(3.0)^{0.116}$$

or

$$t_2 = 590.68°R = 130.68°F$$

Now, the average compressibility for the gas mixture must be obtained. From above the inlet compressibility,

$$z_1 = 0.972$$

Compression ratio,

$$(C_R)_2 = \frac{P_2}{P_c} = \frac{150}{659.20} = 0.228$$

Temperature ratio,

$$(T_R)_2 = \frac{t_2}{t_c} = \frac{560.68}{534.30} = 1.05$$

Using the compression ratio and pressure ratio we determine the outlet compressibility factor from the compressibility charts in Appendix E. Thus,

$$z_2 = 0.93$$

$$\bar{z} = \frac{z_1 + z_2}{2} = \frac{0.972 + 0.93}{2} = 0.95$$

In determining the polytropic head we use Equation 6-33, where

$$P_1 = P_s$$

$$P_2 = P_d$$

and the average ratio of specific heat, k, is k = 1.18 = inlet conditions, which is an approximation. Thus,

$$H = \left(\frac{Rt_s}{mw}\right)\left(\frac{\eta_p k}{k-1}\right)\left[\left(\frac{P_d}{P_s}\right)^{k\eta_p/(k-1)} - 1\right](\bar{z}) \quad (6\text{-}33)$$

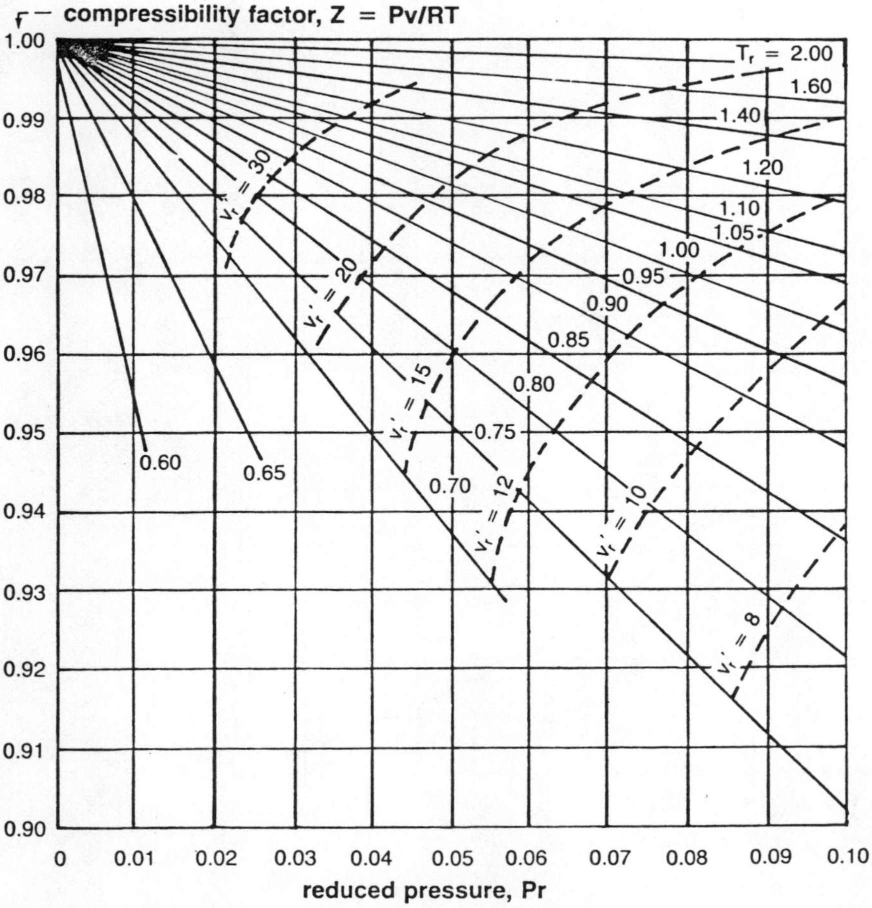

Figure 6-45. Compressibility curves for very low values of reduced pressure. (Reprinted by permission of *Chemical Engineering*, McGraw-Hill Company, July 1954.)

from which

$$H = \left[\frac{(1{,}545)(520)}{31.47}\right] (8.621)\ [(3.0)^{0.116} - 1]$$

$H = 29{,}913.143$ ft-lb$_f$/lb$_m$

The required number of compressor stages is determined by

$$N_{st} = \frac{H_p}{H_{Pm}} \qquad (6\text{-}68)$$

where H_{Pm} = maximum polytropic head per stage, ft-lb$_f$/lb$_m$ (see Figures 6-46 and 6-47)

Using Table 6-5, we have

$$\theta = \left[\frac{(26.1)(mw)}{k_1 z_1 t_1}\right] = \left[\frac{(26.1)(31.47)}{(1.18)(0.972)(520)}\right] \qquad (6\text{-}69)$$

$\theta = 1.377$

From Figure 6-46,

$H_{Pm} = 11{,}000$ ft-lb$_f$/lb$_m$

Thus,

$$N_{st} = \frac{29{,}913.143}{11{,}000} = 2.719 \approx 3$$

The required rpm is

$$N = N_N \left(\frac{H_p}{H_{Pm}\ N_{st}}\right)^{0.5} \qquad (6\text{-}70)$$

$$N = (7{,}700)\left[\frac{29{,}913}{(11{,}000)(3)}\right]^{0.5} = 7{,}331 \text{ rpm}$$

The required shaft power is

$$P_{sf} = \frac{\dot{m}\ H_p}{33{,}000\ \eta_p} = \frac{(3{,}000)(29{,}913)}{(33{,}000)(0.76)}$$

$P_{sf} = 3{,}578.11$ hp

Using Table 6-7 to determine the mechanical losses, L_m, we find that

$L_m = (0.025)(3578.11) = 89.453$ hp

$(P_{sf})_{actual} = P_{sf} + L_m = 3{,}578.11 + 89.453 = 3{,}667.563$ hp

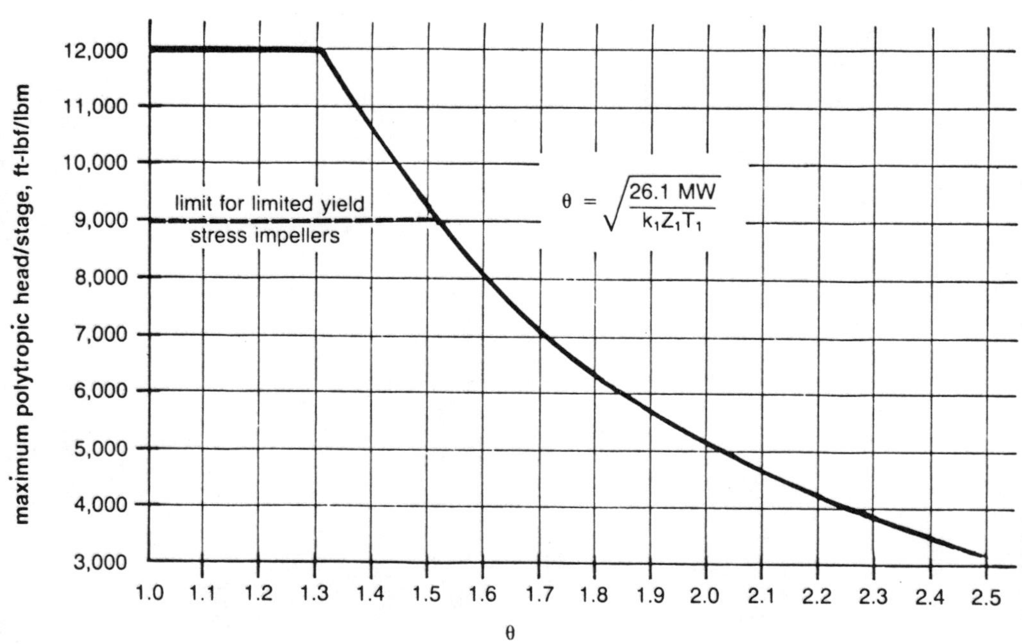

Figure 6-46. Maximum polytropic head per stage—English system [6].

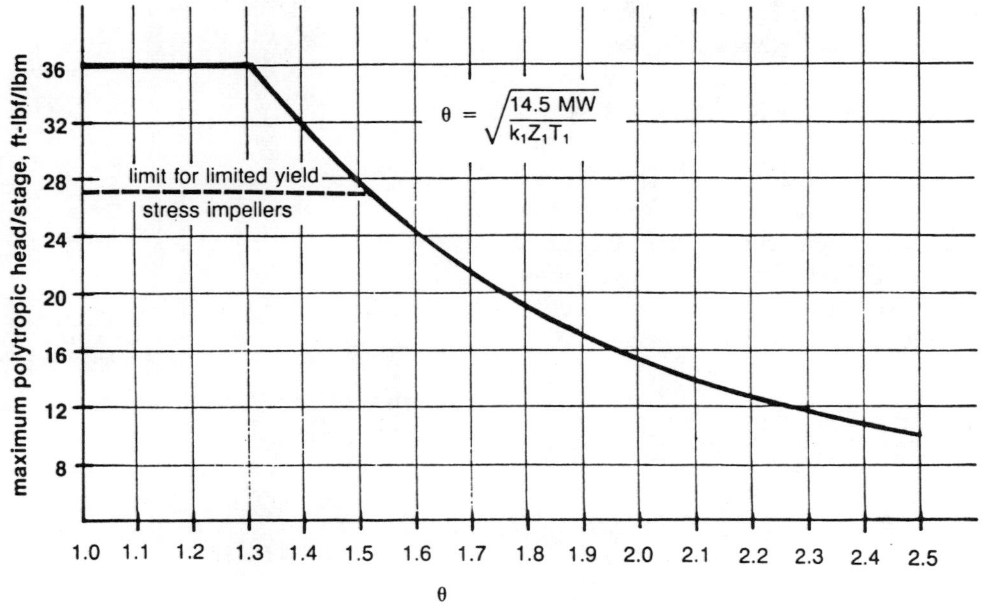

Figure 6-47. Maximum polytropic head per stage—metric system [6].

Table 6-7 [6]
Approximate Mechanical Losses as a Percentage of Gas Power Requirement*

Gas Power Requirement English (hp)	Gas Power Requirement Metric (kW)	Mechanical Losses, L_m (%)
0–3,000	0–2,500	3
3,000–6,000	2,500–5,000	2.5
6,000–10,000	5,000–7,500	2
10,000+	7,500+	1.5

**There is no way to estimate mechanical losses from gas power requirements. This table will, however, ensure that mechanical losses are considered and yieid useful values for estimating purposes.*

The discharge temperature becomes

$$t_2 = t_1(C_R)^{(n-1)/n} = (520)(3.0)^{0.116} = 590.68°R$$

$$t_2 = 130.68°F$$

This example demonstrates how centrifugal compressors are estimated. The reader should be cautioned as when to use inlet values for the values of k and z. The value of k will decrease during the compression process and calculations for the polytropic head and discharge temperature should be made with average values of k, including single stage compressors. Compressor manufacturers use the inlet values at each stage of compression, but the inlet values for each stage will be different. In calculating the polytropic head, the inlet value of k can be used to achieve an approximate value of the head with some error, because the polytropic head is insensitive to the value of k and thus $n/(n - 1)$.

The discharge temperature is much more dependent on the value of k. Using the inlet value of k will yield a conservative value of the discharge temperature, generally 25–50°F in extreme cases.

For a more detailed discussion of the specification and design of centrifugal compressors, the interested reader is referred to Lapina [6].

EXAMPLE 6-4: INSTALLING A COMPRESSOR AT ELEVATION

A reciprocating air compressor is to be installed in a food processing plant, which is at an elevation of 6,562 feet. The desired capacity is 33.3 m³/min. The machine to be used is to be refitted and is of Polish make. From an elevation-barometric conversion chart, such as Figure 6-48, we determine that the atmospheric pressure at the site location is 11.53 psia. The compressor is to compress the air to 7 atmospheres, or 102.87 psia.

Now,

$$V = 33.3\,\frac{m^3}{min}\left(35.314\,\frac{ft^3}{m^3}\right) = 1{,}175.96 \text{ cfm}$$

Compression ratio:

$$P_1 = 11.53 \text{ psia}$$

$$P_f = 102.87 \text{ psia}$$

$$C_R = \frac{102.87}{11.53} = 8.92 > 6, \text{ thus requiring two-stage compression}$$

With an intercooler, you must consider the gas pressure drop across it. The minimum horsepower is developed when the ratios of compression are equal in all cylinders. The ideal case is with no intercooling in which Ludwig [7] suggests

$$\frac{P_2}{P_1} = \frac{P_3}{P_2} = \frac{P_4}{P_3} = \ldots = \frac{P_n}{P_{n-1}} \tag{6-71}$$

and with intercooling,

$$\frac{P_{d_1}}{P_1} = \frac{P_{d_2}}{P'_{d_1}} = \frac{P_{d_3}}{P'_{d_2}} = \ldots \frac{P_{f_n}}{P'_{f_{n-1}}} \tag{6-72}$$

where subscripts 1,2,3, ..., n = gas conditions across a cylinder in which 1 represents the first stage, 2 represents the second stage, etc.

subscript d = interstage discharge pressure condition, directly at the cylinder

prime (′) = represents the actual pressure to the suction of the succeeding cylinder, which is the interstage discharge condition that is reduced by pressure drop over the intercooler system

subscript f = final discharge pressure from a multistage machine

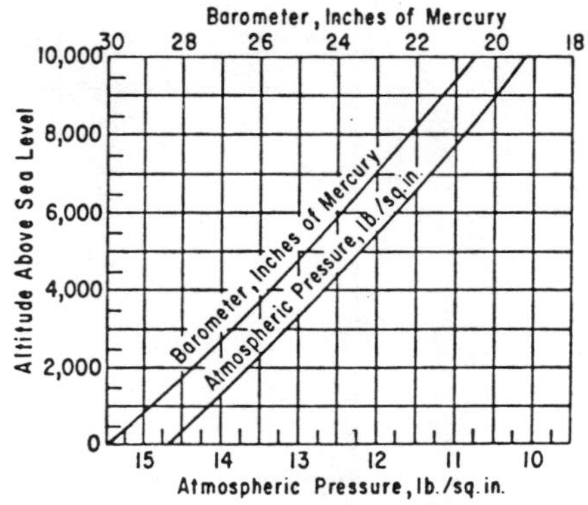

Figure 6-48. Atmospheric and barometric pressures at various altitudes [7].

For a multiple stage unit, the compression ratio is

$$C_{R_1} = C_{R_2} = C_{R_3} = \ldots = C_{R_n} = \left(\frac{P_{f_n}}{P_1}\right)^{1/n} \tag{6-73}$$

where
$$C_{R_1} = \frac{P_{d_1}}{P_1}$$

$$C_{R_2} = \frac{P_{d_2}}{P'_{d\,1}}$$

$$C_{R_3} = \frac{P_{d_3}}{P'_{d\,2}}$$

$$C_{R_n} = \frac{P_{d_n}}{P'_{d\,n-1}}$$

Thus, for two stages,

$$C_{R_1} = C_{R_2} = \left(\frac{P_{f_2}}{P_1}\right)^{0.5} \tag{6-74}$$

Thus, the compression ratio per stage is approximately

$$C_R = (8.92)^{0.5} = 2.99$$

and for the first stage,

$$P_1 = 11.53 \text{ psia}$$

$$P_{d_1} = (2.99)(11.53) + \frac{5}{2} = 36.94 \text{ psia}$$

For second stage,

$$P_{d_1} = (2.99)(11.53) - \frac{5}{2} = 31.97 \text{ psia}$$

$P_{f_2} = 102.87$ psia

The discharge temperature the first stage is by Equation 6-55

$$t_{d_1} = t_1\left(\frac{P_2}{P_1}\right)^{\frac{k-1}{k}}$$

for k = 1.406,

$$t_{d_1} = (85 + 460)(2.99)^{0.289} = 747.94°R$$

or

$$t_{d_1} = 287.94°F$$

The discharge temperature for the second stage is based on the discharge temperature from the intercooler. The intercooler cools the air to 90°F, which is the suction temperature to the second stage. Thus

$$t_{f_2} = t'_{i_1}R_c^{(k-1)/k} = (90 + 460)(2.99)^{0.287} = 754.80°R$$

$$t_{f_2} = 294.80°F$$

Selecting the Reciprocating Compressor

A reliable and quick method to approximate the compressor size is to use the "horsepower per million" curves depicted in Figure 6-49. The "horsepower per million" is the bhp/MMcfd and is used to determine the horsepower per stage by the following relation:

$$hp = \frac{(bhp)}{(MMcfd)}(MMcfd)F_{sg}\left(\frac{z_s + z_d}{2}\right) \quad (6\text{-}75)$$

where F_{sg} is determined in Figure 6-50, converting the acfm to MMcfd we have

$$MMcfd = (1{,}175.96)(60)(24)\left(\frac{11.53}{14.4}\right)\left(\frac{460 + 85}{460 + 60}\right)$$

$$= 1{,}421{,}068.508$$

For the first stage, $F_{sg} = 69.6$

$$bhp = (69.6)\left(\frac{1{,}421{,}068.508}{1.0 \times 10^6}\right) = 98.906 \text{ hp}$$

For the second stage,

$$MMcfd = (1{,}421{,}068.508)\left(\frac{11.53}{14.4}\right)\left(\frac{460 + 90}{460 + 60}\right)$$

$$= 1{,}203{,}486.372$$

$$bhp = (69.6)\left(\frac{1{,}203{,}486.372}{1.0 \times 10^6}\right) = 83.763 \text{ hp}$$

$$\text{Total horsepower} = 98.906 + 83.763 = 182.669 \text{ or } 183 \text{ hp}$$

Equation 6-75 is based on a given compression ratio, C_R, brake horsepower/10^6 ft^3/d at 14.4 psia and suction temperature. F_{sg} is a constant which is a factor for the specific gravity of the gas.

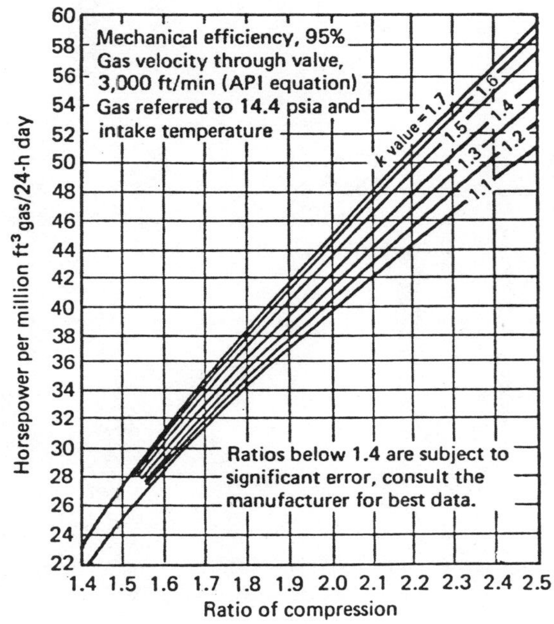

Figure 6-49. Power requirements for reciprocating compressors. (Courtesy of Ingersoll-Rand Company.)

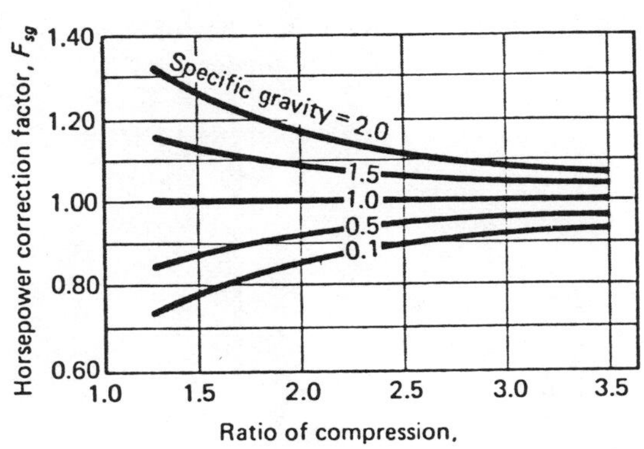

Figure 6-50. Horsepower correction factors for specific gravity [8].

Next, the cylinders must be sized. This can only be done after the interstage temperatures and pressure are defined. Because of the clearance required to allow operation and permit the provision of passages, the piston does not sweep the entire volume of the cylinder.

Thus, the actual cylinder capacity is lower than the displacement of the cylinder. Relating this in terms of volumetric efficiency we have

$$\eta_v = \frac{Q}{C_D} \tag{6-76}$$

where η_v = volumetric efficiency
Q = capacity at inlet conditions, acfm
C_D = cylinder displacement, ft³/min, where

$$C_D = \left(\frac{A_{he} + A_{ce}}{144}\right)\left(\frac{L}{12}\right) N \tag{6-77}$$

where L = piston stroke, in.
A_{he} = area of head end of piston, in.²
A_{ce} = area of crank end piston (A_{he} minus the area of the piston rod), in.²
N = rpm

A convenient formula recommended by Neerken [8] is

$$\eta_v = 0.97 - C_c\left[\frac{(C_R)^{1/k} - 1}{z_d/z_s}\right] \tag{6-78}$$

where C_c = cylinder clearance
C_R = compression ratio
k = ratio of specific heats
z_s, z_d = compressibility factors at the suction and discharge conditions, respectively.

For our machine we have the following design:

L = 220 mm = 8.661 in. = piston stroke

N = 500 rpm

D_1 = 500 mm = 19.685 in. = diameter of first stage cylinder

D_2 = 300 mm = 11.811 in. = diameter of second stage cylinder

For the first stage,

piston rod diameter = 65 mm = 2.559 in.

$$A_{he} = \pi\left(\frac{19.685}{2}\right)^2 = 304.341 \text{ in.}^2$$

$$A_{ce} = 304.341 - \pi\left(\frac{2.559}{2}\right)^2 = 299.198 \text{ in.}^2$$

$$C_s = \left(\frac{304.341 + 299.198}{144}\right)\left(\frac{8.661}{12}\right)(500)$$
$$= 1{,}512.514 \text{ ft}^3/\text{min}$$

For the second stage,

piston rod diameter = 60 mm = 2.362 in.

$$A_{he} = \pi\left(\frac{11.811}{2}\right)^2 = 109.563 \text{ in.}^2$$

$$A_{ce} = 109.563 \text{ in.}^2 - \pi\left(\frac{2.362}{2}\right)^2 \text{ in.}^2 = 105.181 \text{ in.}^2$$

$$C_s = \left(\frac{109.563 + 105.181}{144}\right)\left(\frac{8.661}{12}\right)(500)$$
$$= 538.165 \text{ ft}^3/\text{min}$$

The volumetric efficiency is approximated by Equation 6-76 as

$$\eta_v = 0.97 - (0.13)\left[\frac{(2.99)^{0.711} - 1}{1.0}\right] = 0.817 = 81.7\%$$

This analysis is only a preliminary estimate of what the compressor design is to be, although in this example, data is drawn from an existing unit. The actual selection of a compressor can only be accomplished using the manufacturer's data on such items as piston displacement and the volumetric efficiencies of the cylinders. The manufacturer's data should always be used before attempting a final design. The actual unit in this example is similar to the one shown in Figure 6-51.

A more detailed discussion on how to specify and design reciprocating compressors is given by Chlumsky [5].

EXAMPLE 6-5: NAPHTHA PUMP SYSTEM DESIGN

A cosmetic manufacturer of women's lipstick contracted a chemical company to formulate a chemical that satisfies certain specifications. The chemical process engineers determined that a light cut of naphtha would make an excellent base for the lipstick. The pump in this application can also be used to supply the naphtha to a small chemical company nearby for manufacturing paint thinner. This second application is called the "maximum capacity condition" and will be discussed after the pump is sized for the first application. The pump must be sized for both cases.

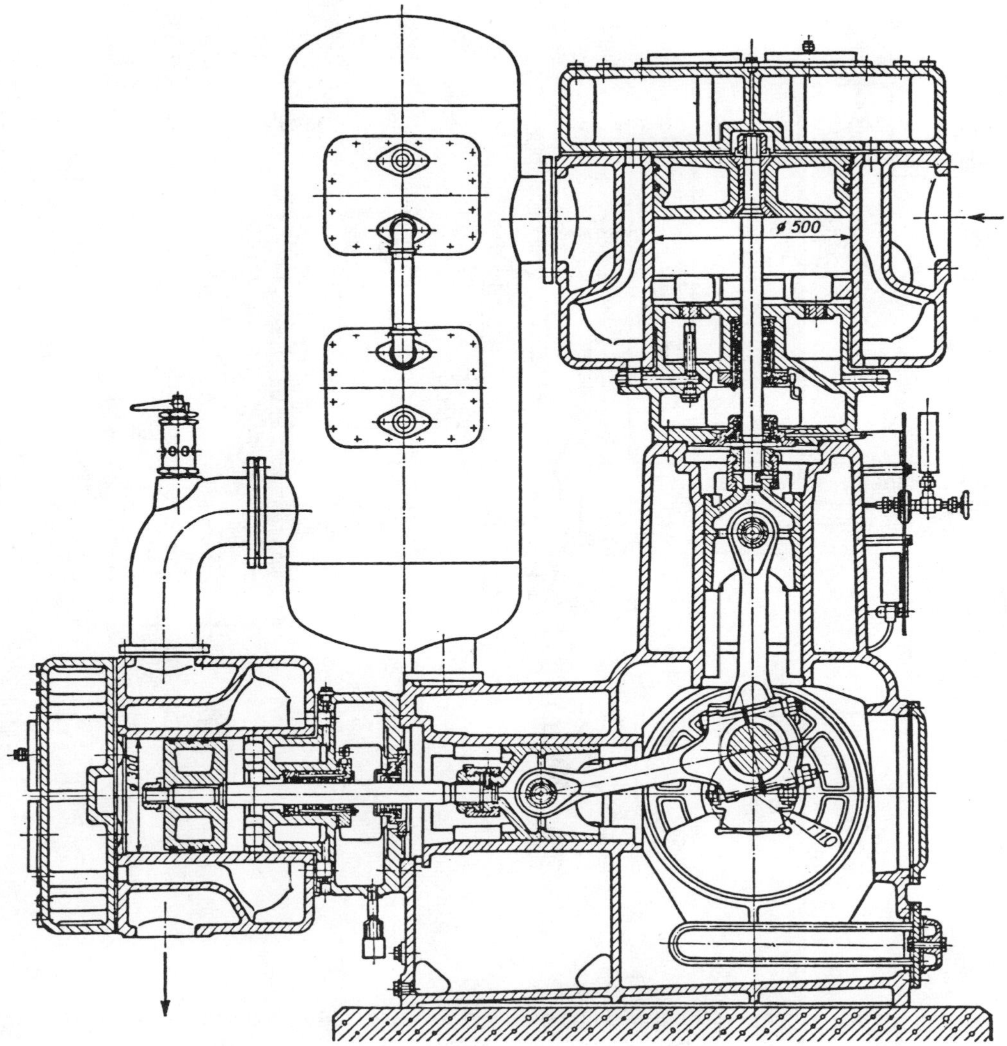

Figure 6-51. Two-stage reciprocating compressor with a shell and tube intercooler. The first stage is achieved with the vertical cylinder and the second stage with the horizontal cylinder. Pistons of the first stage are aluminum and the second stage are cast iron. (Courtesy of Zaklady Budowy Maszyn, Aparatury im Szadkowskiego, Poland.)

In the first case, a rail switcher transports the naphtha to the chemical plant from a nearby refinery. The plant only needs to send one 50,000-gallon railroad tank car once every four months to meet the cosmetic manufacturer's needs. The light naphtha cut is 68°API. The task is to design a pump and hydraulic system that will store and transport the naphtha according to the configuration shown in Figure 6-52.

The reservoir is large enough to consider the fluid as having a constant head. The plant manager estimates that the naphtha head required is 12 feet, but wants to have it evaluated.

The basic process involves the naphtha passing through a scrubber that contains caustic soda (NaOH). The caustic soda removes the straw color in the naphtha, resulting in a colorless liquid. Next, the naphtha is processed through an activated charcoal filter to remove the fuel odor. Finally, the finished process liquid is loaded into the 50,000-gallon tank car.

In the petrochemical industry, the specific gravity of petroleum is given in terms of hydrometer termed °API. The relation for API is as follows:

$$^{\circ}\text{API} = \frac{141.5}{\frac{\gamma_p}{\gamma_w}} = 131.5 \qquad (6\text{-}79)$$

where γ_p = the specific gravity of the petroleum product at 60°F

γ_w = the specific gravity of water at 60°F

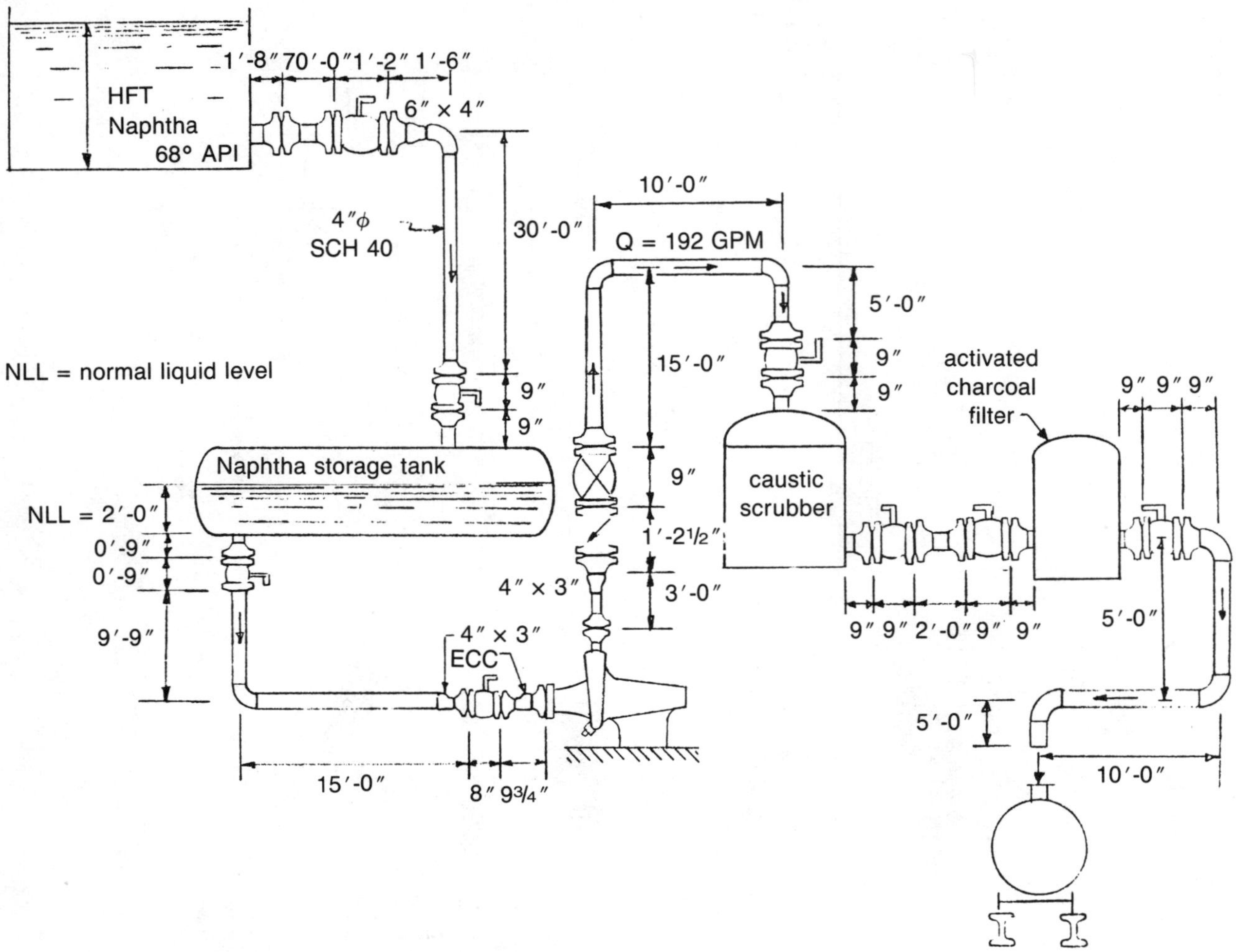

Figure 6-52. Pump-piping scheme of light naphtha cut used to manufacture women's lipstick. (Example 6-5).

The relationship between the °API, temperature is given in Figure 6-53. For our case of 68°API, using Equation 6-79, we have

$$\frac{\gamma_p}{\gamma_w} = \frac{141.5}{199.5} = 0.709$$

in which $\gamma_p = (0.709)(62.4)\text{lb/ft}^3 = 44.26\ \text{lb/ft}^3$ at 60°F

The maximum pumping temperature is controlled at 90°F. The coldest pumping temperature is at 34°F. Since the density is higher at the lower temperature, that is the one used for frictional pressure drop calculations. Thus, referring to Figure 6-53

$$\gamma_p = 0.73$$

and

$$\rho = 45.55\ \text{lb/ft}^3$$

The Flow from the Reservoir to Naphtha Storage Tank

The reservoir is of such large magnitude that the head of liquid is considered constant, because the railroad switch engine delivers the naphtha regularly to the plant.

The flow rate from the reservoir to the storage tank in gallons per minute is determined from the following expression:

$$Q = 19.65\ d^2 \left(\frac{h_L}{K}\right)^{0.5} \tag{6-79}$$

The velocity heads on the line from point ① to point ② are as follows:

Values of f_T are determined from Figure 1-7.

$$\text{Entrance: } K = 0.78 = 0.78$$
$$\text{2–4-in. plug valve: } K = 18\,f_T = 18(0.017)(2) = 0.612$$
$$\text{Exit: } K = 1.00 = 1.00$$
$$\sum K = 2.392$$

$$Q = 19.65(4.026)\left[\frac{12.00}{2.392}\right]^{0.5} = 177.2 \text{ gpm}$$

$$\mu = 1.1 \text{ cp}\left(\frac{2.4 \text{ lb}_m/\text{ft-hr}}{1 \text{ cp}}\right) = 2.64\,\frac{\text{lb}_m}{\text{ft-hr}}$$

$$V = \frac{Q}{A} = \left[\frac{(177.2)\left(\frac{\text{gal}}{\text{min}}\right)\left(\frac{1 \text{ ft}^3}{7.479 \text{ gal}}\right)\left(\frac{1 \text{ min}}{60 \text{ sec}}\right)}{0.0884 \text{ ft}^2}\right]$$

$$= 4.467\,\frac{\text{ft}}{\text{sec}}$$

$$N_{Re} = \frac{DV\rho}{\mu} = \frac{\left(\frac{4.026}{12}\right)\text{ft }(4.467)\times\frac{\text{ft}}{\text{sec}}\,(45.55)\,\frac{\text{lb}_m}{\text{ft}^3}}{2.640\,\frac{\text{lb}_m}{\text{ft-hr}}\left(\frac{1 \text{ hr}}{3{,}600 \text{ sec}}\right)}$$

$$= 93{,}088$$

Using N_{Re} to check the friction factor,

$$f^{-0.5} = -2\log_{10}\left(\frac{\epsilon/d}{3.7} + \frac{2.51}{N_{Re}f^{0.5}}\right) \tag{1-6a}$$

$$(0.03195)^{-0.5} = -2\log_{10}\left[\frac{\left(\frac{0.0018}{0.336}\right)}{3.7} + \frac{2.51}{(93{,}088)(0.17875)}\right]$$

$$f = 0.03198$$

Now,

$$\Delta P_f = \left(\frac{fL}{d} + \sum K_i\right)\frac{\rho V^2}{2g_c} \tag{1-4}$$

$$\Delta P_f = \left[\frac{(0.03198)(105.83)\text{ft}}{\left(\frac{4.026}{12}\right)\text{ft}} + 2.392\right] \times \frac{(45.55)\frac{\text{lb}_m}{\text{ft}^3}(4.467)^2\,\frac{\text{ft}^2}{\text{sec}^2}\,\frac{1 \text{ ft}^2}{144 \text{ in.}^2}}{2(32.2)\,\frac{\text{ft-lb}_m}{\text{sec}^2\text{-lb}_f}}$$

$\Delta P_f = 1.223$ psi

12 ft of naphtha head = (12)(0.325) psi = 3.900 psi

3.90 psi < 10.5 psi nitrogen pad

This pressure differential will cause the naphtha to be forced back into the reservoir. The number of feet required to deliver the liquid to the tank will now be determined. Since we already have 12 ft in the tank, then

$x + 3.90 \geqslant 10.5$ psi

$x = 6.60$ psi $= 20.308$ ft

Adding an additional 26 ft of head we have

38 ft = 12.35 psi − 1.223 psi = 11.127 psi > 10.5 psi

The new flow rate is

$$Q = 19.65(4.026)\left(\frac{38}{2.392}\right)^{0.5} = 315.317 \text{ gpm}$$

$$V = \left[\frac{(315.317)\left(\frac{\text{gal}}{\text{min}}\right)\left(\frac{1 \text{ ft}^3}{7.479 \text{ gal}}\right)\left(\frac{1 \text{ min}}{60 \text{ sec}}\right)}{0.0884 \text{ ft}^2}\right] = 7.948\,\frac{\text{ft}}{\text{sec}}$$

$$N_{Re} = \frac{\left(\frac{4{,}026}{12}\right)\text{ft }(7.948)\times\frac{\text{ft}}{\text{sec}}\,(45.55)\,\frac{\text{lb}_m}{\text{ft}^3}}{2.640\,\frac{\text{lb}}{\text{ft-hr}}\left(\frac{1 \text{ hr}}{3{,}600 \text{ sec}}\right)} = 165{,}633$$

Applying Equation 1-6a, $f = 0.0319$

$$\Delta P_f = \left[\frac{(0.0319)(105.83)\text{ ft}}{\frac{4.026}{12}\text{ ft}} + 2.392\right] \times \frac{(45.55)(7.948)^2\,\frac{\text{ft}^2}{\text{sec}^2}\left(\frac{1 \text{ ft}^2}{144 \text{ in.}^2}\right)}{2(32.2)\frac{\text{ft-lb}_m}{\text{sec}^2\text{-lb}_f}}$$

$\Delta P_f = 3.864$ psi

38 ft − ΔP_f = 12.35 psi − 3.864 psi = 8.486 psi

8.486 psi < 10.5 psi pad

Select a 6-in. ϕ Sch 40 pipe

Repeating the hydraulic analysis we have

$$\text{Entrance and exit: } K = 1.78$$
$$\text{2-6-in. } \phi \text{ plug valve: } K = 18\ f_T = 18(0.015) = 0.27$$
$$\sum K = 2.050$$

$$Q = 19.65(6.065)\left(\frac{38}{2.050}\right)^{0.5} = 513.107 \text{ gpm}$$

$$V = \frac{(513.107)\left(\frac{1}{7.479}\right)\left(\frac{1}{60}\right)}{0.2006} = 5.700\frac{\text{ft}}{\text{sec}}$$

$$N_{Re} = \frac{\left(\frac{6.065}{12}\right)(5.70)(45.55)}{2.640\left(\frac{1}{3,600}\right)} = 178,942$$

$$f = 0.01803$$

$$\Delta P_f = \left[\frac{(0.01803)(105.83)}{\left(\frac{6.065}{12}\right)} + 2.050\right]\frac{(45.55)(5.70)^2\left(\frac{1}{144}\right)}{2(32.2)}$$

$$\Delta P_f = 0.930 \text{ psi}$$

$38 \text{ ft} - \Delta P_f = 12.35 \text{ psi} - 0.930 \text{ psi} = 11.42 \text{ psi} > 10.5 \text{ psi}$

So there is 0.92 psi (11.42 − 10.50) net positive pressure head of naphtha entering the storage tank.

Naphtha Pump Hydraulics

Suction Line

For 4-in. Sch 40 portion of line, L = 23.313 ft

K-Values

$$\text{Entrance and exit: } K = 1.75$$
$$\text{1-4-in. plug valve: } K = 0.306$$
$$\text{4-in.} \times \text{3-in. reducer: } K = 0.163$$
$$\sum K = 2.219$$

For 3-inch Sch 40 portion of line:

K-Values

$$\text{3-in.} \phi \text{ plug valve: } K = (18)(0.018) = 0.324$$
$$\text{4-in.} \times \text{3-in. diffuser: } K = K_1 = 0.055$$
$$\sum K = 0.379$$

To determine the flow rate we must consider what the system is to service. Plant operations dictate that the loading of the tank car must not take longer than four and one-half hours. The rail tank car capacity is 50,000 gallons. We select 4.35 hours, which yields a flow rate of

$$\frac{50,000 \text{ gal}}{4.35 \text{ hrs}}\left(\frac{1 \text{ hr}}{60 \text{ min}}\right) = 191.571 \approx 192 \text{ gpm}$$

We will size a centrifugal pump with 192 gpm capacity. For 192 gpm,

$$V_4 = \frac{(192)\left(\frac{1}{7,479}\right)\left(\frac{1}{60}\right)}{(0.0884)} = 4.84 \text{ ft/sec}$$

$$N_{Re} = \frac{\left(\frac{4.026}{12}\right)(4.84)(45.55)}{2.640\left(\frac{1}{3,600}\right)} = 100,863$$

$f_4 = 0.032$ (from Equation 1-6a)

$$\Delta P_{f_4} = \left[\frac{(0.032)(23.313)}{\left(\frac{4.026}{12}\right)} + 2.219\right]\frac{(45.55)(4.84)^2\left(\frac{1}{144}\right)}{2(32.2)}$$

$$\Delta P_{f_4} = 0.511 \text{ psi}$$

For a 3-in. line, L = 1.0 ft

$$V_3 = \frac{(192)\left(\frac{1}{7.479}\right)\left(\frac{1}{60}\right)}{(0.05130)} = 8.34 \text{ ft/sec}$$

$$N_{Re} = \frac{\left(\frac{3.068}{12}\right)(8.34)(45.55)}{2.640\left(\frac{1}{3,600}\right)} = 132,449$$

$$f = 0.0344$$

$$\Delta P_{f_3} = \left[\frac{(0.0344)(1.0)}{\left(\frac{3.068}{12}\right)} + 0.379\right]\frac{(45.55)(8.34)^2\left(\frac{1}{144}\right)}{2(32.2)}$$

$$\Delta P_{f_3} = 0.175 \text{ psi}$$

The total pressure drop for the suction line = ΔP_s

$$\Delta P_s = \Delta P_{f_4} + \Delta P_{f_3} = 0.686 \text{ psi}$$

Discharge Line

L = 60.708 ft, 4-in. Sch 40

K-Values

1–4-in. swing check: K = (100)(0.017) = 1.70
1–4-in. gate valve: K = (8)(0.017) = 0.136
4–4-in. plug valves: K = (4)(18)(0.017) = 1.224
5–4-in. std 90° elbows: K = (5)(30)(0.017) = 2.550
Entrance: K = 1.0 = 1.00

$$\sum K = 6.610$$

$$V_4 = \frac{(192)\left(\frac{1}{7.479}\right)\left(\frac{1}{60}\right)}{(0.0884)} = 4.84 \text{ ft/sec}$$

$$N_{Re_4} = \frac{\left(\frac{4.026}{12}\right)(4.84)(45.55)}{2.640\left(\frac{1}{3,600}\right)} = 100,863$$

f = 0.032 (from Equation 1-6a)

$$\Delta P_{f_4} = \left[\frac{(0.032)(60.708)}{\left(\frac{4.026}{12}\right)} + 6.610\right]\frac{(45.55)(4.84)^2\left(\frac{1}{144}\right)}{2(32.2)}$$

$$\Delta P_{f_4} = 1.427 \text{ psi}$$

For the 3-in. portion of the discharge line, L = 3 ft. For 3-in. Sch 40 pipe, d_i = 3.068 in.

K-Values

Entrance: K = 0.780
4-in. × 3-in. reducer: K = 0.163

$$\sum K = 0.943$$

$$V_3 = \frac{(192)\left(\frac{1}{7.479}\right)\left(\frac{1}{60}\right)}{(0.0513)} = 8.34 \text{ ft/sec}$$

$$N_{Re_3} = \frac{\left(\frac{3.068}{12}\right)(8.34)(45.55)}{2.640\left(\frac{1}{3,600}\right)} = 132,449$$

$$f_3 = 0.0344$$

$$\Delta P_{f_3} = \left[\frac{(0.0344)(3.0)}{\left(\frac{3.068}{12}\right)} + 0.943\right]\frac{(45.55)(8.34)^2\left(\frac{1}{144}\right)}{2(32.2)}$$

$$\Delta P_{f_3} = 0.460 \text{ psi}$$

The total pressure drop for the discharge line = ΔP_D

$$\Delta P_D = \Delta P_{f_4} + \Delta P_{f_3} = 1.427 + 0.460 = 1.887 \text{ psi}$$

From Figure 6-54, the pump hydraulic design calculation data sheet, it is obvious that the available NPSH is much higher than the required NPSH. This means that the 10.5 psi pressure for the nitrogen pad is excessive. The minimum pad pressure required is

$$\text{ATM. pressure (psia)} + x + \text{static head (psi)} = \left(\begin{array}{c}\text{friction pressure}\\ \text{drop on suction line } (\Delta P_s)\end{array}\right) = \left(\begin{array}{c}\text{liquid vapor}\\ \text{pressure (psia)}\end{array}\right)$$

where x = minimum pad pressure required, psig

$$14.7 + x + 3.557 = 0.511 + 20.85 + 21.361 \text{ psi}$$

select x = 7 psig

Referring to Figure 6-55 and 6-56, we re-evaluate the pump performance. Since the light naphtha cut has a low viscosity

$$\text{bhp} = \frac{QH\gamma}{3,960\eta_c} \quad (6\text{-}2)$$

$$\text{bhp} = \frac{(192)(63.77)(0.73)}{(3,960)(0.61)} = 3.7 \text{ or a 4 hp motor}$$

The Maximum Capacity Condition

The small chemical company nearby that manufactures paint thinner needs the naphtha only about once a year. However, when the naphtha is needed, it must be delivered quickly. Consequently, delivery time is crucial to the client.

Equivalents of Degrees API, Degrees Baumé, Specific Gravity, Weight Density, and Pounds Per Gallon at 60 F/60 F

Degrees on API or Baumé Scale	Values for API Scale — Oil			Values for Baumé Scale — Liquids Lighter Than Water			Values for Baumé Scale — Liquids Heavier Than Water		
	Specific Gravity S	Weight Density, Lb/Ft³ ρ	Pounds per Gallon	Specific Gravity S	Weight Density, Lb/Ft³ ρ	Pounds per Gallon	Specific Gravity S	Weight Density, Lb/Ft³ ρ	Pounds per Gallon
0	...	...	...	...	...	...	1.0000	62.36	8.337
2	...	...	...	...	...	...	1.0140	63.24	8.454
4	...	...	...	...	...	...	1.0284	64.14	8.574
6	...	...	...	...	...	...	1.0432	65.06	8.697
8	...	...	...	...	...	...	1.0584	66.01	8.824
10	1.0000	62.36	8.337	1.0000	62.36	8.337	1.0741	66.99	8.955
12	0.9861	61.50	8.221	0.9859	61.49	8.219	1.0902	67.99	9.089
14	0.9725	60.65	8.108	0.9722	60.63	8.105	1.1069	69.03	9.228
16	0.9593	59.83	7.998	0.9589	59.80	7.994	1.1240	70.10	9.371
18	0.9465	59.03	7.891	0.9459	58.99	7.886	1.1417	71.20	9.518
20	0.9340	58.25	7.787	0.9333	58.20	7.781	1.1600	72.34	9.671
22	0.9218	57.87	7.736	0.9211	57.44	7.679	1.1789	73.52	9.828
24	0.9100	56.75	7.587	0.9091	56.70	7.579	1.1983	74.73	9.990
26	0.8984	56.03	7.490	0.8974	55.97	7.482	1.2185	75.99	10.159
28	0.8871	55.32	7.396	0.8861	55.26	7.387	1.2393	77.29	10.332
30	0.8762	54.64	7.305	0.8750	54.57	7.295	1.2609	78.64	10.512
32	0.8654	53.97	7.215	0.8642	53.90	7.205	1.2832	80.03	10.698
34	0.8550	53.32	7.128	0.8537	53.24	7.117	1.3063	81.47	10.891
36	0.8448	52.69	7.043	0.8434	52.60	7.031	1.3303	82.96	11.091
38	0.8348	52.06	6.960	0.8333	51.97	6.947	1.3551	84.51	11.297
40	0.8251	51.46	6.879	0.8235	51.36	6.865	1.3810	86.13	11.513
42	0.8155	50.86	6.799	0.8140	50.76	6.786	1.4078	87.80	11.737
44	0.8063	50.28	6.722	0.8046	50.18	6.708	1.4356	89.53	11.969
46	0.7972	49.72	6.646	0.7955	49.61	6.632	1.4646	91.34	12.210
48	0.7883	49.16	6.572	0.7865	49.05	6.557	1.4948	93.22	12.462
50	0.7796	48.62	6.499	0.7778	48.51	6.484	1.5263	95.19	12.725
52	0.7711	48.09	6.429	0.7692	47.97	6.413	1.5591	97.23	12.998
54	0.7628	47.57	6.359	0.7609	47.45	6.344	1.5934	99.37	13.284
56	0.7547	47.07	6.292	0.7527	46.94	6.275	1.6292	101.60	13.583
58	0.7467	46.57	6.225	0.7447	46.44	6.209	1.6667	103.94	13.895
60	0.7389	46.08	6.160	0.7368	45.95	6.143	1.7059	106.39	14.222
62	0.7313	45.61	6.097	0.7292	45.48	6.079	1.7470	108.95	14.565
64	0.7238	45.14	6.034	0.7216	45.00	6.016	1.7901	111.64	14.924
66	0.7165	44.68	5.973	0.7143	44.55	5.955	1.8354	114.46	15.302
68	0.7093	44.23	5.913	0.7071	44.10	5.895	1.8831	117.44	15.699
70	0.7022	43.79	5.854	0.7000	43.66	5.836	1.9333	120.57	16.118
72	0.6953	43.36	5.797	0.6931	43.22	5.778	...	...	...
74	0.6886	42.94	5.741	0.6863	42.80	5.722	...	...	...
76	0.6819	42.53	5.685	0.6796	42.38	5.666	...	...	...
78	0.6754	42.12	5.631	0.6731	41.98	5.612	...	...	...
80	0.6690	41.72	5.577	0.6667	41.58	5.558	...	...	...
82	0.6628	41.33	5.526	0.6604	41.19	5.506	...	...	...
84	0.6566	40.95	5.474	0.6542	40.80	5.454	...	...	...
86	0.6506	40.57	5.424	0.6482	40.42	5.404	...	...	...
88	0.6446	40.20	5.374	0.6422	40.05	5.354	...	...	...
90	0.6388	39.84	5.326	0.6364	39.69	5.306	...	...	...
92	0.6331	39.48	5.278	0.6306	39.33	5.257	...	...	...
94	0.6275	39.13	5.231	0.6250	38.98	5.211	...	...	...
96	0.6220	38.79	5.186	0.6195	38.63	5.165	...	...	...
98	0.6166	38.45	5.141	0.6140	38.29	5.119	...	...	...
100	0.6112	38.12	5.096	0.6087	37.96	5.075	...	...	...

Figure 6-53. Relationship between °API and temperature. (Courtesy of Crane Company.)

Pump Hydraulic Design Calculation Sheet

Liquid	Light Naphtha Cut—68° API	
Viscosity at P.T. (Pumping Temp.)	1.1	cp
Vapor pressure at P.T.	20.85	psia
Sp. gr. (γ) at P.T.	0.73	
Flow at ambient temp.	192	gpm
Operating flow at P.T.	192	gpm
Design flow at P.T.	192	gpm

Suction			Discharge		
Source pressure =	25.20	psia	Terminal pressure =	16.7	psia
Static head =	3.559	psi	Static head (lift) =	6.313	psi
$-\Delta P_f$, line loss =	−0.511	psi	$-\Delta P_f$ discharge		
Suction pressure =	28.248	psia	Piping system =	1.887	psi
− Vapor pressure =	−20.85	psia	Other =	20.0	psi
NPSH avail =	7.398	psia	Discharge press. =	44.90	psia
NPSH avail =	23.39	ft	− Suction press. =	28.248	psia
NPSH req'd =	1.3	ft	TDH =	16.652	psia
			TDH =	52.643	feet

bhp at Duty Condition

$$bhp_D = \frac{(gpm)(TDH)(\gamma)}{(3{,}960)(\eta)}$$

bhp at Maximum Capacity Condition

$$bhp_{BP} = \frac{(gpm)(TDH)(\gamma)}{(3{,}960)(\eta)}$$

Figure 6-54. Pump hydraulic design calculation sheet for Example 6-5.

Referring to the pump manufacturer's pump performance curve, Figure 6-55, we see that approximately 400 gpm is the maximum limit. Using this flow rate we re-evaluate the pump for the maximum capacity case.

Suction Line

Referring to previous calculations on the suction side we have the following:

$$V_4 = \left(\frac{400}{192}\right)(4.84) = 10.08\frac{ft}{sec}$$

$$N_{Re} = \left(\frac{10.08}{4.84}\right)(100{,}863) = 210{,}062$$

From Equation 1-6a we obtain

$f = 0.0315$

$$\Delta P_{f_4} = \left[\frac{(0.0315)(23.313)}{\left(\frac{4.026}{12}\right)} + 2.219\right]\frac{(45.55)(10.08)^2\left(\frac{1}{144}\right)}{2(32.2)}$$

$\Delta P_{f_3} = 2.200$ psi

For the 3-in. portion of the suction line,

$$V_3 = \left(\frac{400}{192}\right)(8.34) = 17.375\frac{ft}{sec}$$

$$N_{Re_3} = \left(\frac{400}{192}\right)(132{,}449) = 275{,}935$$

From Equation 1-6a,

$f = 0.03395$

$$\Delta P_{f_3} = \left[\frac{(0.03395)(1.0)}{\left(\frac{3.068}{12}\right)} + 0.379\right]\frac{(45.55)(17.375)^2\left(\frac{1}{144}\right)}{2(32.2)}$$

$\Delta P_f = 0.759$ psi

$\Delta P_s = \Delta P_{f_4} + \Delta P_{f_3}$

$\Delta P_s = 2.20 + 0.759$

$\Delta P_s = 2.959$ psi

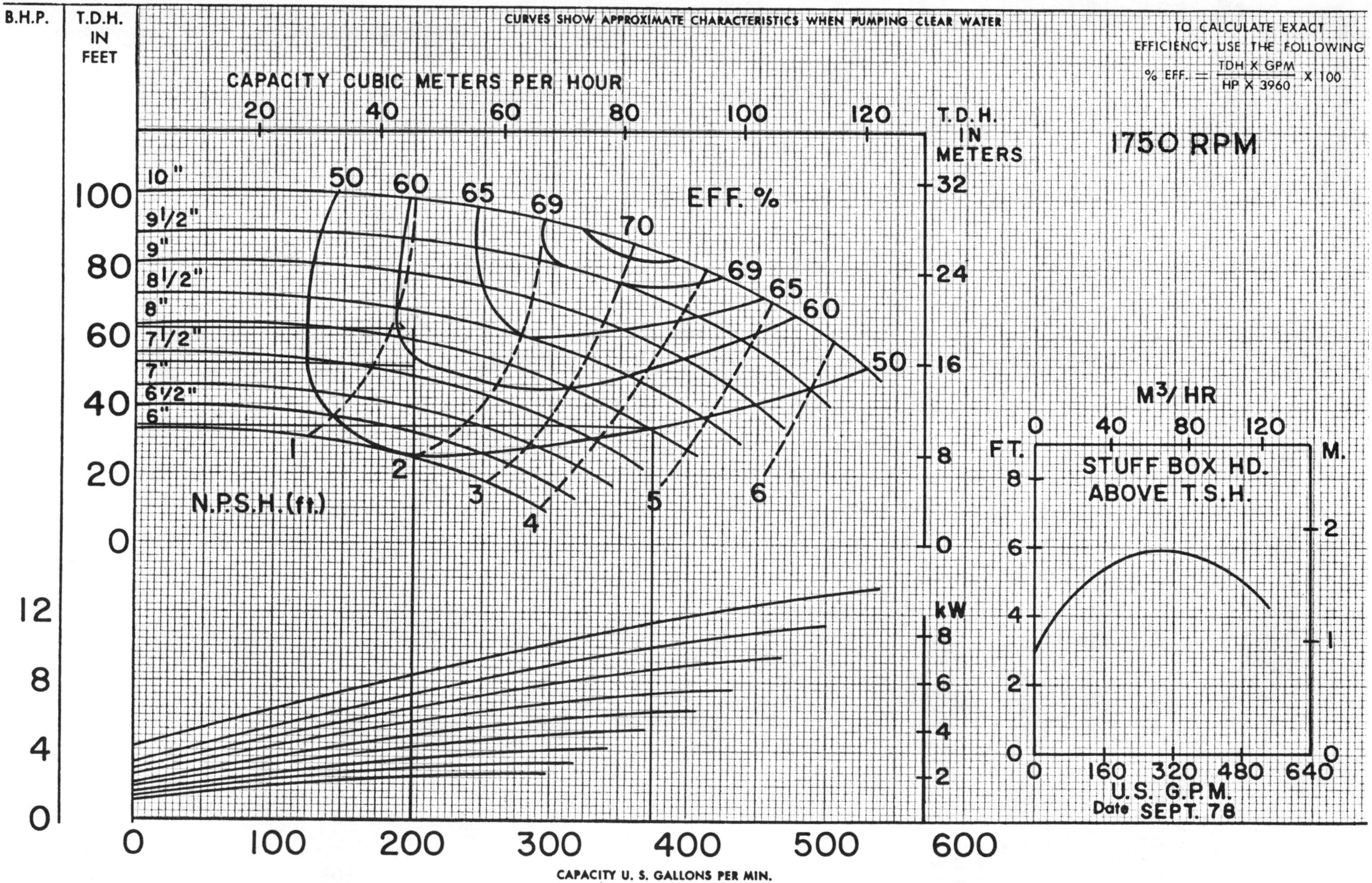

Figure 6-55. Pump performance curve for light naphtha cut. Line 2 is the re-evaluated condition for the naphtha; line 3 is the maximum capacity condition. (Courtesy of the Duriron Company.)

Pump Hydraulic Design Calculation Sheet

Liquid	Light Naphtha Cut—68° API	
Viscosity at P.T. (Pumping Temp.)	1.1	cp
Vapor pressure at P.T.	20.85	psia
Sp. gr. (γ) at P.T.	0.73	
Flow at ambient temp.	192	gpm
Operating flow at P.T.	192	gpm
Design flow at P.T.	192	gpm

Suction				Discharge			
Source pressure	=	21.70	psia	Terminal pressure	=	16.7	psia
Static head	=	3.559	psi	Static head	=	6.313	psi
– ΔP_f, line loss	=	–0.511	psi	– ΔP_f discharge			
Suction pressure	=	24.748	psia	Piping system	=	1.887	psi
– Vapor pressure	=	–20.85	psia	Other	=	20	psi
NPSH avail	=	3.898	psia	Discharge press.	=	44.90	psia
NPSH avail	=	12.3	ft	– Suction press.	=	24.748	psia
NPSH req'd	=	1.3	ft	TDH	=	20.152	psia
				TDH	=	63.77	feet

bhp at Duty Condition

$$bhp_D = \frac{(gpm)(TDH)(\gamma)}{(3{,}960)(\eta)}$$

bhp at Back-Pressure Condition

$$bhp_{BP} = \frac{(gpm)(TDH)(\gamma)}{(3{,}960)(\eta)}$$

Figure 6-56. Re-evaluation of pump hydraulic design calculation sheet of Example 6-5.

Discharge Line

Referring to previous calculations on the discharge side we have the following:

$$V_4 = \left(\frac{400}{192}\right)(4.84) = 10.08\frac{ft}{sec}$$

$$N_{Re} = \left(\frac{400}{192}\right)(100{,}863) = 210{,}062$$

f = 0.0315

$$\Delta P_{f_4} = \left[\frac{(0.0315)(60.708)}{\left(\frac{4.026}{12}\right)} + 6.610\right]\frac{(45.55)(10.08)^2\left(\frac{1}{144}\right)}{2(32.2)}$$

ΔP_{f_4} = 6.143 psi

For 3-in. portion,

$$V_3 = \left(\frac{400}{192}\right)(8.34) = 17.375\frac{ft}{sec}$$

$$N_{Re} = \left(\frac{400}{192}\right)(132{,}449) = 275{,}935$$

f = 0.03395

$$\Delta P_{f_3} = \left[\frac{(0.03395)(3.0)}{\left(\frac{3.068}{12}\right)} + 0.943\right]\frac{(45.55)(17.375)^2\left(\frac{1}{144}\right)}{2(32.2)}$$

ΔP_{f_3} = 1.989

$\Delta P_D = \Delta P_{f_4} + \Delta P_{f_3}$ = 6.143 psi + 1.989 psi

ΔP_D = 8.132 psi

Referring to Figure 6-57, we reevaluate the pump for the maximum capacity condition.

Normally, we would use a 9.5-in. impeller, as indicated on the pump manufacturer's curve, Figure 6-55. In this case, being that the application is infrequent, we keep the 8.0-in. impeller. As the flow rate increases with the same size impeller, the TDH decreases and the required NPSH increases. As we see on Figure 6-55, the available NPSH of 4.589 ft is slightly exceeded at 400

Pump Hydraulic Design Calculation Sheet
Maximum Capacity Condition Reevaluation

Liquid	Light Naphtha Cut—68° API	
Viscosity at P.T. (Pumping Temp.)	1.1	cp
Vapor pressure at P.T.	20.85	psia
Sp. gr. (γ) at P.T.	0.73	
Flow at ambient temp.	400	gpm
Operating flow at P.T.	400	gpm
Design flow at P.T.	400	gpm

Suction			**Discharge**		
Source pressure =	21.70	psia	Terminal pressure =	16.70	psia
Static head =	3.559	psi	Static (lift) =	6.313	psi
$-\Delta P_f$, line loss =	−2.959	psi	$-\Delta P_f$ discharge		
Suction pressure =	22.30	psia	Piping system =	8.132	psi
− Vapor pressure =	−20.85	psia	Other =	20.00	psi
NPSH avail =	1.45	psia	Discharge press. =	51.145	psia
NPSH avail =	4.589	ft	− Suction press. =	22.300	psia
NPSH req'd =	3.75	ft	TDH =	28.845	psia
			TDH =	91.282	feet

bhp at Duty Condition

$$bhp_D = \frac{(gpm)(TDH)(\gamma)}{(3{,}960)(\eta)}$$

bhp at Maximum Capacity Condition

$$bhp_{BP} = \frac{(gpm)(TDH)(\gamma)}{(3{,}960)(\eta)}$$

Figure 6-57. Maximum capacity re-evaluation of pump hydraulic design calculation sheet of Example 6-5.

gpm. It is suggested that a flow rate of 375 gpm be used to avoid cavitation. From Figure 6-55 the actual TDH is

$$TDH = 34 \text{ ft}$$

The required brake horsepower is

$$bhp = \frac{(375)(34.0)(0.73)}{(3{,}960)(0.65)} = 3.62 \text{ hp}$$

A 4-hp motor is sufficient for normal and maximum capacity operations.

Re-evaluation of Reservoir Line

Since the nitrogen pad on the naphtha storage tank was decreased from 10.5 psi to 7.0 psi, we must reconsider the line size.

With 38 feet of head in the reservoir, we incurred a pressure drop of 3.9 psi, yielding an entry pressure of 8.5 psi. In the back-pressure condition, we need a flow rate of 375 gpm. The new presure drop in the line connecting the reservoir to the storage tank, considering the pipe to be 4-in. schedule 40, is as follows:

$$V = \frac{(375)\left(\frac{gal}{min}\right)\left(\frac{1\ ft^3}{7.479\ gal}\right)\left(\frac{1\ min}{60\ sec}\right)}{0.0884\ ft^2} = 9.453\ \frac{ft}{sec}$$

$$N_{Re} = \frac{\left(\frac{4.026}{12}\right) ft\ (9.453) \times \frac{ft}{sec}\ (45.55)\ \frac{lb_m}{ft^3}}{2.640\ \frac{lb_m}{ft\text{-}hr}\left(\frac{1\ hr}{3{,}600\ sec}\right)} = 196{,}992$$

From Equation 1-6a,

$$f^{-0.5} = -2\log_{10}\left(\frac{\epsilon/d}{3.7} + \frac{2.51}{N_{Re}\ f^{0.5}}\right)$$

$$f^{-0.5} = -2\log_{10}\left[\frac{\left(\frac{0.0018}{0.336}\right)}{3.7} + \frac{2.51}{(196{,}992)(f)^{0.5}}\right]$$

$$f = 0.0315$$

From Equation 1-4 we have

$$\Delta P_f = \left(\frac{fL}{d} + \sum K\right) \frac{\rho V^2}{2g_c}$$

$$\Delta P_f = \left[\frac{(0.0315)(105.83)\text{ft}}{\left(\frac{4.026}{12}\right)\text{ft}} + 2.392\right] \times \frac{(45.55)\ \frac{\text{lb}_m}{\text{ft}^3}\ (9.453)^2\ \frac{\text{ft}^2}{\text{sec}^2}\left(\frac{1\ \text{ft}^2}{144\ \text{in.}^2}\right)}{2(32.2)\frac{\text{ft-lb}_m}{\text{sec}^2\text{-lb}_f}}$$

$\Delta P_f = 5.411$ psi

With 38 feet of head in the reservoir we have an entry pressure to the storage tank of

38 ft = 12.008 psi

Entry pressure = 12.008 psi − 5.411 psi = 6.597 psi

Because 6.597 psi < 7.00 psi pad, we keep the 6-in. schedule 40 pipe. The 6-in. line was evaluated for 513 gpm, so it is adequate for the 375 gpm in the 4-in. line. The system is now completely designed for hydraulics, using a 4-in. × 3-in. horizontal centrifugal pump.

NOTATION

acfm = actual cubic feet per minute, ft^3/min
bhp = brake horsepower, hp
$\bar{C}$ = clearance volume, $in.^3$
C_p = specific heat at constant pressure, Btu/lb_m-mole-°F
C_r = compression ratio
C_v = specific heat at constant volume, Btu/lb_m-mole-°F
D = diameter of impeller or rotor, in.
D_s = specific diameter, dimensionless
ghp = gas horsepower = horsepower delivered to gas, hp
H = head = energy per pound of mass, ft-lb_f/lb_m, or better known as feet of head, ft
icfm = actual cubic feet per minute at compressor inlet, ft^3/min
J = mechanical equivalent of heat = 778 ft-lb_f/Btu
k = ratio of specific heats = C_p/C_v, dimensionless
m = mass, lb_m and re-expansion polytropic exponent
$\dot{m}$ = mass flow rate, lb_m/hr
m_o = moles of gas = m/mw
mw = molecular weight
n = polytropic exponent
N = speed, rpm
N_s = specific speed, dimensionless
NPSH = net positive suction head, feet or psia
P = pressure, psi
Q = flow rate, gpm or ft^3/sec
R = $\bar{R}$/mw = gas constant of a particular gas
$\bar{R}$ = universal gas constant = 1545 ft-lb_f/lb_m mole-°R
scfm = standard cubic feet per minute, ft^3/min—see discussion under standard volumetric flow
t = temperature, °F
Δt = temperature differential, °F
V = volume of gas or cylinder, ft^3
v = specific volume of gas, ft^3/lb_m
W_w = weight of fluid
whp = water horsepower, hp
y = constant = (k−1)/k
z = compressibility factor, dimensionless

Greek Symbols

γ = specific gravity, dimensionless
η = efficiency, expressed as percent
ϵ = ratio of clearance volume to the volume swept by the piston stroke
ρ = density, lb_m/ft^3

REFERENCES

1. Buchter, H. Hugo, *Industrial Sealing Technology,* John Wiley & Sons, New York, N.Y., 1979.
2. Dimoplon, William, "What Process Engineers Need to Know About Compressors," *Compressor Handbook for the Hydrocarbon Processing Industries,* Gulf Publishing Co., Houston, Tx., 1979.
3. Balje, O. E., "A Study on Design Criteria and Matching of Turbo-machines-Part B," Trans. ASME, *J. Eng. Power,* Jan. 1962.

4. Tezekjian, E. A., "How to Control Centrifugal Compressors," *Compressor Handbook for the Hydrocarbon Processing Industries,* Gulf Publishing Co., Houston, Tx., 1979.
5. Chlumsky, Vladimir, *Reciprocating and Rotary Compressors,* Publishers of Technical Literature, Prague, Czechoslovakia, 1965.
6. Lapina, Ronald P., *Estimating Centrifugal Compressor Performance, Volume 1,* Gulf Publishing Company, Houston, Tx., 1982.
7. Ludwig, Ernest E., *Applied Process Design for Chemical and Petrochemical Plants,* Volume 3, 2nd Edition, Gulf Publishing Company, Houston, Tx., 1983.
8. Neerken, R. F., "Keys to Compressor Selection," *Chemical Engineering,* Jan. 25, 1975, New York.

7

The Mechanical Design of Shell and Tube Heat Exchangers

A heat exchanger in process systems allows the transfer of energy as heat from one source to another. Without this essential piece of equipment most industrial processes would be impossible.

There are various types of heat exchangers, each of which is designed to accommodate the requirements of the specific needs at hand. Shell and tube heat exchangers are by far the most common because of their relative simplicity and ability to handle the largest variety of fluids. Plate fin heat exchangers have become quite popular in cryogenic gas services and have largely replaced shell and tube exchangers in gas processing plants. Finned-tube exchangers are used for gas-gas heat transfer, such as in waste heat recovery units, and have gained popularity in the past few years because of emphasis on cogeneration to satisfy energy needs.

FUNDAMENTALS OF SHELL AND TUBE HEAT EXCHANGERS

A shell and tube heat exchanger is a cylindrical vessel housing a set of tubes (called the tube bundle) containing a fluid at some temperature and immersed in another fluid at a different temperature. The transfer of heat occurs between the fluid flowing over the tubes and the fluid flowing inside the tubes. The fluid flow inside the tubes is said to be "tube side" and the fluid flow external to the tube bundle is said to be "shell side."

The simplest type of shell and tube heat exchanger is the type shown in Figure 7-1, where warm kerosene enters on the top shell side. The kerosene's flow path is guided between the tubes by baffle plates and exits at the bottom shell-side nozzle cooled to the desired temperature. The tube bundle is supported between two tubesheets with baffle plates spaced at intervals to support and brace the tubes. In this figure the tube-side flow enters the tube bundle on the bottom left side and exits on the top left side with a horizontal baffle plate separating the two tube-side flows. This type of arrangement is called a 1-2 exchanger, one shell-side pass and two tube-side passes. The various configurations of exchangers will be discussed shortly.

Figure 7-2 shows a reboiler in which isobutane vapor is formed by heating liquid isobutane. This type of reboiler is called a "kettle" type reboiler because of the excess area above the tube bundle that is provided for vapor separation. Figure 7-3 shows another type of reboiler where the shell and tube exchanger is mounted vertically alongside a process tower. Here the heat energy of steam is used to separate the propane and propylene liquid into a gas-liquid two-phase mixture. This type of arrangement is common in the gas processing industry and, as will be discussed later, one must be very careful in designing the support(s) for such an exchanger, because of the tubes' thermal expansion.

All shell and tube heat exchangers are exposed to internal pressures, tube-side and shell-side. Thus, in the United States the ASME Section VIII Division I Pressure Vessel Code governs the vessel design of such exchangers. The detailed design of shell and tube exchangers is governed by TEMA (Tubular Exchanger Manufacturing Association), whose published standard classifies exchangers by the severity of process requirements. The three classes are Class "R," Class "C," and Class "B" exchangers. Before discussing these classes, we must clarify heat exchanger design types and terminology (see Figure 7-4).

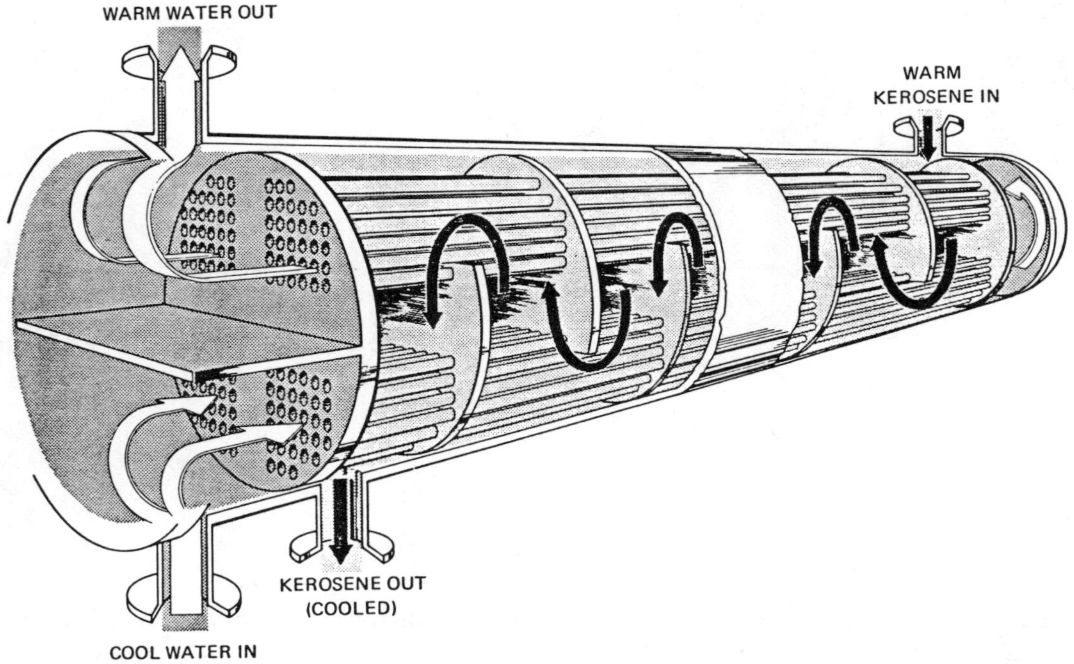

Figure 7-1. An example of a fixed tubesheet heat exchanger. (Courtesy of Howell Training Company.)

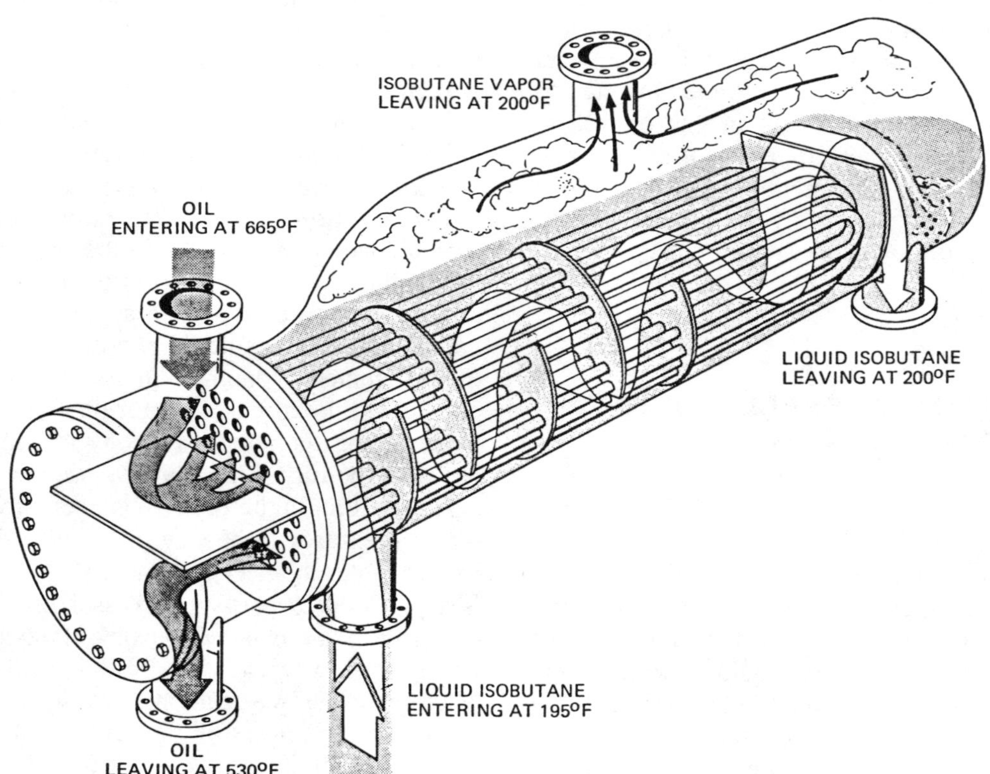

Figure 7-2. This U-tube exchanger represents a kettle type reboiler. (Courtesy of Howell Training Company.)

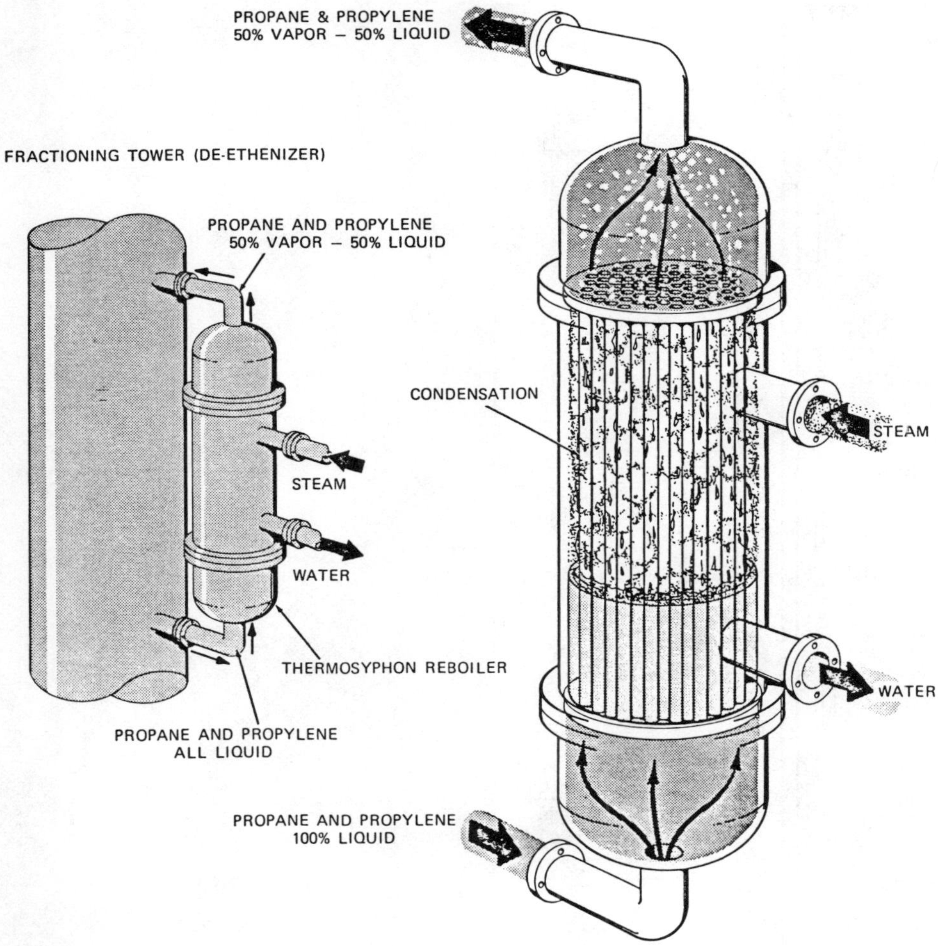

Figure 7-3. Illustration of a thermosyphon reboiler. (Courtesy of Howell Training Company.)

Design Classifications of Heat Exchangers

Typical shell and tube heat exchangers and their functions are as follows:

Reboiler—transfers heat to a liquid to produce a two-phase, gas-liquid mixture used in a distillation column.

Thermosiphon Reboiler—provides natural circulation of the boiling fluid by a static liquid head shown in Figure 7-3.

Forced Circulation Reboiler—a reboiler in which a pump is used to force the liquid through the heat exchanger (reboiler) into the distillation column.

Condenser—a heat exchanger to condense vapors by removing heat from a gas.

Partial Condenser—only partially condenses a gas to provide heat to another medium to satisfy a process condition. The residual gas is recirculated through a heater and recycled. A common application is using excess steam to heat up a process fluid. A typical application of a partial condenser on a distillation column is to condense only enough liquid for the reflux when the overhead product is vapor.

Final Condenser—an exchanger where all the gas is condensed and all the heat is transferred to the other medium.

Steam Generator—a device that generates steam, such as a boiler, to provide energy for process requirements. The most classic example is the old steam locomotive, which is a shell and tube exchanger "mounted on wheels" with the steam used to power the locomotion. (This unit is a fired vessel and is not covered by ASME Section VIII Division.)

Vaporizer—an exchanger that fully or partially vaporizes a liquid.

Chiller—an exchanger in which a process medium is cooled by evaporating a refrigerant, or by cooling and heating with little or no phase change.

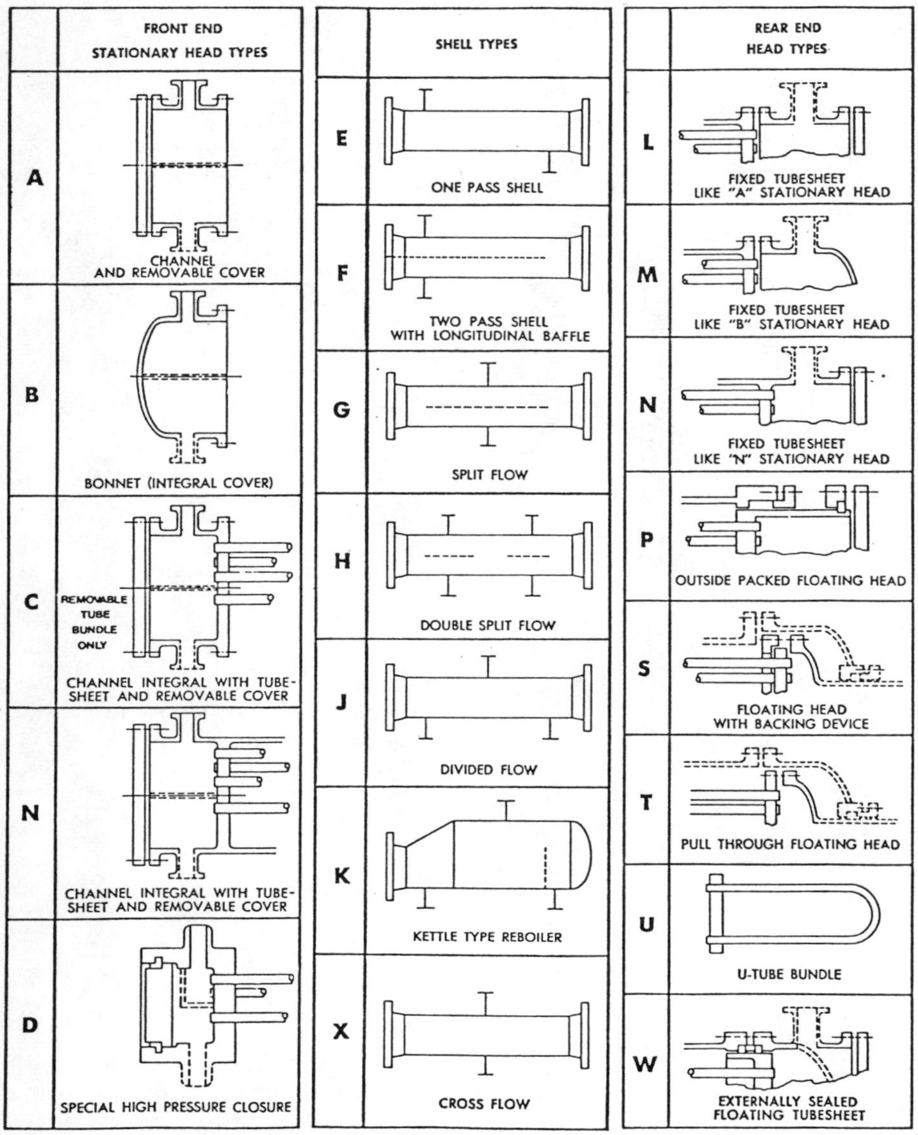

Figure 7-4. Nomenclature of shell and tube heat exchangers. (©1978 by Tubular Exchanger Manufacturers Association.)

These classifications are the major types of services that shell and tube exchangers provide in the process industries.

Process requirements dictate the type of design to be used. Figure 7-4 shows some of the major types of construction. The standard TEMA classification of exchangers is to use the shell identification and number with the exchanger designation type. For example, an 18–150 BEM is an exchanger having an 18-in. shell with 150 tubes, a bonnet (integral) cover with a fixed tubesheet at one end (B in Figure 7-4), a fixed tubesheet and a stationary head at the other end (M), and a one-pass shell between both ends (E).

Fixed Tubesheet Shell and Tube Heat Exchangers

Fixed tubesheet shell and tube heat exchangers are the simplest of the shell and tube designs. They consist of a tube bundle attached to a tubesheet on each side of the tube bundle. The tubesheets are welded to the shell providing an absolute seal to prevent the shell-side fluid from leakage. Often the tubesheets extend beyond the shell diameter and have flange bolt holes that allow the tube heads to be bolted to the tubesheets.

In fixed tubesheet exchangers, tubes can fill the entire shell to achieve maximum heat exchange (of course, this also increases shell-side fluid pressure drop) such that tolerances between tubes are minimum. However, this factor limits the shell-side fluid to a relatively clean service, because the exterior of the closely-packed tubes cannot be mechanically cleaned or inspected. Another limitation to the design is that there is no allowance for thermal growth of the tubes, except if an external expansion joint is used, which is quite common for this type of exchanger. Normally, single convoluted bellows are used since the maximum temperature differential is 200°F and the cyclic loading is insignificant.

Tube-side headers, channel covers, and internals of tubes can be cleaned quite easily and the shell side can be cleaned only by circulating a cleaning fluid or backwashing.

U-Tube Shell and Tube Heat Exchangers

U-tube shell and tube heat exchangers consist of one tubesheet with tubes bent in a U-shape attached to the single tubesheet. This type of exchanger is used for large temperature differentials where there is a lot of tube growth. This type of design allows for easy access to the shell side of the tubes and removal of the tube bundle. The inside of tubes must be cleaned with special tools and then only when the bending radius is fairly large. This type of design is also very suitable for chemical cleaning.

The maximum number of tubes per tubesheet is less than the fixed tubesheet design because of the minimum bending radius required to form the U-shape. The U-tube design is also very applicable to high-pressure services.

Floating Head Shell and Tube Heat Exchangers

This type of shell and tube heat exchanger has a floating head that is designed to accommodate thermal expansion of the tubes and to provide access to the tube-side and shell-side exchanger components. This type of design is expensive and its use should be considered against other possible designs.

Packed Lantern Ring Exchanger (Figure 7-5a). This construction is normally limited to design tempera-

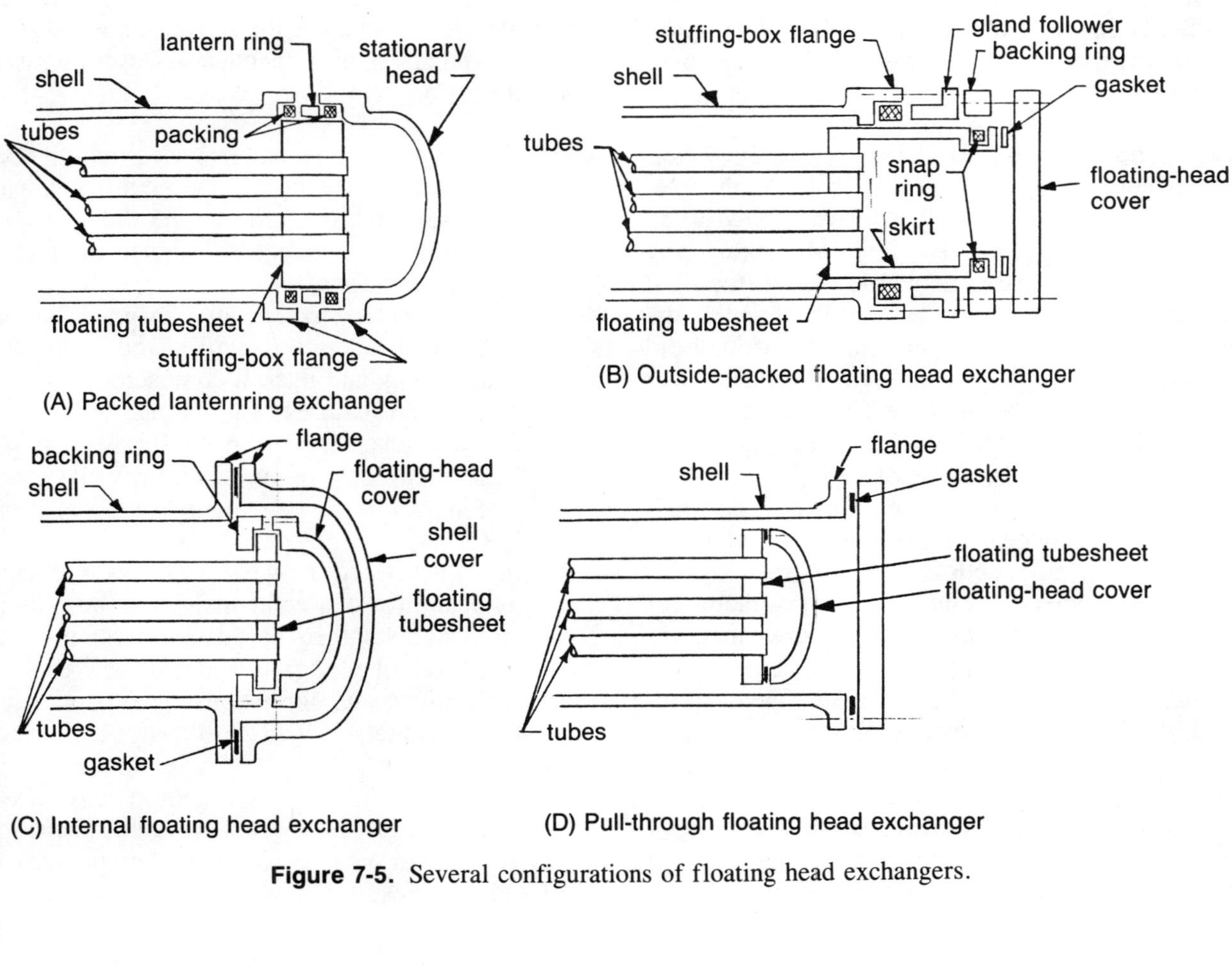

Figure 7-5. Several configurations of floating head exchangers.

tures < 370°F and design pressures < 300 psig. This type of design is used only for mild services, such as steam, air, low viscous oils. In this design the shell-side and tube-side fluids are sealed by separate packings which, in turn, are separated by a lantern ring. The lantern ring fits between the packings that separate the shell and tube-side fluids and normally contains weep holes that accommodate any leakage through the packing. Such leakage, which is passed to the outside and drops to the foundation below, will not cause shell and tube-side fluids to mix.

The tubesheet must be designed such that it is large enough in diameter to encompass the packing-lantern-ring ensemble and differential thermal expansion of the tubes. Occasionally, a skirt is attached to a thin tubesheet to act as a bearing surface for the packing-lantern-ring ensemble.

Outside-Packed Floating Head Exchanger (Figure 7-56). Rings of packing contain the shell-side fluid, which is compressed by a gland follower that is guided by a tube sheet skirt. The skirt is integral to the floating tubesheet. This removable-bundle construction allows for differential expansion between the shell and tubes. This design is normally limited to 600°F and 600 psig, which is one reason why it is the most commonly used removable-bundle type exchanger in the petroleum-chemical industry, even though usage has decreased over recent years.

Internal Floating-Head Exchanger (Figure 7-5c). This design consists of an internal floating tubesheet held by an internal backing ring, which is bolted to an internal floating head cover. The internal backing ring and internal shell cover are beyond the end of the shell containing the tubes. To remove the tube bundle, the shell cover, split backing ring, and internal floating head cover must be removed. The internal floating head cover acts as a return cover for the tube fluid with an even number of tube-side passes. With an odd number of tube-side passes, a nozzle must be extended from the internal floating-head cover through the outside shell cover. Clearances between the shell and the outermost tubes are 1 1/8 in. for pipe shells and 1 7/16 in. for medium-sized rolled plate shells. This design is more suitable for higher shell-side temperatures and pressures than for pull-through bundle types of construction. This design has been used extensively in the petroleum-chemical industry, but there has been a decline of use over the past few years.

Pull-Through Bundle Floating-Head Exchanger (Figure 7-5d). This design consists of a floating head directly bolted to an internal floating head cover. The tube bundle can be removed without removing either internal floating head cover or shell cover when bundle is pulled out an opposite end of shell cover facing internal floating head. This feature reduces down and maintenance time during inspection and repair.

The clearance between the outside of the tubes and shell inside must be sufficient to allow space for both the gasket and bolting at the internal floating head cover. This clearance is usually twice that required for the split ring design used in the internal floating head in the previous section. This type of design is normally limited to services where leakage of the internal gasket is tolerable. With an odd number of tube-side passes, a nozzle must extend from the internal floating-head cover through the shell cover. The number of tube-side passes is simply limited by the number of tubes. This design is generally suited for lower temperatures and pressures than that of the internal floating head exchanger described earlier.

General TEMA Exchanger Classes—R, C, and B

There are three basic categories of shell and tube heat exchangers in TEMA—Class R, Class C, and Class B. The difference in class is the degree of severity of service the exchanger will encounter. Descriptions of the three classes are as follows:

Class R includes heat exchangers specified for the most severe service in the petroleum-chemical processing industry. Safety and durability are required for exchangers designed for such rigorous conditions.

Class C includes heat exchangers designed for the generally moderate services and requirements. Economy and overall compactness are the two essential features of this class.

Class B are exchangers specified for general process service. Maximum economy and optimum compactness are the main criteria of design.

Rubin [3] described the TEMA classes of exchangers in terms of the various components and how they vary from one class to another. This data is given in Table 7-1. Ludwig [4] described various types of heat exchangers, their applications and limitations, which include shell and tube exchangers as well as other types. This data is presented in Table 7-2.

Tables 7-1 and 7-2 provide a comprehensive view of the various types of heat exchangers and their applications, so we can now focus on the components of the shell and tube design.

Table 7-1
Comparison of TEMA Classes R, C and B Exchangers [3]

Paragraph	Topic	R	C	B
1.12	Definition	for the generally severe requirements of petroleum and related processing applications.	for the generally moderate requirements of commercial and general process applications.	for general process service.
1.51	Corrosion allowance on carbon steel	1/8 inch	1/16 inch	1/16 inch
2.2	Tube diameters	3/4, 1, 1 1/4, 1 1/2, and 2 inch od	R + 1/4, 3/8, 1/2, and 5/8	R + 5/8
2.5	Tube pitch and minimum cleaning lane	1.25 × tube od. 1/4 inch lane.	R + 5/8 tubes may be located 1.2 × tube od	R + lane may be 3/16 inch in 12 inch and smaller shells for 5/8 and 3/4 tubes.
3.3	Minimum shell diameter	8 inch tabulated	6 inch tabulated	6 inch tabulated.
4.42	Longtudinal baffle thickness	1/4 inch minimum	1/8 inch alloy, 1/4 inch CS	1/8 inch alloy, 1/4 inch carbon steel
4.71	Minimum tie rod diameter	3/8 inch	1/4 inch in 6–15 inch shells	1/4 inch 6-15 inch shells.
5.11	Floating head cover cross-over area	1.3 times tube flow area	Same as tube flow area	Same as tube flow area
5.31	Lantern ring construction	375°F maximum. 300 psi up to 24 inch diam shell 150 psi for 25–42 inch shells 75 psi for 43–60 inch shells	600 psi maximum.	(same as TEMA R)
6.2	Gasket materials	Metal jacketed or solid metal for (a) internal floating head cover. (b) 300 psi and up. (c) all hydrocarbons.	Metal jacketed or solid metal (a) internal floating head. (b) 300 psi and up. Asbestos permitted for 300 psi and lower pressures.	(same as TEMA C)
6.32	Peripheral gasket contact surface	Flatness tolerance specified.	No tolerance specified.	No tolerance specified.
7.131	Minimum tubesheet thickness with expanded tube joints	Outside diameter of the tube.	0.75 × tube od for 1 inch and smaller. 7/8 inch for 1 1/4 od 1 inch for 1 1/2 od 1.25 inch for 2 od	(same as TEMA C)
7.44	Tube Hole Grooving	Two grooves	Above 300 psi design pressure: above 350°F design temp.-2 grooves	(same as TEMA R)
7.51	Length of expansion	Smaller of 2 inch or tubesheet thickness	Smaller of 2 × tube od or 2″	(same as TEMA R)
7.7	Tubesheet pass partition grooves	3/16 inch deep grooves required	Over 300 psi 3/16 inch deep grooves required or other suitable means for retaining gaskets in place	(same as TEMA C)
9.3	Pipe Tap Connections	6000 psi coupling with bar stock plug	3000 psi coupling	3000 psi coupling with bar stock plug
9.32	Pressure Gauge Connections	required in nozzles 2 inch & up.	(shall be specified by purchaser)	(same as TEMA R)
9.33	Thermometer Connections	required in nozzles 4 inch & up.	(shall be specified by purchaser)	(same as TEMA R)
9.1	Nozzle construction	no reference to flanges	same as TEMA R	All nozzles larger than one inch must be flanged.
10.1	Minimum bolt size	3/4 inch	1/2 inch recommended, smaller bolting may be used	5/8 inch

Table 7-2
Selection Guide Heat Exchanger Types [4]

Type Designation	Significant Feature	Applications Best Suited	Limitations	Relative Cost in Carbon Steel Construction
Fixed Tube Sheet	Both tube sheets fixed to shell	Condensers; liquid-liquid; gas-gas; gas-liquid; cooling and heating, horizontal or vertical, reboiling	Temperature difference at extremes of about 200° F. due to differential expansion	1.0
Floating Head or Tube Sheet (Removable and non-removable bundles)	One tube sheet "floats" in shell or with shell, tube bundle may or may not be removable from shell, but back cover can be removed to expose tube ends.	High temperature differentials, above about 200° F. extremes; dirty fluids requiring cleaning of inside as well as outside of shell, horizontal or vertical.	Internal gaskets offer danger of leaking. Corrosiveness of fluids on shell side floating parts. Usually confined to horizontal units.	1.28
U-Tube; U-Bundle	Only one tube sheet required. Tubes bent in U-shape. Bundle is removable.	High temperature differentials which might require provision for expansion in fixed tube units. Clean service or easily cleaned conditions on both tube side and shell side. Horizontal or vertical.	Bends must be carefully made or mechanical damage and danger of rupture can result. Tube side velocities can cause erosion of inside of bends. Fluid should be free of suspended particles.	1.08
Kettle	Tube bundle removable as U-type or floating head. Shell enlarged to allow boiling and vapor disengaging.	Boiling fluid on shell side, as refrigerant, or process fluid being vaporized. Chilling or cooling of tube side fluid in refrigerant evaporation on shell side.	For horizontal installation. Physically large for other applications.	1.2–1.4
Double Pipe	Each tube has own shell forming annular space for shell side fluid. Usually use externally finned tube.	Relatively small transfer area service, or in banks for larger applications. Especially suited for high pressures in tube above 400 psig.	Services suitable for finned tube. Piping-up a large number often requires cost and space.	0.8–1.4
Pipe Coil	Pipe coil for submersion in coil-box of water or sprayed with water is simplest type of exchanger.	Condensing, or relatively low heat loads on sensible transfer.	Transfer coefficient is low, requires relatively large space if heat load is high.	0.5–0.7
Open Tube Sections (Water cooled)	Tubes require no shell, only end headers, usually long, water sprays over surface, sheds scales on outside tubes by expansion and contraction. Can also be used in water box.	Condensing, relatively low heat loads on sensible transfer.	Transfer coefficient is low, takes up less space than pipe coil.	0.8–1.1
Open Tube Sections (Air Cooled) Plain or finned tubes	No shell required, only end headers similar to water units.	Condensing, high level heat transfer.	Transfer coefficient is low, if natural convection circulation, but is improved with forced air flow across tubes.	0.8–1.8
Plate and Frame	Composed of metal-formed thin plates separated by gaskets. Compact, easy to clean.	Viscous fluids, corrosive fluids slurries, High heat transfer.	Not well suited for boiling or condensing; limit 350-500°F by gaskets. Used for Liquid-Liquid only; not gas-gas.	0.8-1.5
Spiral	Compact, concentric plates; no bypassing, high turbulence.	Cross-flow, condensing, heating.	Process corrosion, suspended materials.	0.8-1.5
Small-tube Teflon	Chemical resistance of tubes; no tube fouling.	Clean fluids, condensing, cross-exchange.	Low heat transfer coefficient.	2.0-4.0

Basic Components of Shell and Tube Heat Exchangers

There are various components to a shell and tube heat exchanger, but the following are the essential ones:

1. Tubes
2. Baffles
3. Tie rods
4. Tubesheets

Tubes

There are basically two types—finned tubes and bare tubes. Finned tubes have external fins mounted by various mechanical means. The necessity of having external fins mounted on tubes is to provide more heat transfer area and thus more heat influx to the tube fluid. Finned tubes are most common where there is a gas-liquid or gas-gas transfer of heat with the gas always being external to the tubes. Typical applications of finned tubes are waste heat recovery exchangers, waste heat boilers, gas turbine regenerators, and air-cooled exchangers. Examples of some finned tube designs are shown later.

Plain or bare tubes are the most common in shell and tube design. These tubes come in two basic types—solid wall construction and duplex construction. The duplex design consists of a tube within a tube in which the outer tube is mechanically drawn over the inner tube. The solid wall tube is what the name implies, a simple tube of solid wall construction. Tubing is available in almost as many materials as piping and is available in standard gauge sizes listed in Table 7-3, along with diameters and section properties.

In applying the U-tube exchanger design, tubes must be bent 180°. Table 7-4 lists the recommended minimum bend radii.

Baffles

Baffles serve several functions and consequently the design of each is dependent on its purpose. Baffles can act as:

1. Structural supports for the tubes.
2. Dampers against vibration.
3. Devices to control and direct flow patterns of the shell-side liquid.

Baffles as Tube Structural Supports. Like piping, tubes behave as structural beams and consequently will develop excessive deflection, or sag, if left unsupported. Baffles act as the structural supports in the shell and tube exchanger. Another structural function of baffles is to add stiffness to the tubes so that each tube, in effect, is constrained at each baffle. Thus, the hole in the baffle, being larger by varying amounts than the outside tube diameter, acts as a limit stop for the tube. In piping mechanics (see Chapter 2) a limit stop is a restraint that limits the amount of pipe (in this case, tube) movement to the distance between the hole diameter and the outside diameter of the tube. In other words, the tube can translate in the lateral direction perpendicular to the tube axis only by the amount of clearance between the tube OD and the hole diameter. Translation is mentioned instead of rotation because even though the tube rotates, it is insignificant. Thus, the baffle hole acts as a limit stop and prevents lateral buckling of the tubes when they are induced to thermal expansion by temperature differentials. In this sense the tubes are much stiffer and stronger than they would be without the baffle supports. The consequences of strengthened tubes affect the integrity of tube joint connections in the tubesheets and this will be discussed shortly. We see from this discussion that the baffle plates act as both structural supports and as buckling stabilizers.

Baffles as Tube Vibration Dampers. Figure 7-6 shows baffles of circular rings with rods that run vertically in the first two rings and horizontally in the second two rings, thus damping vibration much in the same way as helical vortex strakes on stacks (Chapter 5). The rods break up forming vortices that induce vibrations, a phenomenon discussed in Chapters 4 and 5 called vortex shedding. The rods also reduce turbulence to below resonant levels of the natural frequency of the tubes and they reduce fluid elastic vibration.

Baffles Control and Direct the Flow Pattern of the Shell-Side Fluid. There are various types of baffles that direct and/or control the flow of the shell side fluid. Figures 7-1 and 7-2 are examples of baffles guiding or directing the flow in the vertical direction. Figure 7-7 shows baffles diverting flow in the horizontal direction. The flow direction is a function of the orientation of the baffles and their respective geometries and is dependent upon process requirements. The arrangement in Figure 7-7 is said to be vertically cut and the arrangements in Figures 7-1 and 7-2 are said to be horizontally cut.

Often, process conditions require the shell-side fluid to flow horizontally, parallel to the longitudinal axis of the exchanger. This arrangement, called a longitudinal baffle, is shown in Figure 7-8. Figure 7-8a shows a two-pass shell-side arrangement and Figure 7-8b shows a four-pass shell-side arrangement. The baffles control the flow in the sense that both the direction and flow rate are dependent on orientation and number of passes, respectively. With the same inlet flow rate, the fluid velocity

Table 7-3
Characteristics of Tubing [2]

Tube O.D. Inches	B.W.G. Gage	Thickness Inches	Internal Area Sq. Inch	Sq. Ft. External Surface Per Foot Length	Sq. Ft. Internal Surface Per Foot Length	Weight Per Foot Length Steel Lbs.*	Tube I.D. Inches	Moment of Inertia Inches[4]	Section Modulus Inches[3]	Radius of Gyration Inches	Constant C**	O.D./I.D.	Transverse Metal Area Sq. Inch
1/4	22	.028	.0295	.0655	.0508	.066	.194	.00012	.00098	.0792	46	1.289	.0195
1/4	24	.022	.0333	.0655	.0539	.054	.206	.00011	.00083	.0810	52	1.214	.0159
1/4	26	.018	.0360	.0655	.0560	.045	.214	.00009	.00071	.0824	56	1.168	.0131
1/4	27	.016	.0373	.0655	.0570	.040	.218	.00008	.00064	.0829	58	1.146	.0117
3/8	18	.049	.0603	.0982	.0725	.171	.277	.00068	.0036	.1164	94	1.354	.0502
3/8	20	.035	.0731	.0982	.0798	.127	.305	.00055	.0029	.1213	114	1.233	.0374
3/8	22	.028	.0799	.0982	.0835	.104	.319	.00046	.0025	.1227	125	1.176	.0305
3/8	24	.022	.0860	.0982	.0867	.083	.331	.00038	.0020	.1248	134	1.133	.0244
1/2	16	.065	.1075	.1309	.0969	.302	.370	.0022	.0086	.1556	168	1.351	.0888
1/2	18	.049	.1269	.1309	.1052	.236	.402	.0018	.0072	.1606	198	1.244	.0694
1/2	20	.035	.1452	.1309	.1126	.174	.430	.0014	.0056	.1649	227	1.163	.0511
1/2	22	.028	.1548	.1309	.1162	.141	.444	.0012	.0046	.1671	241	1.126	.0415
5/8	12	.109	.1301	.1636	.1066	.602	.407	.0061	.0197	.1864	203	1.536	.177
5/8	13	.095	.1486	.1636	.1139	.537	.435	.0057	.0183	.1903	232	1.437	.158
5/8	14	.083	.1655	.1636	.1202	.479	.459	.0053	.0170	.1938	258	1.362	.141
5/8	15	.072	.1817	.1636	.1259	.425	.481	.0049	.0156	.1971	283	1.299	.125
5/8	16	.065	.1924	.1636	.1296	.388	.495	.0045	.0145	.1993	300	1.263	.114
5/8	17	.058	.2035	.1636	.1333	.350	.509	.0042	.0134	.2016	317	1.228	.103
5/8	18	.049	.2181	.1636	.1380	.303	.527	.0037	.0118	.2043	340	1.186	.089
5/8	19	.042	.2298	.1636	.1416	.262	.541	.0033	.0105	.2068	358	1.155	.077
5/8	20	.035	.2419	.1636	.1453	.221	.555	.0028	.0091	.2089	377	1.126	.065
3/4	10	.134	.1825	.1963	.1262	.884	.482	.0129	.0344	.2229	285	1.556	.260
3/4	11	.120	.2043	.1963	.1335	.809	.510	.0122	.0326	.2267	319	1.471	.238
3/4	12	.109	.2223	.1963	.1393	.748	.532	.0116	.0309	.2299	347	1.410	.220
3/4	13	.095	.2463	.1963	.1466	.666	.560	.0107	.0285	.2340	384	1.339	.196
3/4	14	.083	.2679	.1963	.1529	.592	.584	.0098	.0262	.2376	418	1.284	.174
3/4	15	.072	.2884	.1963	.1587	.520	.606	.0089	.0238	.2410	450	1.238	.153
3/4	16	.065	.3019	.1963	.1623	.476	.620	.0083	.0221	.2433	471	1.210	.140
3/4	17	.058	.3157	.1963	.1660	.428	.634	.0076	.0203	.2455	492	1.183	.126
3/4	18	.049	.3339	.1963	.1707	.367	.652	.0067	.0178	.2484	521	1.150	.108
3/4	20	.035	.3632	.1963	.1780	.269	.680	.0050	.0134	.2532	567	1.103	.079
1	8	.165	.3526	.2618	.1754	1.462	.670	.0392	.0784	.3009	550	1.493	.430
1	10	.134	.4208	.2618	.1916	1.237	.732	.0350	.0700	.3098	656	1.366	.364
1	11	.120	.4536	.2618	.1990	1.129	.760	.0327	.0654	.3140	708	1.316	.332
1	12	.109	.4803	.2618	.2047	1.037	.782	.0307	.0615	.3174	749	1.279	.305
1	13	.095	.5153	.2618	.2121	.918	.810	.0280	.0559	.3217	804	1.235	.270
1	14	.083	.5463	.2618	.2183	.813	.834	.0253	.0507	.3255	852	1.199	.239
1	15	.072	.5755	.2618	.2241	.714	.856	.0227	.0455	.3291	898	1.167	.210
1	16	.065	.5945	.2618	.2278	.649	.870	.0210	.0419	.3314	927	1.149	.191
1	18	.049	.6390	.2618	.2361	.496	.902	.0166	.0332	.3366	997	1.109	.146
1	20	.035	.6793	.2618	.2435	.360	.930	.0124	.0247	.3414	1060	1.075	.106
1 1/4	7	.180	.6221	.3272	.2330	2.057	.890	.0890	.1425	.3836	970	1.404	.605
1 1/4	8	.165	.6648	.3272	.2409	1.921	.920	.0847	.1355	.3880	1037	1.359	.565
1 1/4	10	.134	.7574	.3272	.2571	1.598	.982	.0741	.1186	.3974	1182	1.273	.470
1 1/4	11	.120	.8012	.3272	.2644	1.448	1.010	.0688	.1100	.4018	1250	1.238	.426
1 1/4	12	.109	.8365	.3272	.2702	1.329	1.032	.0642	.1027	.4052	1305	1.211	.391
1 1/4	13	.095	.8825	.3272	.2775	1.173	1.060	.0579	.0926	.4097	1377	1.179	.345
1 1/4	14	.083	.9229	.3272	.2838	1.033	1.084	.0521	.0833	.4136	1440	1.153	.304
1 1/4	16	.065	.9852	.3272	.2932	.823	1.120	.0426	.0682	.4196	1537	1.116	.242
1 1/4	18	.049	1.042	.3272	.3016	.629	1.152	.0334	.0534	.4250	1626	1.085	.185
1 1/4	20	.035	1.094	.3272	.3089	.456	1.180	.0247	.0395	.4297	1707	1.059	.134
1 1/2	10	.134	1.192	.3927	.3225	1.955	1.232	.1354	.1806	.4853	1860	1.218	.575
1 1/2	12	.109	1.291	.3927	.3356	1.618	1.282	.1159	.1546	.4933	2014	1.170	.476
1 1/2	14	.083	1.398	.3927	.3492	1.258	1.334	.0931	.1241	.5018	2181	1.124	.370
1 1/2	16	.065	1.474	.3927	.3587	.996	1.370	.0756	.1008	.5079	2299	1.095	.293
2	11	.120	2.433	.5236	.4608	2.410	1.760	.3144	.3144	.6660	3795	1.136	.709
2	12	.109	2.494	.5236	.4665	2.201	1.782	.2904	.2904	.6697	3890	1.122	.647
2	13	.095	2.573	.5236	.4739	1.934	1.810	.2586	.2586	.6744	4014	1.105	.569
2	14	.083	2.642	.5236	.4801	1.699	1.834	.2300	.2300	.6784	4121	1.090	.500

* Weights are based on low carbon steel with a density of 0.2833 lbs./cu. in. For other metals multiply by the following factors:

Aluminum	0.35	Nickel-Chrome-Iron	1.07
Titanium	0.58	Admiralty	1.09
A.I.S.I. 400 Series Stainless Steels	0.99	Nickel and Nickel-Copper	1.13
A.I.S.I. 300 Series Stainless Steels	1.02	Copper and Cupro-Nickels	1.14
Aluminum Bronze	1.04		
Aluminum Brass	1.06		

** Liquid Velocity $= \dfrac{\text{Lbs. Per Tube Per Hour}}{\text{C x Sp. Gr. of Liquid}}$ in feet per sec. (Sp. Gr. of Water at 60° F. = 1.0)

Table 7-4
Minimum Tube Bend Radii [4]

Tube Outside Dia. (in.)	Bend Radius (in.)	Center-to-Center Distance (in.)
Duplex, all sizes	3 times Tube O.D.	6 times Tube OD
*Plain: 5/8	13/16	1 5/8
3/4	1	2
1	1 3/16	2 3/8

**For bends this sharp, the tube wall on the outer circumference of the tube may thin down $1^1/2$ to 2 gauge thicknesses, depending on condition and specific tube material. More generous radii will reduce this thinning. TEMA presents a formula for calculating the minimum wall thickness.*

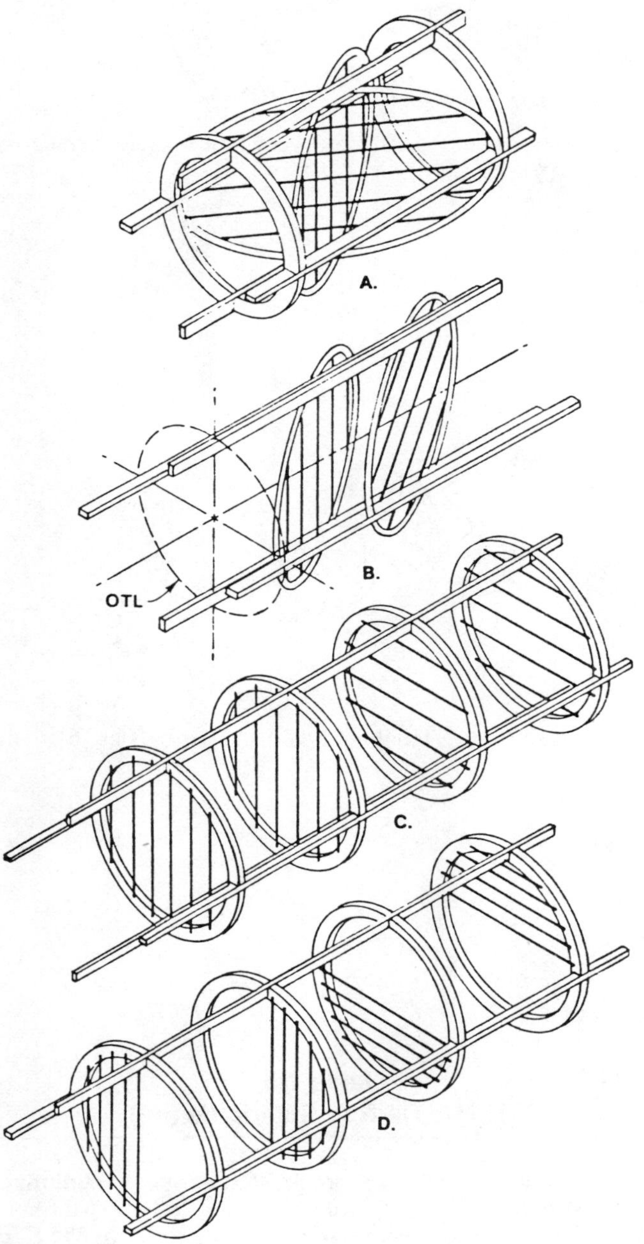

Figure 7-6. Although complex, this design eliminates tube vibration. To use this configuration, one must be cognizant of pressure data [5]. (Courtesy of *Heat Transfer Engineering*, Hemisphere Publishing Corporation, New York, Washington, D.C.)

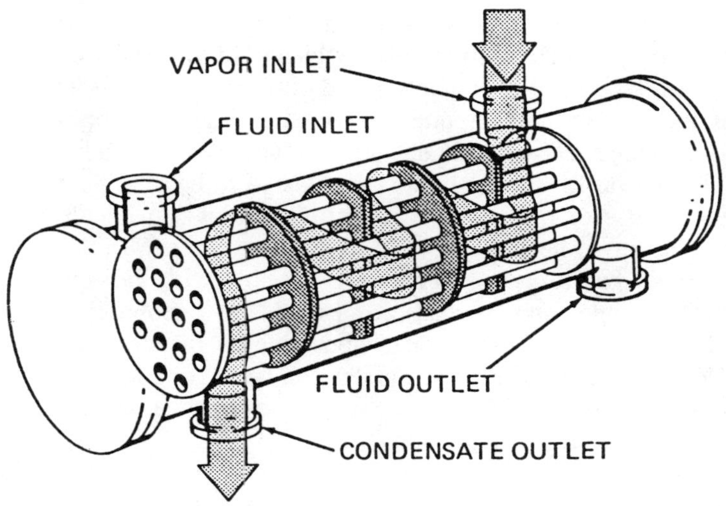

Figure 7-7. Baffles can divert flow horizontally. (Courtesy of Howell Training Company.)

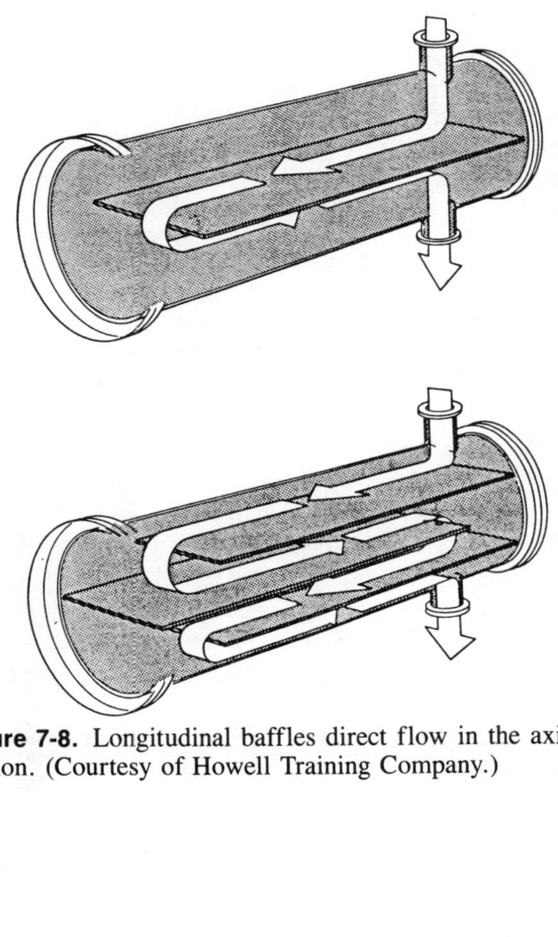

Figure 7-8. Longitudinal baffles direct flow in the axial direction. (Courtesy of Howell Training Company.)

increases as the flow area decreases, that is, the velocity increases with an increase in the number of passes.

The control of flow in exchangers is accomplished as well with orifice baffles. Figure 7-9 shows an annular orifice baffle. To utilize this type of design a very clean shell-side fluid is required, since the fluid must flow in the annular space between the tube outside diameter and the hole in the baffle forming the orifice. The flow at the orifice is very turbulent and the pressure drop through an orifice-baffle arrangement is very high. Consequently, these baffles are not used often in industry. Also, since the orifice baffle requires a very clean fluid, non-Newtonian fluids are completely ruled out. We will see later in the chapter that the plate fin type of exchanger is superior to the shell and tube design for many clean services. The reason for the shell and tube design to be dominant is because of the wider variety of fluids it can handle versus any other design.

Other baffle arrangements are possible with varying baffle shapes and orientations. Figure 7-10 shows baffles in disc and doughnut shapes, which disperse the flow in a radial direction. Baffles can be cut to allow for horizontal or vertical flow in varying amounts as shown in Figure 7-11.

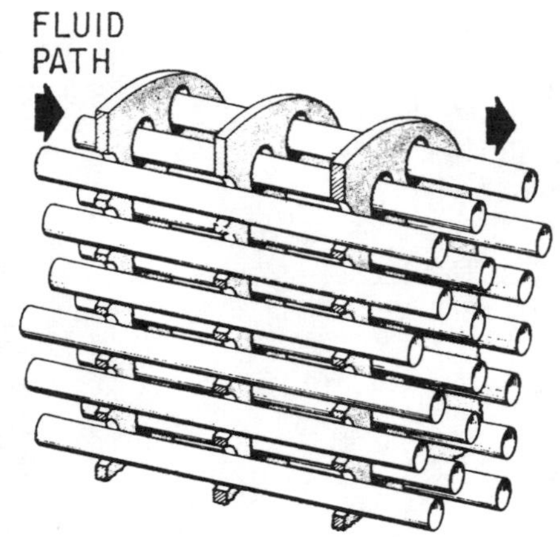

Figure 7-9. Annular orifices between tube outside surface and hole in baffle plate [6].

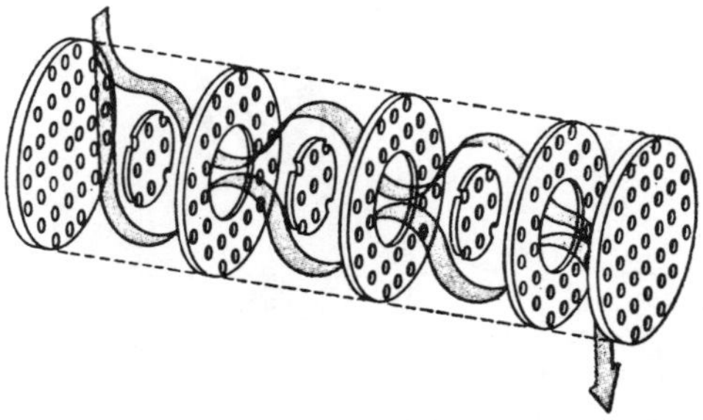

Figure 7-10. Doughnut and disc type baffles [6].

Tie Rods

These are structural rods that run parallel to the exchanger tubes through the outer perimeter of the baffles, fastened to the tubesheets such that they space and support the baffles. Tie rods, being attached to the baffle plates, also prevent them from vibrating and damaging the tubes. Table 7-5 lists what TEMA recommends as a minimum number of tie rods and rod diameters for a set of shell diameters.

Tubesheets

These are the structured plates in which the tubes are connected at each end of the exchanger. Tubesheets come in two basic types—single and double. Double tubesheets consist of two tubesheets mounted together at each end of the tubes with a clearance between the two sheets. The reason for using two tubesheets at each end is to reduce the possibility of a leak of the tube-side fluid. Double tubesheets are quite common with highly toxic services, where a leak cannot be tolerated.

Single tubesheets are much more common than double tubesheets because of process applications and economy. Typical tube-tubesheet connections are shown in Figure 7-12.

Of great immediate concern in tubesheet design is the loading induced by the tubes thermal movement, which

Table 7-5
TEMA Tie Rod Standards (in.)

Nominal "R" Exchanger Shell Diameter	"R" Exchanger Tie Rod Diameter	"C" & "B" Exchanger Tie Rod Diameter	Minimum Number of Tie Rods
8–15	3/8	1/4	4
16–27	3/8	3/8	6
28–33	1/2	1/2	6
34–48	1/2	1/2	8
49–60	1/2	1/2	10

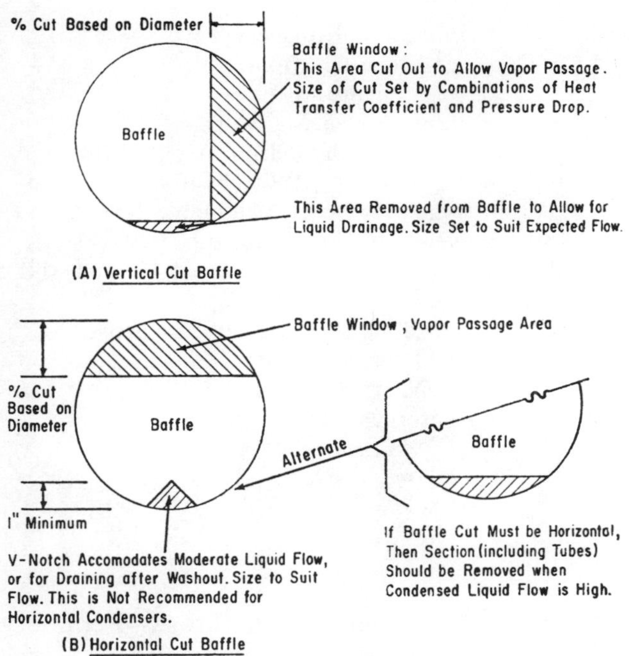

Figure 7-11. Baffle details [4].

is a definite problem in fixed tubesheet exchangers. TEMA gives two equations for determining the compressive stress induced on tubesheets for all three types of exchangers—Classes R, C, and B:

$$\sigma_c = \frac{\pi^2 E_t}{2(k\ell/r)^2} \quad \text{when } C_c \leqslant k\ell/r \tag{7-1}$$

$$\sigma_c = \frac{\sigma_y}{2}\left[1 - \frac{(k\ell/r)}{2C_c}\right] \text{ when } C_c > k\ell/r \tag{7-2}$$

$$\text{where } C_c = \left[\frac{2\pi^2 E_t}{\sigma_y}\right]^{0.5}$$

σ_y = minimum yield stress of tube material of design temperature

r = radius of gyration of tube

= $0.25[d_o^2 + (d_o - 2t_t)^2]^{0.50}$, in.

t_t = tube wall thickness, in.

$k\ell$ = equivalent effective unsupported length of the tube, in.

= unsupported tube span, in.

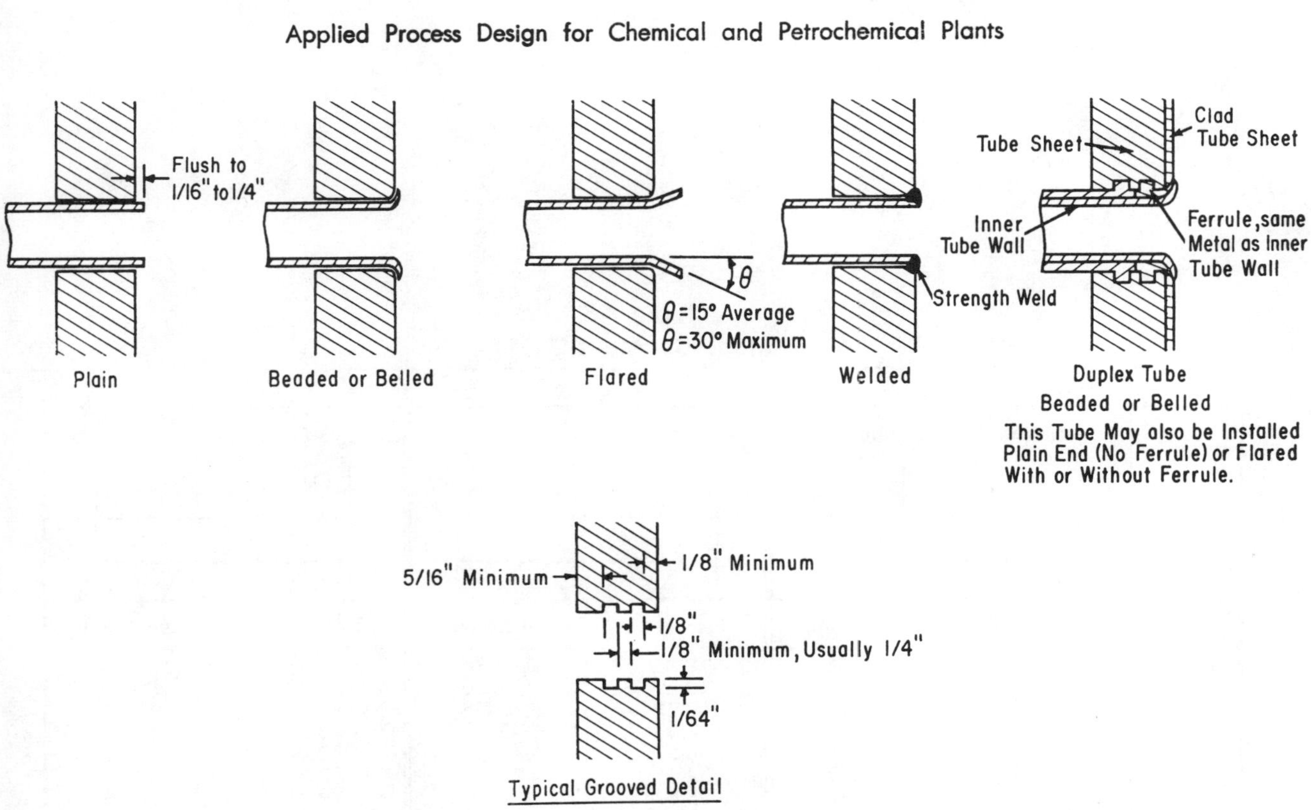

Figure 7-12. Typical tubesheet-tube connections [4].

$$k = \begin{cases} 0.6 & \text{for unsupported tube lengths between two tubesheets} \\ 0.8 & \text{for unsupported tube lengths between a tubesheet and a baffle} \\ 1.0 & \text{for unsupported tube lengths between two baffles} \end{cases}$$

E_t = modulus of elasticity of tube material at mean tube metal temperature, psi
d_o = outside diameter of tubes, in.
σ_c = allowable tube compressive stress, psi, for the tubes at the outer periphery of the tube bundle

Equation 7-1 is based on Euler's column equation and Equation 7-2 is based on the short column formula developed by Professor J. B. Johnson during the nineteenth century.

Other TEMA formulations are summarized in the following sections. The reader is urged to be familiar with the TEMA standard and follow its guidelines in designing a shell and tube heat exchanger.

TEMA Formulations

Baffles and Support Plates

Natural Frequencies of Straight Tubes on Multiple Equal Spans

$$f_n = \frac{3.36C}{l^2}\sqrt{\frac{EI}{W}}$$

where
f_n = tube natural frequency, Hz
C = mode constant from Table 7-6
l = span length, in.
E = modulus of elasticity, psi
I = moment of inertia, in.[4] (Table 7-3)
W = $W_t + W_{fi} + MW_{fo}$, lbs/ft
W_t = weight of empty tube (Table 7-3)
W_{fi} = weight of fluid inside tube $0.00545\ \rho_i d_i^2$
W_{fo} = weight of fluid displaced by tube $0.00545\ \rho_o d_o^2$
M = added mass coefficient from Table 7-6
ρ = fluid density, lbs/ft[3]
d = diameter of tube, in.
subscripts:
i = inside
o = outside

Allowable Tube Compressive Stress—Periphery of Bundle. The allowable tube compressive stress, psi, for the tubes at the periphery of the bundle is given by:

$$S_c = \frac{\pi^2 E_t}{2(k\ell/r)^2} \qquad \text{when } C_c \leq k\ell/r \text{ or}$$

$$S_c = \frac{S_y}{2}\left[1 - \frac{(k\ell/r)}{2C_c}\right] \qquad \text{when } C_c > k\ell/r$$

$$\text{where } C_c = \sqrt{\frac{2\pi^2 E_t}{S_y}}$$

Table 7-6
Mode Constant-C [2]

No. of Spans	Extreme Ends Supported 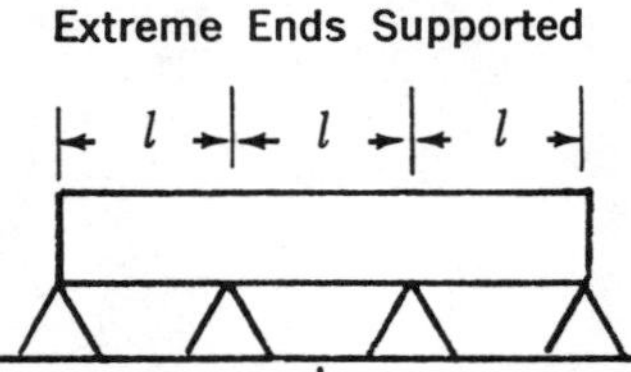		Extreme Ends Clamped 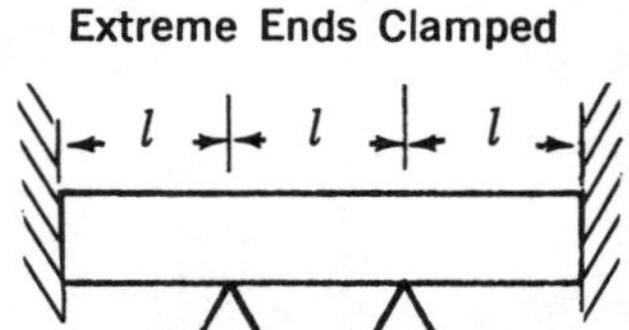		Extreme Ends Clamped-Supported 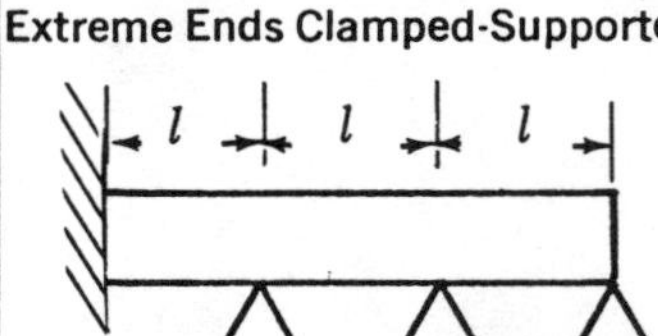	
	1st Mode	2nd Mode	1st Mode	2nd Mode	1st Mode	2nd Mode
1	31.73	126.94	72.36	198.34	49.59	160.66
2	31.73	49.59	49.59	72.36	37.02	63.99
3	31.73	40.52	40.52	59.56	34.32	49.59
4	31.73	37.02	37.02	49.59	33.02	42.70
5	31.73	34.99	34.99	44.19	33.02	39.10
6	31.73	34.32	34.32	40.52	32.37	37.02
7	31.73	33.67	33.67	38.40	32.37	35.66
8	31.73	33.02	33.02	37.02	32.37	34.99
9	31.73	33.02	33.02	35.66	31.73	34.32
10	31.73	33.02	33.02	34.99	31.73	33.67

S_y = yield stress, psi, of tube material at design metal temperature used.
r = radius of gyration of tube

$= 0.25\sqrt{d_o^2 + (d_o - 2t_t)^2}$, in. (Table 7-3)

$k\ell$ = equivalent unsupported buckling length of the tube, inches. Use the largest value considering unsupported tube spans.
ℓ = unsupported tube span, in.

$$k = \begin{cases} 0.6 & \text{for unsupported spans between two tubesheets.} \\ 0.8 & \text{for unsupported spans between a tubesheet and a baffle.} \\ 1.0 & \text{for unsupported spans between two baffles.} \end{cases}$$

Note: The value of S_c shall not exceed the Code allowable tensile stress of the tube material at design metal temperature used.

Effect of Longitudinal Tube Stress

$$f_{nP} = f_n \sqrt{1 + \frac{Pl^2}{EI\pi^2}}$$

where f_{nP} = tube natural frequency in stressed condition, Hz
P = axial force, lbs (positive for tensile, negative for compressive)

Natural Frequencies of Straight Tubes on Unequal Multiple Spans

$$f_n = 10.83\ k^2 \sqrt{\frac{EI}{W}}$$

For a tube on multiple unequal spans with the extreme ends fixed and simply supported at the intermediate supports, kl can be obtained by solving the following characteristic determinant for an n span system.

Natural Frequencies of U-Tubes. It must be recognized that each tube is a continuous beam that has a single fundamental frequency. This frequency may be largely governed by the lowest "stand alone" frequency of either the longest straight span or the U-bend. It is suggested that both be calculated and that the lower value be used, keeping in mind the approximate and somewhat conservative nature of the result. The straight span frequency may be determined from Table 7-6 using the appropriate mode constant. The U-bend out-of-plane frequency, assuming simple supports and for the first mode only, may be calculated as follows:

$$f_{nu} = \frac{2.74C_u}{R^2}\sqrt{\frac{EI}{W}}$$

where f_{nu} = U-tube natural frequency, Hz
C_u = mode constant for U-bend
R = bend radius, in.

Note: For other than simple support conditions the calculated frequency may be estimated by multiplying the above value for f_{nu} by the appropriate ratio of mode constants from Table 7-6 using single span values.

ASME Tube Joint Load Criteria

The ASME Section VIII Division 1 pressure vessel code lists formulations in evaluating tube forces exerted on tubesheets. Referring to Figure 7-13 and Table 7-7 the formulas for the maximum tube force are as follows:

For joint types a, b, c, d, e:

$$F_t = A_t\sigma_{all}f_r \tag{7-3}$$

For joint types f, g, h, i, j, k:

$$F_t = A_t\sigma_{all}f_rf_ef_y \tag{7-4}$$

where F_t = maximum tube joint force, lb_f
A_t = cross-sectional metal area of tube, in.^{2}
σ_{all} = ASME maximum allowable stress, psi
f_r = joint reliability factor
f_r (no test) = maximum value without test given in Table 7-7
f_r (test) = maximum value with test as specified in the ASME Section VIII Division 1 code, per section UA-002

Figure 7-14 shows how the tube joint load varies for various tube gauges of various process conditions. Naturally, as the tube wall increases, the tube stiffens and, consequently, the force exerted by the tube on the tubesheet joint increases. The engineer should evaluate the tube loads with the various process conditions possible and use the worst for determining the maximum tube joint force, as shown in Figure 7-14. The TEMA standard gives the formulations to determine the tube joint forces and the user is referred to this standard for these expressions.

The buckling of exchanger tubes can be a problem if thermal expansion is not properly accounted for in de-

Table 7-7
Reliability Factors, f_r [7]

Type Joint	Description[8]	Notes	f_r (test)	f_r (no test)
a	Welded only, $a \geq 1.4t$	(1)(7)(8)	1.00	0.80
b	Welded only, $t \leq a < 1.4t$	(1)(2)	0.70	0.55
c	Brazed, examined	(1)(3)	1.00	0.80
d	Brazed, not fully examinable	(1)(6)	0.50	0.40
e	Rolled, welded, $a \geq 1.4t$	(1)(7)(8)	1.00	0.80
f	Rolled, two or more grooves, and welded, $a < 1.4t$	(1)(4)(5) (7)	0.95	0.75
g	Rolled, single-groove, and welded, $a < 1.4t$	(1)(4)(5) (7)	0.85	0.65
h	Rolled, no grooves, and and welded, $a < 1.4t$	(1)(4)(5) (7)	0.70	0.50
i	Rolled, two or more grooves	(1)(4)(5)	0.90	0.70
j	Rolled, single groove	(1)(4)(5)	0.80	0.65
k	Rolled, no grooves	(1)(4)(5)	0.60	0.50

Notes:
(1) The use of f_r (test) factor requires qualification in accordance with UA-003 and UA-004.
(2) For welds where a is less than t, f_r (no test) = 0. Tubes with Type (b) joints where $a<t$ may be considered as acting as stays and contributing to the strength of the tubesheet only when the joint is tested in accordance with UA-003 and UA-004.
(3) A value of 1.00 for f_r (test) or .80 for f_r (no test) can be applied only to joints in which visual examination assures that the brazing filler metal has penetrated the entire joint [see UB-14(a)] and the depth of penetration is not less than three times the nominal thickness of the tube wall.
(4) When the ratio of O.D. to I.D., using nominal tube dimensions, is less than 1.05 or greater than 1.410, qualification in accordance with UA-003 and UA-004 is required.
(5) The nominal pitch used in the design of tubesheets for roller expanded joints shall not be less than the following:

$$P = d_o + 0.165\,(d_o + 2t)$$

where P = nominal pitch (center-to-center distance of adjacent tube holes), in.
d_o = tube O.D., in.
t = nominal thickness of average wall tube, in.

except that:
(a) nominal pitch shall not be less than $d_o + 2t$ unless the joint is qualified in accordance with UA-003 and UA-004; and
(b) 96% of the ligaments between tube holes throughout the thickness of the machined tubesheet shall not be less than 0.85 $(P-d_o)$. Ligaments which do not meet this requirement shall be evaluated and corrections made as may be necessary.
(6) A value of .50 for f_r (test) or .40 for f_r (no test) shall be used for joints in which visual examination will not provide proof that the brazing filler metal has penetrated the entire joint [see US-14(b)].
(7) The value of f_r (no test) applies only to material combinations as provided for under Section IX. For material combinations not provided for under Section IX, f_r must be determined by test in accordance with UA-003 and UA-004.
(8) For joint types involving more than one fastening method, the sequence used in the joint descriptions does not necessarily indicate the order in which the operations are performed.

sign. One such formulation to predict the critical buckling load is as follows:

$$F_{cr} = \frac{\pi^2 E_t I_t}{\left(\frac{L_{tube}}{NB + 1}\right)^2}(0.5216) \quad (7\text{-}5)$$

where L_{tube} = total length of tubes between tubesheets
NB = number of baffles

Equation 7-5 is based on the Euler column formula. In situations where there are several baffles, such that the effective length, L_0, divided by the radius of gyration, k, is between 30 and 120, exclusive, then the Johnson short column equation is more accurate. For a tube to be considered as a series of short columns constrained by fixed ends, one must be certain that the baffles constraining the tubes allow practically no translational or rotational movement. The stiffness of the baffle plate should be analyzed, as small translational and rotational tube movement allowed by the baffle plate could considerably alter the buckling characteristics of the tube. The evaluation of a baffle plate containing several tubes can be a somewhat detailed analysis, and it may be faster to consider the tube as a continuous beam in determining buckling characteristics.

For further details on the mechanical design of exchangers, the reader is referred to TEMA. We will discuss tube vibrations shortly.

PROCESS EVALUATION OF SHELL AND TUBE EXCHANGERS

We are concerned here only with any particular heat exchanger and determining whether it can transfer heat energy as required. How the unit affects process conditions of the entire system is not our concern here, because we are interested only in the proper performance of the unit. Evaluating the exchanger in relation to the process system is the primary concern of the chemical engineer. The thermal evaluation of the exchanger is one area where chemical and mechanical engineering overlap; just as in Chapters 2 and 4 we saw how civil and mechanical engineering coincide. Thus, the mechanical engineer must be cognizant of process evaluation of heat exchangers in order to design these units.

A thermal evaluation of shell and tube heat exchangers concerns primarily two modes of heat transfer—conduction and convection.

In Chapter 3 we considered heat transfer through piping and vessel components as well as jacketed systems. As described in Chapter 3, the basic expressions used in convection are as follows:

$$q = \dot{m}C_p\Delta t \quad (3\text{-}24)$$

$$q = UA(LMTD) \quad (3\text{-}26)$$

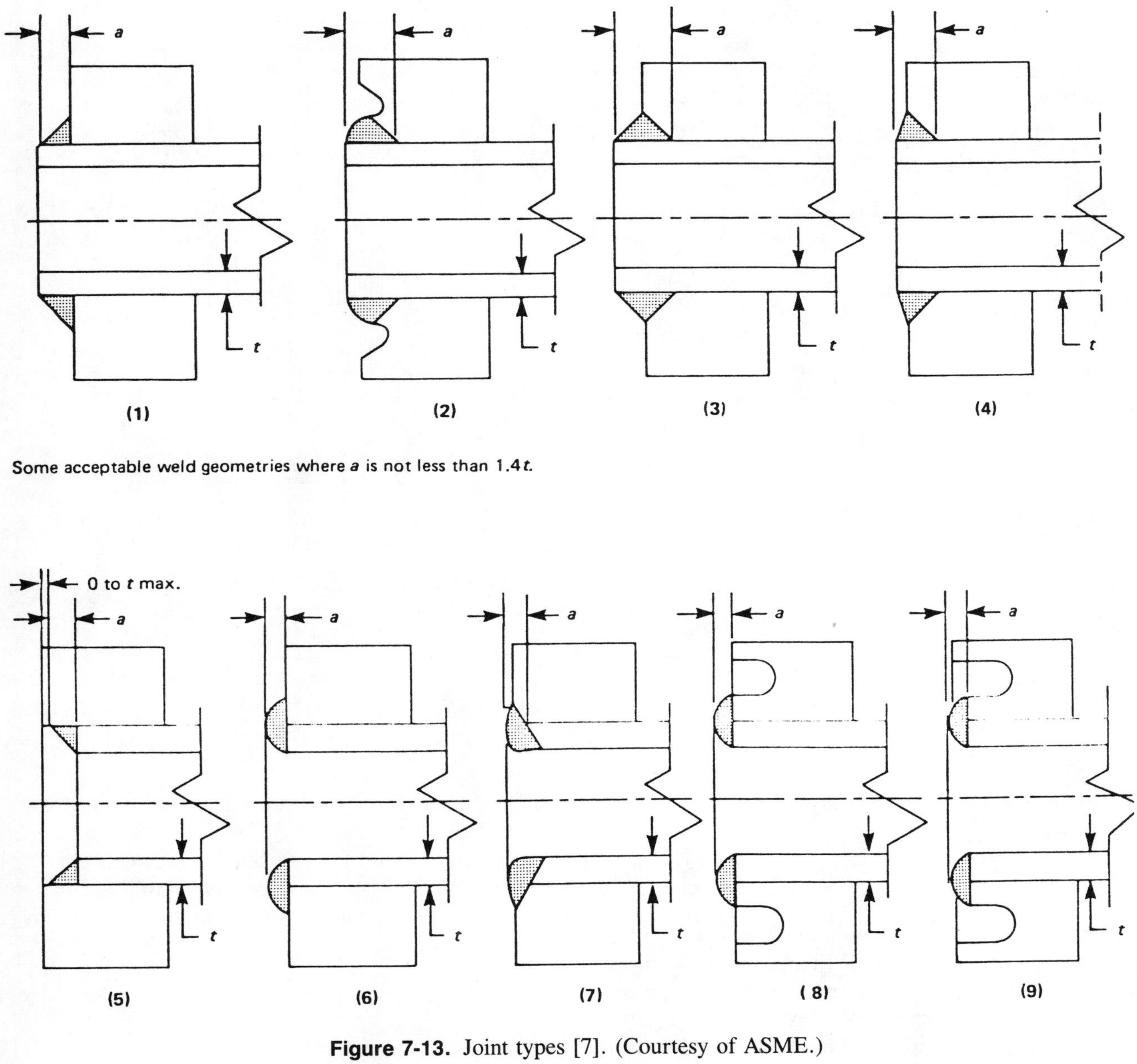

Figure 7-13. Joint types [7]. (Courtesy of ASME.)

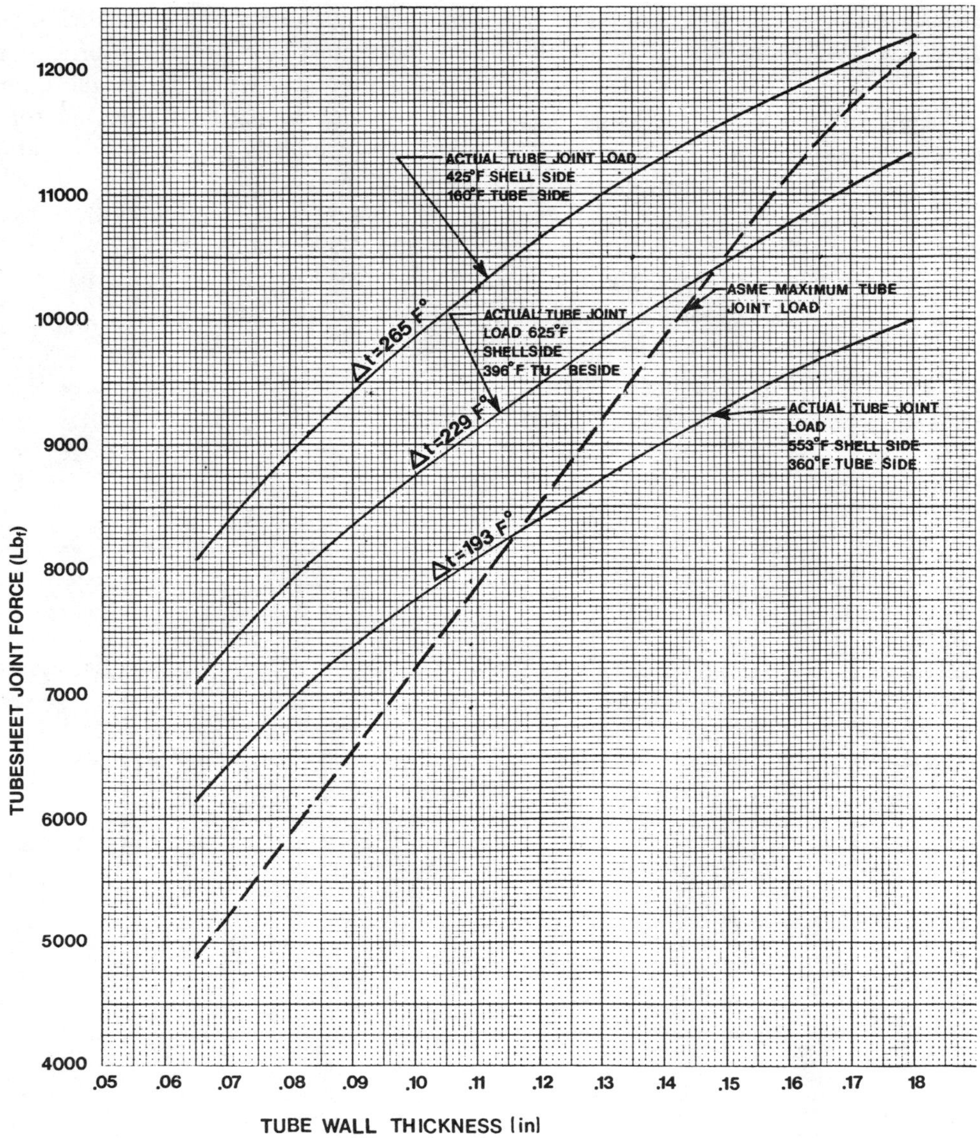

Figure 7-14. Tube joint loads.

Equation 3-9 is a variant of Fourier's heat law of conduction in which,

$$q = KA\Delta t \tag{7-6}$$

The treatment of shell and tube exchangers requires the same basic theory for use in Chapter 3, but a different application. In these types of exchangers we are primarily concerned with the heat duty or heat load required in the same general sense as the jacketed vessels in Chapter 3. Process requirements are the criteria used to determine the heat duty. The two basic components of heat transfer in the shell and tube exchanger are sensible heat and latent heat. These concepts are described mathematically with the use of Equation 3-24. Using this relation we have:

$$q = \dot{m}C_p(\Delta t) \tag{7-7}$$

$$q = \dot{m}h_{fg} \tag{7-8}$$

Equation 7-7 determines sensible heat change and Equation 7-8 determines latent heat change and is a form of Equation 3-13. You will recall that sensible heat is the amount of heat energy required to either heat or cool a given mass (solid, liquid, or gas—without a phase change) to a measurable degree. Thus, if we have a kettle full of water and we heat the water mass at atmospheric pressure to 212°F, adding additional heat to the kettle will not raise the temperature. Thus, the sensible heat is the amount of heat required to raise the water temperature to 212°F. The additional amount of heat required to convert the water to steam is called latent heat. Sensible heat can be detected by the human senses and thus physically measured, as with a thermometer. Latent heat is heat energy that cannot be detected by the human senses and is more intrinsic to the basic process—the hot steam plus the additional heat provided to the kettle to convert the water to steam.

While these terms are rudimentary, their importance is fundamental to heat exchangers. Combining Equations 7-7 and 7-8 we arrive at the total heat duty of the exchanger as:

$$q = \dot{m}C_p(\Delta t) = \dot{m}h_{fg} \tag{7-9}$$

The first term on the right side of Equation 7-9 represents the sensible (cooling or heating) heat and the second right-hand term represents the latent (condensing or boiling) heat.

In Chapter 3 we discussed the LMTD and the reader can use Figure 3-10 to quickly determine this parameter. The difference between that value used in Chapter 3 and the application here is that the LMTD will *vary* from one pass to another in a *multipass* exchanger. Figure 7-15 illustrates how a different LMTD occurs for each pass, showing that using the inlet and outlet of the tubeside and shellside would not produce an accurate LMTD value. Thus, the value obtained from Figure 3-10 must be multiplied by a correction factor, F, as provided in Figure 7-16. These correction factors were tabulated by TEMA and are determined by the "P" and "R" parameters shown in Figure 7-16. For a true counterflow or parallel exchanger a correction factor is not necessary. Use of these tables is demonstrated in the examples which follow.

It should be pointed out that the straight-line exchanger curve in Figure 7-15 is for a single-component fluid. Most industrial applications are multicomponent mixtures. Chemical engineers do vapor-liquid equilibrium calculations that show vapor and liquid compositions in multicomponent mixtures to be different *and* changing within the exchanger. Thus, real curves are not straight, but often can be approximated as such in multicomponent flow. Assuming a straight line is a source of many design errors.

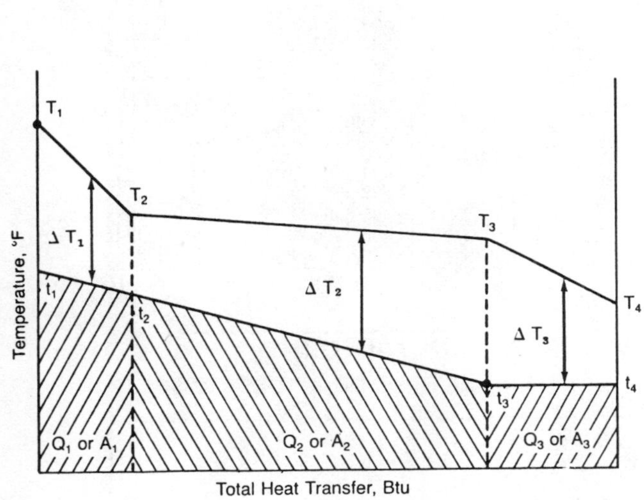

Figure 7-15. The amount of heat transferred varies from one pass to another [4].

Tube Wall Temperature and Caloric Temperature

Chapter 3 dealt with the LMTD in the computation of heat transfer problems. One of the assumptions used in defining the LMTD is that the overall heat transfer coefficient, U, remains constant. With multipass exchangers this is certainly not always the case, as indicated by Figure 7-15. Even with counterflow exchangers, as the cold fluid gets hotter, the viscosity decreases, changing the overall U-value. In many instances, the U value will vary more than the inside tube coefficient.

Colburn [8] addressed this problem by assuming that the U value will change linearly with temperature and deriving an expression for the temperature differential. Colburn thus elected to obtain a single overall coefficient, U_x, at which all heat transfer surfaces can be assumed to transfer heat at the computed LMTD. In this way the overall U-value, U_x, can be expressed as

$$\frac{Q}{A} = U_x\left(\frac{GTTD - LTTD}{L_n\left(\frac{GTTD}{LTTD}\right)}\right) \tag{7-10}$$

Such a value of U_x exists at a temperature, t_c, known as the caloric temperature. It is this value of t_c at which the inside and outside tube film coefficients, h_i and h_o, respectively, are evaluated. The caloric temperature is found by multiplying the respective hot and cold temper-

(text continued on page 122)

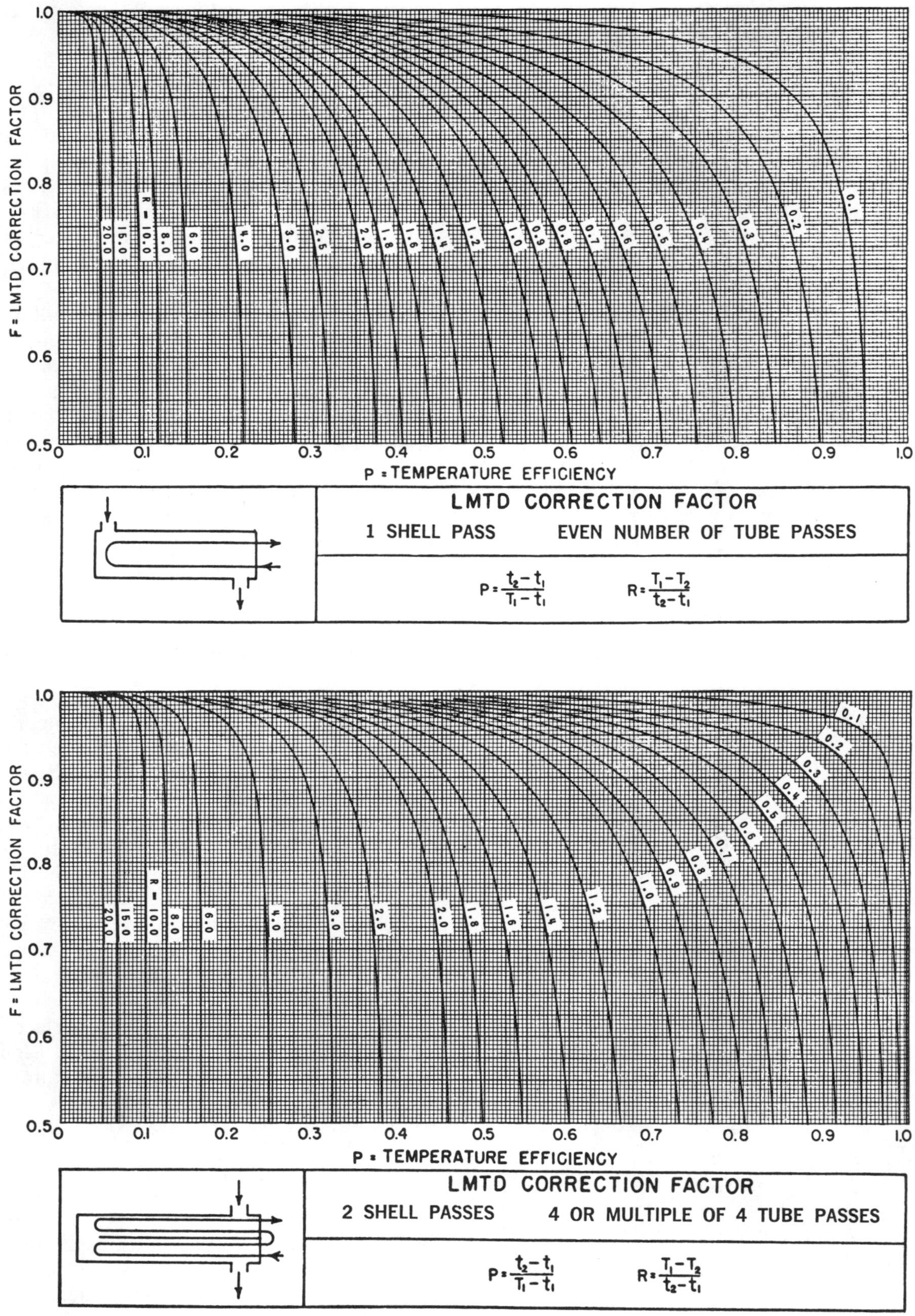

Figure 7-16. LMTD correction factor. (©1978 Tubular Exchanger Manufacturers Association.)

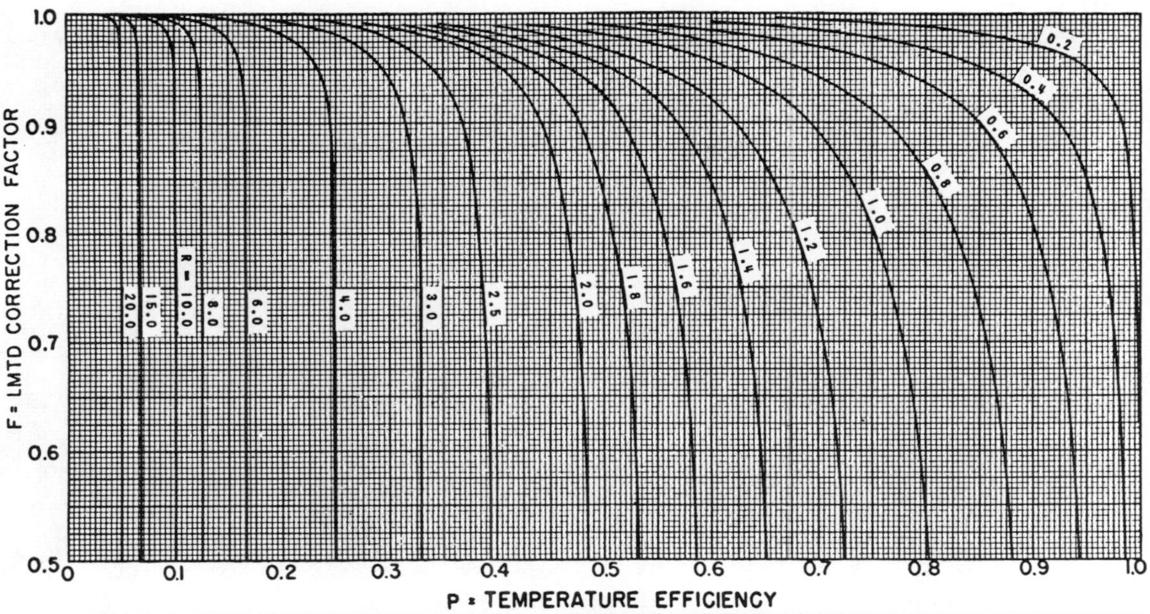

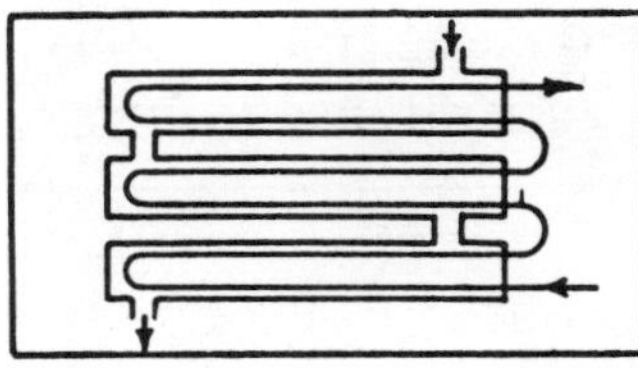

LMTD CORRECTION FACTOR

3 SHELL PASSES 6 OR MORE EVEN NUMBER OF TUBE PASSES

$$P = \frac{t_2 - t_1}{T_1 - t_1} \qquad R = \frac{T_1 - T_2}{t_2 - t_1}$$

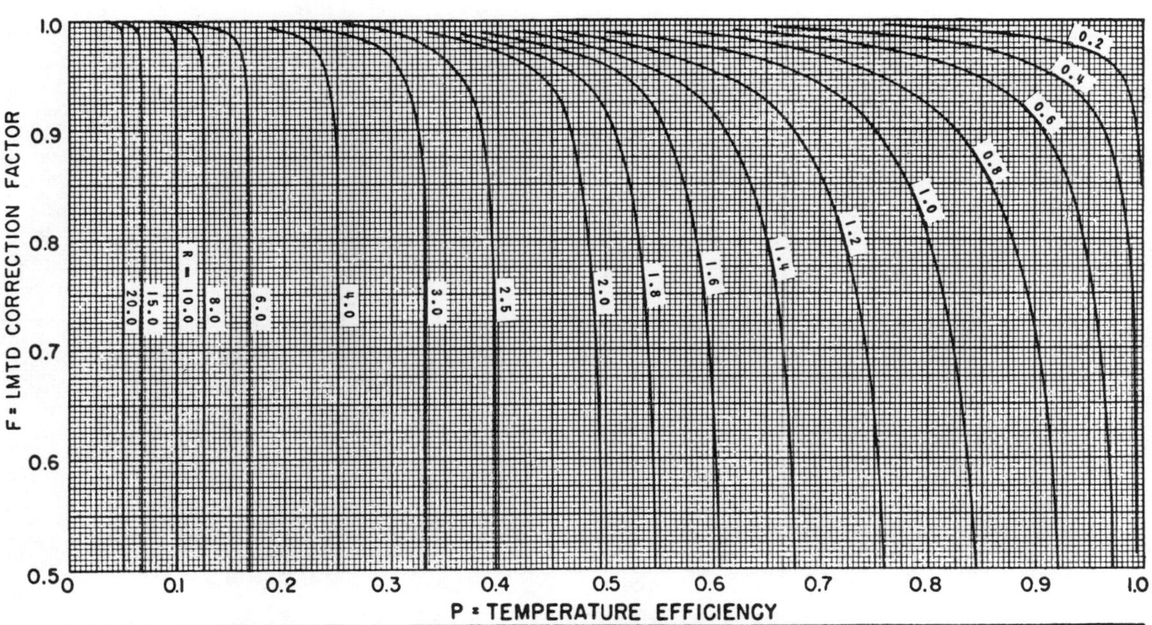

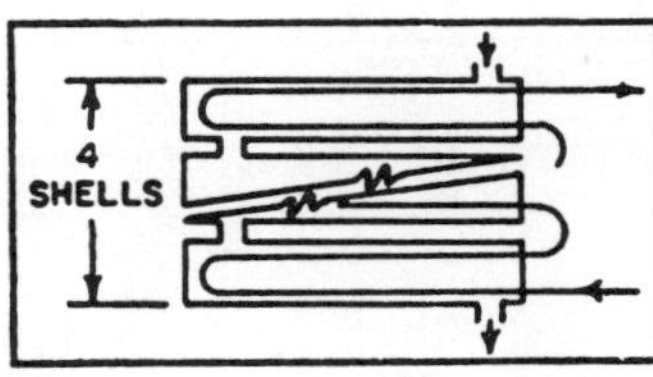

LMTD CORRECTION FACTOR

4 SHELL PASSES 8 OR MULTIPLE OF 8 TUBE PASSES

$$P = \frac{t_2 - t_1}{T_1 - t_1} \qquad R = \frac{T_1 - T_2}{t_2 - t_1}$$

Figure 7-16. Continued.

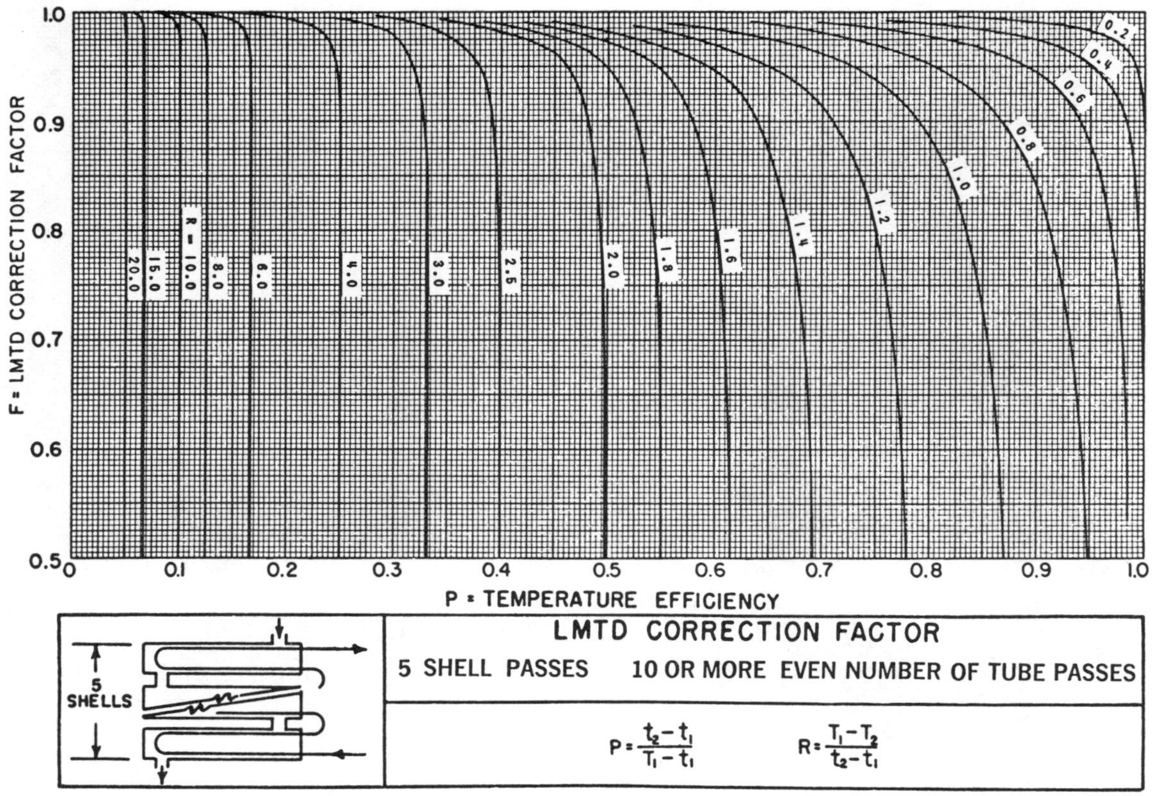

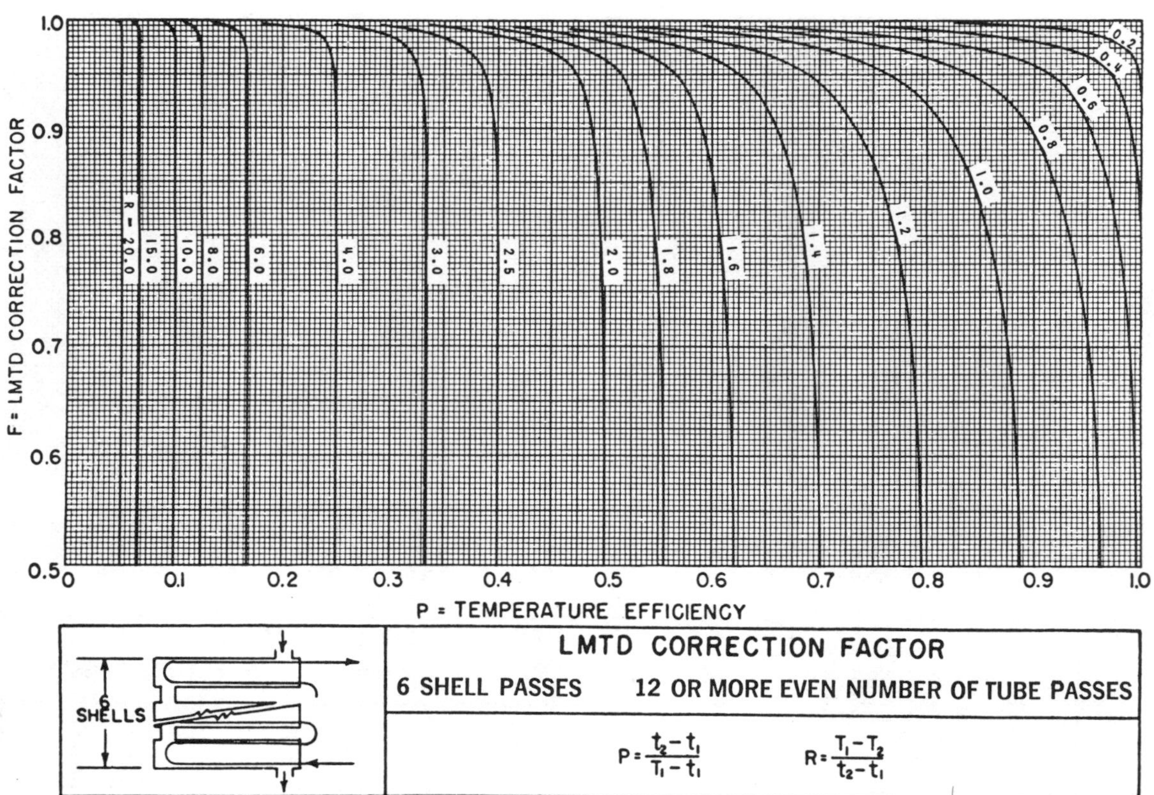

Figure 7-16. Continued.

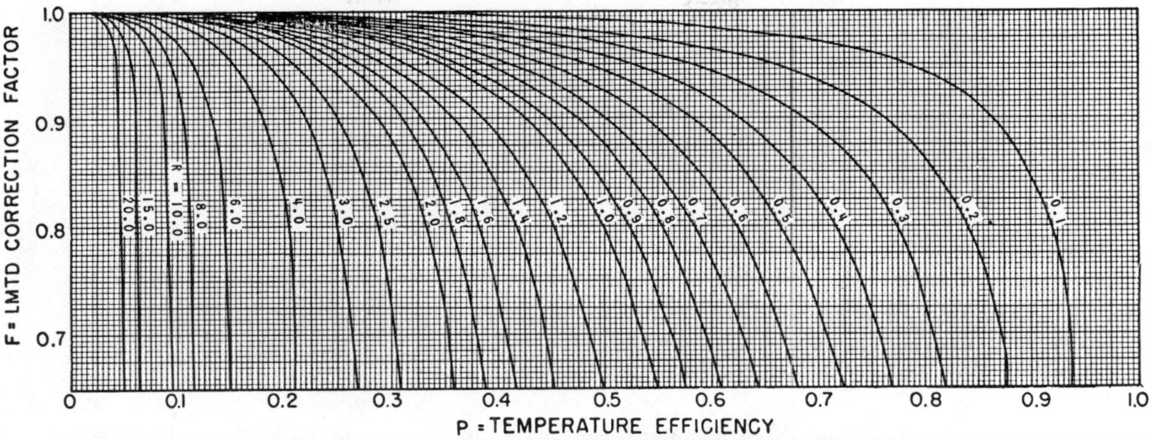

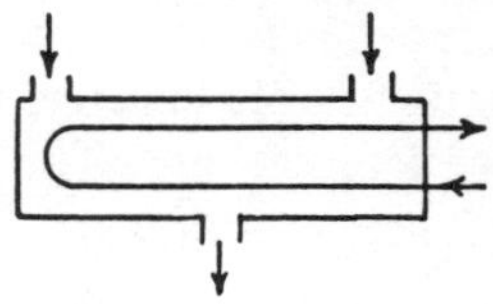

LMTD CORRECTION FACTOR

1 DIVIDED FLOW SHELL PASS **EVEN NUMBER OF TUBE PASSES**

$P = \frac{t_2 - t_1}{T_1 - t_1}$ $R = \frac{T_1 - T_2}{t_2 - t_1}$

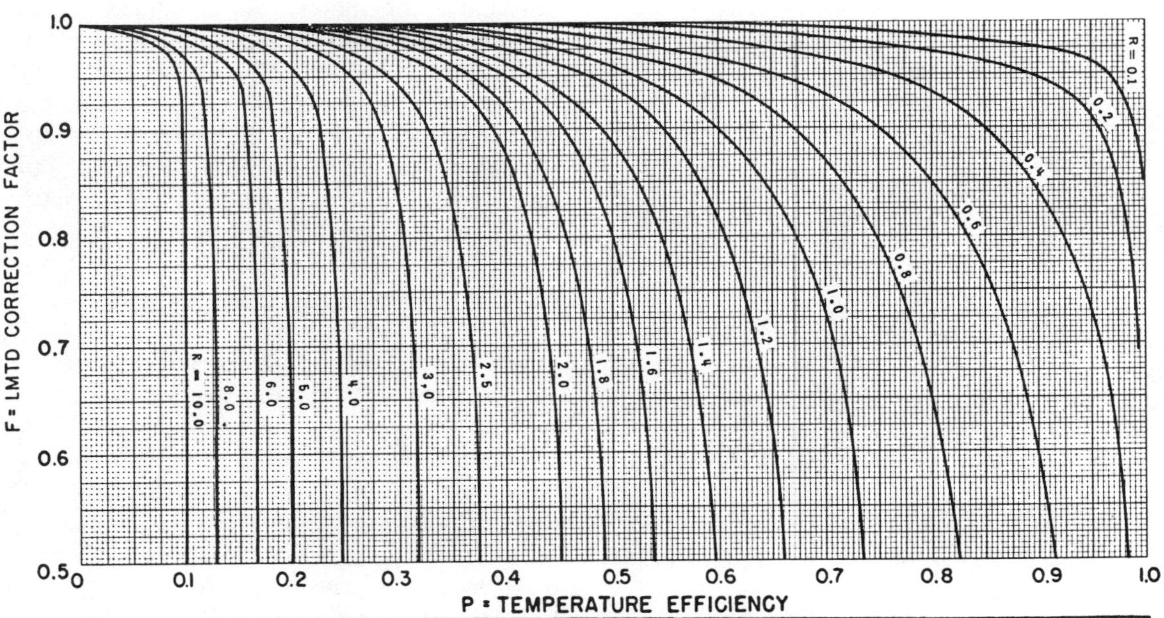

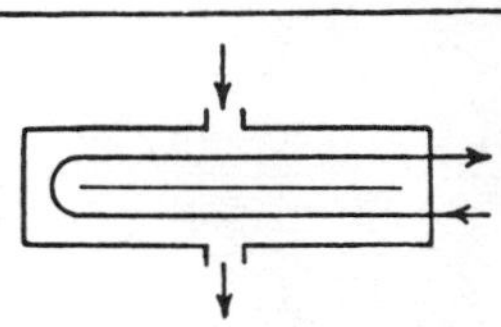

LMTD CORRECTION FACTOR

SPLIT FLOW SHELL **2 TUBE PASSES**

$P = \frac{t_2 - t_1}{T_1 - t_1}$ $R = \frac{T_1 - T_2}{t_2 - t_1}$

Figure 7-16. Continued.

ature differentials by a caloric fraction, F_c. That is, the fraction, F_c, is multiplied by the temperature rise of the controlling stream and adding the resulting rise to the lower terminal temperature of the stream. Figure 7-17 helps in determining the controlling stream. Colburn [8] correlated the data for the insert in the top left-hand corner of Figure 7-17.

The fluid stream, either shell-side or tube-side, that has the largest U-value corresponds to the controlling heat transfer film coefficients, h_i and h_o, which are used to determine U_x. Basically, this implies that we must find the values of U_h and U_c for both separate streams, and the stream that has the largest c value, based on Figure 7-17 is the one used to compute h_o and h_i in computing U_x, where U_h is the overall heat transfer coefficient at the hot end of each respective stream and U_c is the overall heat transfer coefficient at the cold end of each respective stream. In equation form we express the caloric temperatures as derived by Kern [9] as follows:

The hot fluid caloric temperature is

$$t_{ch} = t_{h_o} + F_c(t_{h_i} - t_{h_o}) \tag{7-11}$$

The cold fluid caloric temperature is

$$t_{cc} = t_{c_i} + F_c(t_{c_o} - t_{c_i}) \tag{7-12}$$

where t_{ch} = caloric temperature of hot fluid, °F
t_{h_o} = outlet hot fluid temperature, °F
t_{h_i} = inlet hot fluid temperature, °F
t_{cc} = caloric temperature of cold fluid, °F
t_{c_i} = inlet cold fluid temperature, °F
t_{c_o} = outlet cold fluid temperature, °F
F_c = correction factor determined from Figure 7-16

Once the hot and cold caloric temperatures are determined we can now compute the tube wall temperature, because the caloric temperatures represent the true mean values for the varying values of U_x, h_i, and h_o. Kern [9] has expressed the tube wall temperatures in the following forms:

Outside tube wall temperature with hot fluid external to tube is

$$t_w = t_{ch} - \frac{h_{i_o}}{h_{i_o} + h_o}(t_{ch} - t_{cc}) \tag{7-13}$$

or

$$t_w = t_{ch} - \frac{h_o}{h_{i_o} + h_o}(t_{ch} - t_{cc}) \tag{7-14}$$

Outside tube wall temperature for the hot fluid on the inside of the tube is

$$t_w = t_{cc} + \frac{h_{i_o}}{h_{i_o} + h_o}(t_{ch} - t_{cc}) \tag{7-15}$$

or

$$t_w = t_{ch} + \frac{h_o}{h_{i_o} + h_o}(t_{ch} - t_{cc}) \tag{7-16}$$

Thus, the pipe wall temperature can be computed when the caloric temperature values are determined. The temperature difference across the tube wall is customarily assumed to be negligible because the entire tube is at the outside surface temperature. The terms in Equations 7-13 through 7-16 are as follows:

t_w = tube (outside surface) wall temperature, °F
h_{i_o} = inside film coefficient of tube using outside surface temperature, Btu/hr-ft²-°F
h_o = outside film coefficient of tube, using outside tube surface temperature, Btu/hr-ft²-°F

The tube wall temperatures are to be used in computing thermal movements of the tubes and all other mechanical computations. As explained previously, these tube wall temperatures are based on the caloric temperature values that truly reflect the mean value of the varying values of h_i, h_o, and U_x. It is a common oversight in exchanger design to use the arithmetic mean rather than the caloric values. Equation 3-26 should not be used because Equation 7-10 more accurately describes the true mean values in the exchanger.

Overall Heat Transfer Coefficient

The thermal duty of an exchanger cannot be discussed without first defining the overall heat transfer coefficient, or "U-value." This parameter is referred to in the previous discussion and it is noted that this value can vary with various types of configurations. Now, after discussing how a variable U-value is handled on multipass units using the caloric temperature, we are ready to treat the overall coefficient in detail. The overall heat transfer coefficient, denoted as U, is defined as follows:

$$U = \frac{1}{\frac{1}{h_i} + \frac{T_{fi}}{K_{fi}} + \frac{T_w}{k_w} + \frac{1}{h_o} + \frac{T_{fo}}{k_{fo}}} \tag{7-17}$$

where h_i = inside tube film coefficient, Btu/hr-ft²-°F
T_{fi} = thickness of inside tube deposits, ft

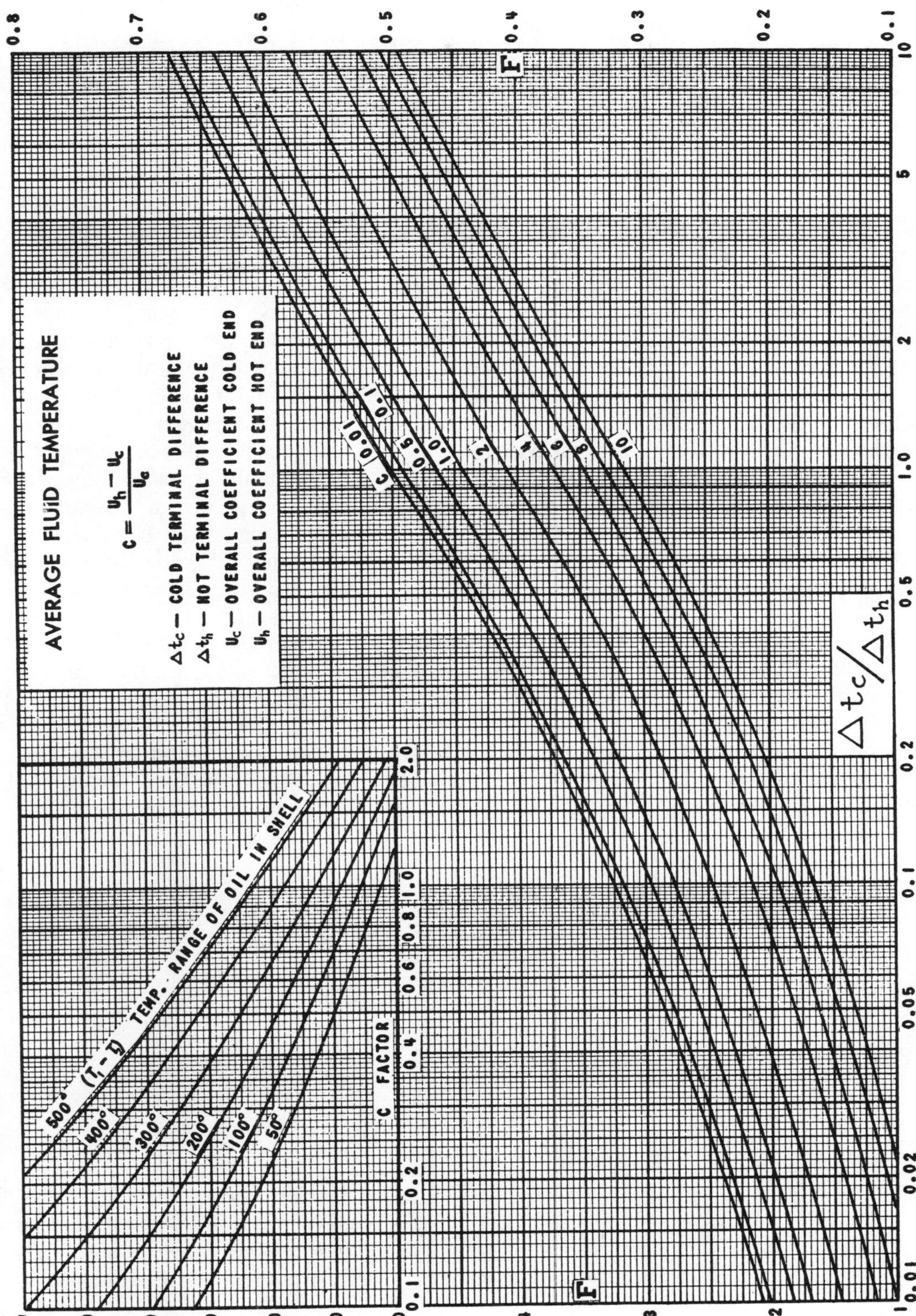

Figure 7-17. Average or caloric temperature. (©1959 Tubular Exchanger Manufacturers Association.)

k_{fi} = thermal conductivity of foreign deposits on inside of tube, Btu/hr-ft^2-°F
T_w = tube wall thickness, ft
k_w = thermal conductivity of tube wall, Btu/hr-ft^2-°F
h_o = outside tube film coefficient, Btu/hr-ft^2-°F
T_{fo} = thickness of outside tube deposits, ft
k_{fo} = thermal conductivity of deposits on outside of tube, Btu/hr-ft^2-°F

The terms in Equation 7-17, 1/h, T_f/k_f, and T_w/k_w, are known as film resistance, fouling resistance (we will refer to this as fouling factors), and tube wall resistance, respectively. These parameters represent the resistance to heat flow through the fluid film, foreign deposits, and the tube wall. This is shown in Figure 7-18 where the temperature is shown varying through the various resistance zones. This figure is a conceptualization of the temperature profile, as the degree of gradient change in temperature is a function of the flow conditions (laminar versus turbulent) and on the type and amount of foreign deposits. To understand Equation 7-17 we will discuss each resistance separately.

Fouling of Inside and Outside Tube Surfaces

Fouling occurs when deposits are made on the walls by particles contained in the fluid medium or by the fluid itself forming a layer on the tube walls. This can occur two ways, either by adhesive characteristics of the deposited matter or by the foreign material being bonded to the tube surface by thermal gradients between the tube wall and the foreign material, so that the latter changes phases when it contacts tube surface, resulting in a coating effect. Thus, the depositing of foreign material adds to the resistance of heat flow from the tube and shellside flows. Fouling can occur inside and outside of tube surfaces. The complexity of fouling and how it occurs does not easily allow this phenomenon to be treated analytically. There are far too many variables involved for one to accurately compute fouling factors. Thus, this phenomenon is treated in a more subjective light, using experience as a guideline. Years of experience with various services have resulted in the use of accurate fouling factors.

Fouling factors are very important in the design of shell and tube heat exchangers. Bare or plain tubes, which are almost always used, generate low U-values when compared to those generated by tubes with fin attachments. Finned tubes, especially those with fairly high fins, experience very little fouling unless the deposits cover an appreciable portion of the fin height. With the normally accepted long periods between tube cleaning in plants, fouling certainly must be considered in the calculation of the U-value. One must be aware of the shell- and tube-side fluids and select those fouling factors that best reflect the optimum fouling that will affect thermal duty.

The fouling factor in Equation 7-17 is T_f/k_f. This term is the inverse of the thermal conductance of heat through the foreign matter, denoted by k_f/T_f. Thus, the *reciprocal* of the thermal conductivity of the foreign material is known as the fouling factor. Fouling can exist on both or one side of the tube. Typical values for fouling factors for common services are given in Table 7-8.

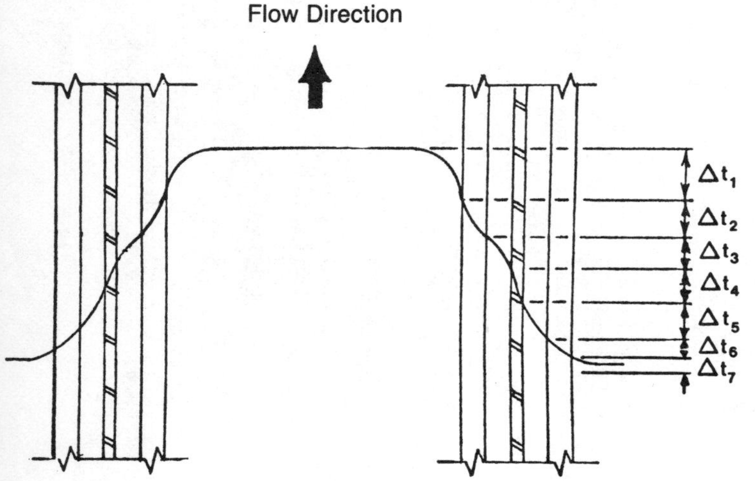

Δt_1 = Temperature drop through inside turbulent boundary layer
Δt_2 = Temperature drop through laminar boundary layer inside tube
Δt_3 = Temperature drop through fouling layer inside tube
Δt_4 = Temperature drop through tube wall
Δt_5 = Temperature drop through outside fouling layer
Δt_6 = Temperature drop through outside laminar boundary layer
Δt_7 = Temperature drop through outside turbulent boundary layer

Figure 7-18. Temperature profile through tube wall.

Table 7-8
Recommended Minimum Fouling Resistances

Gases and Vapors	Fouling Factor
Centrifugal compressor exhaust	0.001
Reciprocating compressor exhaust	0.01
Reciprocating compressor refrigerant vapor	0.0025
Centrifugal compressor refrigerant vapor	0.0015
Oil-free and clean high-quality steam	0.0003
Oil-free and clean low-quality steam	0.0005
Oil-bearing steam	0.001
Compressed air	0.002
Acid gas	0.001
Solvent vapors	0.001
Natural gas	0.001
Liquids	
Bay water	0.0025
Distilled water	0.0005
Hard well water	0.0033
Untreated cooling tower water	0.0033
Treated cooling tower water	0.0015
Engine jacket water	0.0012
Treated boiler feed water	0.0015
Fuel oils	0.006
Clean organic solvents	0.001
Vegetable oils	0.004
Refrigerant liquids	0.001
Industrial heat transfer oils	0.001
Hydraulic fluid	0.001
Natural gasoline and liquefied petroleum gases	0.001
Rich oil	0.001
Lean oil	0.002

Tube Film Coefficients

In the region close to the tube wall the fluid becomes stagnant and forms a film around the tube surface—on inside and outside surfaces. This stagnant region is called the "boundary layer" because it forms at the boundary of the tube wall. The size and properties of the particular boundary layer are a function of the fluid properties itself and whether the fluid flow is either turbulent or laminar. Turbulent fluid motion always leads to greater heat transfer between the shell-side and tube-side fluids, resulting in higher film coefficients. For a more detailed discussion of boundary layer theory one is referred to a basic text on heat transfer, such as the *Handbook of Heat and Mass Transfer,* Volumes 1 and 2, N. P. Cheremisinoff, ed., Gulf Publishing Co., Houston, Texas, 1986.

Inside Tube Coefficients. In Example 3-5 we used correlations to determine film coefficients inside tubes. Here, we will give a more comprehensive treatment of film coefficients inside tubes.

In laminar flow there is more fluid stagnation around the tubes because in the boundary layer itself the flow is laminar, whereas in a turbulent boundary layer the sub-laminar boundary layer is only a small percentage of the total layer. Thus, fluid properties at the walls in laminar flow must be evaluated. The laminar film coefficient is largely dependent upon the viscosity and, thus, the temperature (which controls the viscosity) of the wall controls the value of h. The most commonly used correlation for laminar flow inside pipes is that given by Sieder and Tate as follows:

$$N_{Nu_o} = \frac{hd_i}{K} = 1.86(N_{Re})^{1/3}(N_{Pr})^{1/3}\left(\frac{d_i}{L}\right)^{1/3}\left(\frac{\mu}{\mu_w}\right)^{0.14} \tag{7-18}$$

where
d_i = inside pipe or tube diameter, ft
K = coefficient of thermal conductivity of fluid inside tube, Btu/hr-ft²-°F
N_{Nu_o} = Nusselt number (see Chapter 3)
N_{Re} = Reynolds number
N_{Pr} = Prandtl number (see Chapter 3)
L = total tube length, ft
μ = fluid viscosity at bulk temperature, lb_m/ft-hr or cp
μ_w = fluid viscosity at wall, lb_m/ft-hr or cp

The viscosity at the wall, μ_w, should be evaluated using either Equation 7-15 or Equation 7-16.

The bulk temperature of the fluid is in practice the average of the inlet and outlet tube fluid temperatures.

For turbulent flow inside tubes wall effects can play a role in the film coefficient value. The correlation that is widely used is the one developed by Sieder and Tate, which is as follows:

$$N_{Nu_o} = 0.027(N_{Re})^{0.8}(N_{Pr})^{1/3}\left(\frac{\mu}{\mu_w}\right)^{0.14} \tag{7-19}$$

for $0.7 < N_{Pr} < 17{,}000$

As for Equation 7-18, the value of μ_w should be determined using temperature values calculated using Equations 7-15 or 7-16.

McAdams [10] suggests that for temperature differences between the bulk fluid temperature and the pipe wall surface temperature the following expression can be used:

$$N_{Nu} = 0.023(N_{Re})^{0.8}(N_{Pr})^{n} \tag{7-20}$$

where n = 0.4 for heating
n = 0.3 for cooling

And the temperature differences are as follows:

Δt = pipe surface temp—bulk fluid temp
$\Delta t \leqslant 10°F$ for liquids
$\Delta t \leqslant 100°F$ for gases

Outside Tube Film Coefficients. Forced convection around immersed bodies is a complex subject, especially when a bundle of tubes is involved. We will only give a rather brief discussion of how one can obtain a general order of magnitude of film coefficients. The reader should be aware that process design is not addressed. Thus, for solving problems dealing with condensation, nucleate boiling, and film boiling—to name a few—the reader should consult other sources that treat process design in detail [4,8].

For gases flowing normal to circular cylinders a simple relationship is contrived by M. Jakob [1] using an *average* Nusselt number for the gas. An empirical version of this expression is given by

$$N_{Nu} = \frac{\bar{h}d_t}{k_f} = C(N_{Re})^n \tag{7-21}$$

where $\bar{h}$ = average film coefficient for gas, Btu/hr-ft²-°F
d_t = tube diameter, ft
k_f = gas coefficient of thermal conductivity, Btu/hr-ft-°F
C and n = parameters from Table 7-9

A variant of Equation 7-21 is widely used for forced convection of air normal to a cylinder is given by the following:

$$N_{Nu} = \frac{\bar{h}d_t}{k_f} = C(P_r)^{1/3}(N_{Re})^n \tag{7-22}$$

Forced convection normal to tube bundles is much more complex than that of a single tube. The size of the bundle and how the tubes are oriented (tube pitch arrangements) in the bundle are of prime importance. First, we will discuss an approach to determining the film coefficients for bundles and then discuss the merits of arranging tubes in various geometries.

There are four basic types of tube arrangements—triangular pitch, in-line triangular pitch, in-line square pitch, and diamond-square pitch. These four geometries are shown in Figure 7-19. Tubes arranged in bundles are more complex than a single tube because the flow vortices formed by the flow around the first tubes affect the flow around the tubes farther inside the bundle. Mose researchers agree that this transient effect is substantially dampened after the flow passes over the first ten tubes. Numerous research studies have been made that analyzed flow effects on tube bundles. E. D. Grimson [12] concluded from several studies that for tube bundles at least 10 tubes in depth the following expression can be used to predict the film coefficient:

$$\frac{\bar{h}d_t}{k_f} = B(\rho V d_o/\mu_f)^n \tag{7-23}$$

where μ_f = absolute viscosity, lb_m/ft-hr
V = velocity of air, ft/hr
N_{Re} = Reynolds number at maximum fluid velocity, V_{max}
$\bar{h}$ = average film coefficient, Btu/hr-ft²-°F
ρ = air density, lb_m/ft^3
k_f = thermal conductivity of fluid, Btu/hr-ft-°F
B and n = constants given in Table 7-10
d_o = tube outside diameter

The Reynolds number in Equation 7-23 is evaluated at the maximum fluid velocity. This velocity is obtained at the minimum flow passage between the tubes. This minimum distance is shown in Figure 7-19. The minimum distance is expressed in terms of the tube bundle geometry for each of the four configurations, as follows (see Figure 7-19):

(a) Triangular pitch, $d_{min} = \frac{W}{2} - d_t$
(b) In-line triangular pitch, $d_{min} = W - d_t$
(c) In-line square pitch, $d_{min} = W$
(d) Diamond square pitch, $d_{min} = P\cos 45° - D = 0.707p - D$

Table 7-9
Parameters for Fluid Flow Normal to Circular Cylinders

Range of Reynolds Numbers	C	n
$0.40 \leqslant N_{Re} \leqslant 4.0$	0.989	0.330
$4 \leqslant N_{Re} \leqslant 40$	0.911	0.385
$40 \leqslant N_{Re} \leqslant 4000$	0.683	0.466
$4000 \leqslant N_{Re} \leqslant 40,000$	0.193	0.618
$40,000 \leqslant N_{Re} \leqslant 400,000$	0.027	0.805

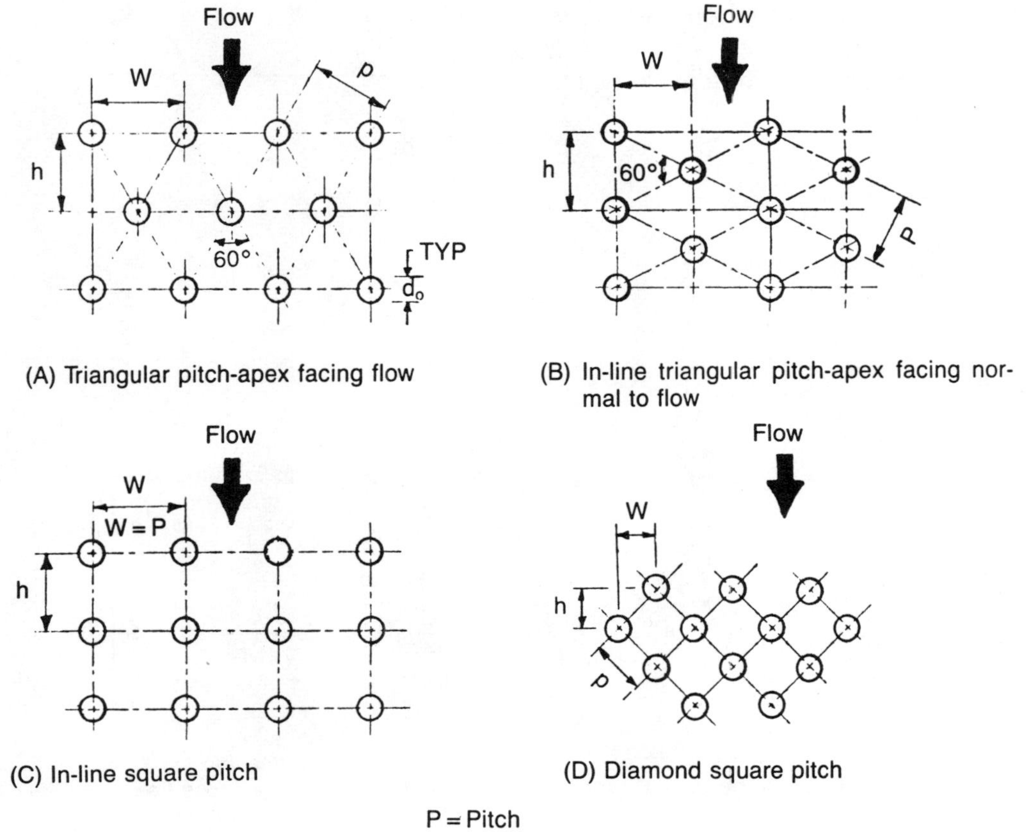

Figure 7-19. Tube bundle arrangements.

Table 7-10
Grimson Constants for Tube Bundles Containing 10 or More Tubes

Tube Bank Geometry	W/d_o (see Figure 7-19)	W/d_o Ratio 1.25 B	1.25 n	1.5 B	1.5 n	2.0 B	2.0 n	3.0 B	3.0 n
In-line	1.25	.348	.592	.275	.608	.100	.704	.063	.752
	1.50	.367	.586	.250	.620	.101	.702	.068	.744
	2.00	.418	.570	.299	.602	.229	.632	.198	.648
	3.00	.290	.601	.357	.584	.374	.581	.286	.608
Staggered	0.600	—	—	—	—	—	—	.213	.636
	0.900	—	—	—	—	.446	.571	.401	.581
	1.000	—	—	.497	.558	—	—	—	—
	1.125	—	—	—	—	.478	.565	.518	.560
	1.250	.518	.556	.505	.554	.519	.556	.522	.562
	1.500	.451	.568	.460	.562	.452	.568	.488	.568
	2.000	.404	.572	.416	.568	.482	.556	.449	.570
	3.000	.310	.592	.356	.580	.440	.562	.421	.574

The cross-flow are for various types of tube bundles is shown in Figure 7-20.

From the concept of continuity, where for two points along a flow path, or streamline,

$$V_1A_1 = V_2A_2 \tag{7-24}$$

where V_1 = velocity of fluid at point 1, ft/sec
A_1 = cross-sectional area, ft^2

we can deduce

$$V_2 = V_1 \frac{A_1}{A_2}$$

With all tubes being placed at a constant pitch and $V_1 = V_f$ = fluid velocity, we have

$$V_{max} = V_f \left(\frac{w}{d_{min}}\right) \tag{7-25}$$

Equation 7-25 represents the fluid velocity that would be used in Equation 7-23.

For tube bundles containing less than 10 tubes, values of the film coefficient in Equation 7-23 must be multiplied by the correction factors in Table 7-11.

Each tube pitch arrangement has its own advantages and disadvantages. A listing of these facts is given in Table 7-12. Whatever the tube arrangement selected, the tube arrangement in the tubesheet should be made very carefully. Clearances, which could be such items as impingement baffles, channel and head baffle lanes, must be considered. Table 7-13 is a compilation of various industrial standards for tube sheet layouts. Figure 7-21 shows a typical tube sheet layout.

One of the easiest and most common methods used to calculate shell-side film coefficients is that proposed by Kern [9]. The Kern correlation, which is used for all fluids, is as follows:

$$\frac{h_oD_e}{k} = 0.36\left(\frac{D_eG_s}{\mu}\right)^{0.55}\left(\frac{C_p\mu}{k}\right)^{1/3}\left(\frac{\mu}{\mu_w}\right)^{0.14} \tag{7-26}$$

or

$$\frac{h_oD_e}{k} = 0.36(N_{Re})^{0.55}(N_{Pr})^{1/3}\left(\frac{\mu}{\mu_w}\right)^{0.14}$$

Equation 7-26 is divided into two components, j_H and N_{Re} in which

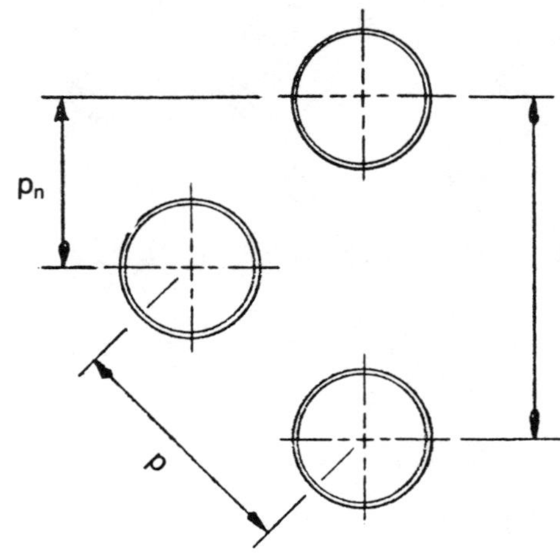

For staggered and in-line tube arrays,

$$A_F = \frac{B}{144}\left[D_s - D_{t_o} + \frac{D_{t_o} - d_t}{P_n}(p - d_t)\right], ft^2$$

For triangular layouts,

$$A_F = \frac{B}{144}\left[D_s - D_{t_o} + \frac{D_{t_o} - d_t}{P}(p - d_t)\right], ft^2$$

where, D_{t_o} = OD of tube bundle
D_s = ID of shell
d_t = OD of tube
B = baffle spacing
A_F = flow area-cross-flow area for one section between two baffles

Figure 7-20. Tube bundle cross-flow area.

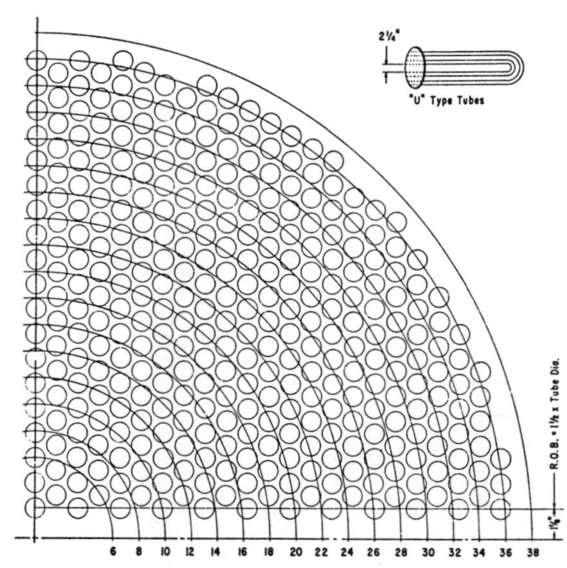

Figure 7-21. Typical tubesheet layouts.

Table 7-11
Kays and London Constants for Tube Bundles Containing 9 Tubes or Fewer

	Number of Tubes								
	1	2	3	4	5	6	7	8	9
In-line	0.64	0.80	0.87	0.90	0.92	0.94	0.96	0.98	0.99
Staggered	0.68	0.75	0.83	0.89	0.92	0.95	0.97	0.98	0.99

Table 7-12
Pros and Cons of Various Tube Arrangements

Tube Pitch Arrangement	Advantage	Disadvantage
(a) Triangular Pitch	Yields higher film coefficients than in-line square pitch. More tubes can be contained in shell because of compact arrangement.	Medium to high pressure drop. Cannot be used in fouling services. Can only have chemical cleaning.
(b) In-line Triangular Pitch	Film coefficients are not as high as triangular pitch, but greater than in-line square pitch. Suitable for fouling conditions.	Medium to high pressure drop. Can only have chemical cleaning.
(c) In-line Square Pitch	Good for conditions requiring low pressure drop. Arrangement allows for easy access of tubes for mechanical cleaning. Good for fouling service.	Relative low film coefficients.
(d) Diamond Square Pitch	Better film coefficients than in-line square pitch, but not as good as triangular or in-line pitch. Easy access for mechanical cleaning. Good for fouling service.	Relative low film coefficients. Does not have as low-pressure drop as the in-line square pitch arrangement.

$$j_H = \frac{h_o D_e}{k}\left(\frac{C_p \mu}{k}\right)^{-1/3}\left(\frac{\mu}{\mu_w}\right) \quad (7\text{-}27)$$

where h_o = outside tube bundle film coefficient, Btu/hr-ft-°F
G_s = mass flow rate of fluid, lb_M/hr
k = thermal conductivity of shell-side fluid, Btu/hr-ft-°F
D_e = shell-side equivalent tube diameter, in.
C_p = specific heat of fluid, Btu/lb-°F

For a square pitch tube arrangement,

$$D_e = \frac{4(p^2 - \pi d_{t_o}^2)/4}{\pi d_{t_o}} \text{ in.} \quad (7\text{-}28)$$

For a 60° equilateral triangular arrangement,

$$D_e = \frac{4(0.43p^2 - 0.5\pi d_{t_o}^2/4)}{\frac{\pi d_{t_o}}{2}} \text{ in.} \quad (7\text{-}29)$$

where p = tube pitch, in.
d_{t_o} = ID of shell, in.
a_s = flow area of tube bundle, ft^2

$$= \frac{D_s(cB)}{p(144)} \text{ ft}^2 \quad (7\text{-}30)$$

D_s = ID of shell, in.
c = clearance between tubes measured along tube pitch, in.
B = baffle spacing, in.
G_s = mass flow rate of fluid, lb_m/hr

$$G_s = \frac{\dot{m}}{a_s}$$

μ = viscosity of the shell-side fluid at the caloric temperature, lb/ft-hr
μ_w = viscosity of the shell-side fluid at the tube wall temperature, lb/ft-hr

The parameter j_H is plotted against N_{Re} in Figure 7-22a. The value of j_H is determined from the figure after the Reynolds number is calculated. Then from Equation 7-27 the film coefficient is determined.

The use of baffles is extremely important in directing the shell-side flow, tube support, and controlling the shell-side flow rate. As the number of baffles is increased, the flow rate increases. Likewise with an increased flow rate, the pressure drop increases substantially with an increasing number of baffles, with the film coefficient increasing as well. Ludwig [4] reports that for a constant flow rate, the velocity across the bundle is doubled with an increase in the film coefficient of approximately 44%.

(text continued on page 139)

Table 7-13
Tube Count for 3/8 in. OD Tubes on 13/16-in. △ Pitch

	TEMA Type L or M				TEMA Type P				TEMA Type S				TEMA Type U			
	Fixed Tubesheet				Outside Packed Floating Head				Inside Floating Head				U-Tube			
Shell	No. of Passes				No. of Passes				No. of Passes				No. of Passes			
ID in.	1	2	4	6	1	2	4	6	1	2	4	6	2	4	6	8
5.047	22	20	16		19	18	12		19	14	12					
6.065	37	30	28		31	26	24		31	26	16					
7.981	68	66	60		61	52	48		56	52	44					
10.02	110	106	96		104	98	84		96	90	76					
12.00	170	164	148		151	142	128		151	138	128					
13.25	212	196	188		178	168	156		187	184	160					
15.25	283	270	252		241	232	220		258	242	224					
17.25	364	348	332		316	298	292		336	326	304					
19.25	454	440	420		396	388	352		421	412	392					
21.25	562	554	524		490	484	456		526	502	480					
23.25	668	646	612		588	570	548		608	598	556					
25.00																
27.00	922	902	868		812	808	764		868	836	804					
29.00																
31.00	1230	1212	1172		1106	1092	1040		1152	1124	1088					
33.00																
35.00	1590	1560	1516		1438	1430	1336		1496	1468	1424					

Tube Count for 5/8 in. OD Tubes on 7/8 in. △ Pitch

	TEMA Type L or M				TEMA Type P				TEMA Type S				TEMA Type U			
	Fixed Tubesheet				Outside Packed Floating Head				Inside Floating Head				U-Tube			
Shell	No. of Passes				No. of Passes				No. of Passes				No. of Passes			
ID in.	1	2	4	6	1	2	4	6	1	2	4	6	2	4	6	8
5.047	22	18	16		19	14	12		14	14	12					
6.065	31	30	24		26	26	16		22	20	16					
7.981	61	52	48		55	48	44		51	48	40					
10.02	96	94	80		88	82	76		85	76	72					
12.00	151	138	132		130	124	112		130	120	112					
13.25	187	176	168		151	148	132		163	152	144					
15.25	241	232	224		206	196	184		216	214	196					
17.25	313	302	292		270	266	252		288	282	264					
19.25	396	384	352		336	334	312		358	350	340					
21.25	482	472	456		418	416	396		450	436	416					
23.25	568	554	536		506	492	472		526	506	484					
25.00																
27.00	792	780	752		704	700	660		724	720	696					
29.00																
31.00	1062	1030	1008		946	930	896		994	978	948					
33.00																
35.00	1356	1346	1304		1234	1220	1180		1288	1252	1220					

Table 7-13
Continued

Tube Count for 5/8-in. OD Tubes on 7/8-in. □ Pitch

Shell ID in.	TEMA Type L or M				TEMA Type P				TEMA Type S				TEMA Type U		
	Fixed Tubesheet				Outside Packed Floating Head				Inside Floating Head				U-Tube		
	No. of Passes				No. of Passes				No. of Passes				No. of Passes		
	1	2	4	6	1	2	4	6	1	2	4	6	2	4	6
5.047	21	16	16		13	12	12		12	12	12				
6.065	26	26	24		22	22	16		21	16	16				
7.981	52	52	52		45	44	44		45	38	32				
10.02	89	82	80		76	76	68		74	70	68				
12.00	128	124	120		109	104	92		109	92	88				
13.25	158	158	148		137	128	120		138	136	128				
15.25	213	208	208		177	176	164		188	184	176				
17.25	277	266	264		241	236	220		246	244	240				
19.25	344	332	336		293	284	284		316	308	304				
21.25	420	404	400		366	364	360		394	388	384				
23.25	502	476	488		436	432	432		456	448	444				
25.00															
27.00	694	668	664		612	608	608		640	636	624				
29.00															
31.00	922	910	908		828	812	812		862	848	840				
33.00															
35.00	1181	1166	1160		1070	1064	1048		1122	1112	1100				

Tube Count for 3/4-in. OD Tubes on 15/16-in. △ Pitch

Shell ID in.	TEMA Type L or M				TEMA Type P				TEMA Type S				TEMA Type U			
	Fixed Tubesheet				Outside Packed Floating Head				Inside Floating Head				U-Tube			
	No. of Passes				No. of Passes				No. of Passes				No. of Passes			
	1	2	4	6	1	2	4	6	1	2	4	6	2	4	6	8
5.047	19	14	12		14	14	8		10	10	4		6	4		
6.065	27	26	20	18	22	18	16		19	18	12		14	8		
7.981	55	48	40	38	42	40	36	32	42	40	32	30	32	24		
10.02	85	76	72	68	74	72	60	54	73	66	60	54	56	52		
12.00	126	118	104	98	109	106	96	86	109	106	92	86	92	80		
13.25	151	148	140	134	130	124	112	108	140	138	124	108	114	104		
15.25	206	196	180	176	174	168	156	152	187	184	168	152	83	74		
17.25	268	266	240	230	241	222	216	210	253	242	224	210	220	204		
19.25	340	330	320	302	288	282	264	260	320	294	280	260	290	268		
21.25	416	408	392	384	384	368	344	338	400	380	352	338	360	340		
23.25	499	480	460	456	469	449	430	418	454	436	416	410	220	210		
25.00	576	558	530	516	544	529	500	490	514	498	471	465	506	488		
27.00	675	661	632	596	643	616	600	575	607	587	560	558	614	580		
29.00	790	773	736	720	744	732	704	695	707	690	663	657	720	684		
31.00	896	875	858		859	835	812	800	816	797	769	760	830	804		
33.00	1018	1011	976		973	959	926	900	931	910	876	870	944	916		
35.00	1166	1137	1098		1118	1093	1054	1010	1062	1039	998	993	1076	940		
37.00	1307	1277	1242		1253	1224	1184	1150	1200	1177	1135	1124	1218	1184		
39.00	1464	1425	1386		1392	1359	1318	1286	1341	1318	1282	1264	1366	1324		
42.00	1688	1669	1618		1616	1602	1552	1482	1558	1554	1502	1482	1600	1552		
45.00	1943	1912	1878		1870	1833	1800	1770	1875	1834	1736	1708	1854	1800		
48.00	2229	2189	2134		2145	2107	2060	2025	2132	2100	1998	1964	2122	2064		
51.00	2513	2489	2432		2411	2395	2344	2305	2431	2392	2286	2250	2410	2356		
54.00	2823	2792	2752		2737	2683	2642	2612	2730	2684	2574	2536	2732	2668		
60.00	3527	3477	3414		3400	3359	3294	3220	3395	3346	3228	3196	3398	3336		

Table 7-13
Continued

Tube Count for ¾-in. OD Tubes on 1-in. △ Pitch

	TEMA Type L or M				TEMA Type P				TEMA Type S				TEMA Type U			
	Fixed Tubesheet				Outside Packed Floating Head				Inside Floating Head				U-Tube			
Shell	No. of Passes				No. of Passes				No. of Passes				No. of Passes			
ID in.	1	2	4	6	1	2	4	6	1	2	4	6	2	4	6	8
5.047	14	14	12		10	10	8		10	10	4		6	4		
6.065	22	20	16	22	19	18	16		19	14	12		10	8		
7.981	42	40	36	40	38	36	32	28	37	32	28	26	28	24		
10.02	73	72	64	64	64	62	60	58	61	60	48	46	56	44		
12.00	109	86	80	98	96	94	84	78	96	94	80	78	86	72		
13.25	139	134	124	122	121	110	100	98	121	118	104	98	106	96		
15.25	187	180	168	164	151	146	140	138	163	164	144	140	148	136		
17.25	241	232	220	212	208	196	188	160	216	214	196	158	200	184		
19.25	296	290	280	270	258	242	232	230	276	270	260	235	254	240		
21.25	372	354	344	330	320	316	296	298	338	338	324	300	314	300		
23.25	434	420	404	404	380	372	364	335	396	396	376	339	388	368		
25.00	507	489	476	482	475	466	452	430	460	440	420	414	452	432		
27.00	604	594	568	582	530	526	508	495	558	554	536	494	538	524		
29.00	689	679	660	672	653	642	620	610	624	605	589	581	632	612		
31.00	808	804	772		724	696	688	669	756	744	716	669	732	708		
33.00	906	891	860		859	848	818	805	818	797	783	771	838	808		
35.00	1030	1026	1000		946	922	904	880	980	978	944	880	950	916		
37.00	1152	1134	1090		1106	1081	1054	996	1047	1039	1001	996	1074	1040		
39.00	1273	1259	1222		1218	1208	1174	1125	1172	1164	1130	1125	1200	1164		
42.00	1485	1461	1434		1426	1399	1376	1306	1367	1350	1322	1306	1406	1364		
45.00	1721	1693	1650		1652	1620	1586		1635	1608	1536	1504	1632	1584		
48.00	1968	1941	1902		1894	1861	1820		1887	1842	1768	1740	1870	1832		
51.00	2221	2187	2134		2142	2101	2060		2143	2104	2019	1992	2122	2076		
54.00	2502	2465	2414		2417	2379	2326		2399	2366	2270	2244	2396	2340		
60.00	3099	3069	3010		2990	2957	2906		2981	2940	2932	2800	2992	2936		

Tube Count for ¾-in. OD Tubes on 1-in. □ Pitch

	TEMA Type L or M				TEMA Type P				TEMA Type S				TEMA Type U			
	Fixed Tubesheet				Outside Packed Floating Head				Inside Floating Head				U-Tube			
Shell	No. of Passes				No. of Passes				No. of Passes				No. of Passes			
ID in.	1	2	4	6	1	2	4	6	1	2	4	6	2	4	6	8
5.047	12	12	12		12	12	4		9	8	4		6	4		
6.065	21	16	16		16	16	12		16	16	12		8	8		
7.981	38	38	32		37	32	32	12	32	32	32	12	24	20		
10.02	61	60	52		57	56	52	24	56	52	52	24	44	40		
12.00	97	90	88		89	82	76	56	89	82	80	56	68	68		
13.25	117	116	112		97	94	88	80	104	104	96	80	90	88		
15.25	158	158	148		137	128	120	114	145	140	140	114	128	120		
17.25	216	208	188		177	176	164	160	188	184	180	160	176	168		
19.25	262	256	244		224	216	208	198	238	236	232	198	112	108		
21.25	316	316	308		274	270	268	260	304	292	284	260	138	134		
23.25	370	372	368		333	332	316	308	344	332	332	308	340	332		
25.00	442	432	428		414	406	392	344	398	386	366	344	400	388		
27.00	524	524	500		464	456	448	424	484	472	468	424	472	460		
29.00	602	596	580		570	562	548	496	554	532	510	496	554	544		
31.00	698	692	688		628	620	612	576	650	648	640	576	640	624		

Table 7-13
Continued

Tube Count for 3/4-in. OD Tubes on 1-in. □ Pitch

Shell ID in.	TEMA Type L or M, Fixed Tubesheet, No. of Passes 1	2	4	6	TEMA Type P, Outside Packed Floating Head, No. of Passes 1	2	4	6	TEMA Type S, Inside Floating Head, No. of Passes 1	2	4	6	TEMA Type U, U-Tube, No. of Passes 2	4	6	8
33.00	782	768	768		742	732	732	668	730	712	682	668	724	720		
35.00	894	892	880		816	812	804	760	848	828	824	760	836	812		
37.00	1004	978	964		952	931	928	872	937	918	882	872	940	924		
39.00	1102	1096	1076		1062	1045	1026	972	1048	1028	996	972	1048	1040		
42.00	1283	1285	1270		1232	1222	1218	1140	1224	1200	1170	1140	1222	1204		
45.00	1484	1472	1456		1424	1415	1386	1336	1421	1394	1350	1336	1420	1400		
48.00	1701	1691	1670		1636	1634	1602	1536	1628	1598	1548	1536	1624	1604		
51.00	1928	1904	1888		1845	1832	1818	1764	1862	1823	1779	1764	1852	1820		
54.00	2154	2138	2106		2080	2066	2044	1992	2096	2048	2010	1992	2084	2064		
60.00	2683	2650	2636		2582	2566	2556	2476	2585	2552	2512	2476	2596	2564		

Tube Count for 3/4-in. OD Tubes on 1-in. ◇ Pitch

Shell ID in.	TEMA Type L or M, Fixed Tubesheet, No. of Passes 1	2	4	6	TEMA Type P, Outside Packed Floating Head, No. of Passes 1	2	4	6	TEMA Type S, Inside Floating Head, No. of Passes 1	2	4	6	TEMA Type U, U-Tube, No. of Passes 2	4	6	8
5.047	12	10	8		12	10	8						4	4		
6.065	21	18	16		16	12	8						10	8		
7.981	37	32	28		32	28	24						24	20		
10.02	61	54	48		52	46	40						42	36		
12.00	97	90	84		81	74	68						66	64		
13.25	113	108	104		97	92	84						86	80		
15.25	156	146	136		140	134	128						124	116		
17.25	208	196	184		188	178	168						174	164		
19.25	256	244	236		241	228	216						218	202		
21.25	314	299	294		300	286	272						272	260		
23.25	379	363	352		359	343	328						334	320		
25.00	448	432	416		421	404	392						390	380		
27.00	522	504	486		489	472	456						468	452		
29.00	603	583	568		575	556	540						550	532		
31.00	688	667	654		660	639	624						626	608		
33.00	788	770	756		749	728	708						720	700		
35.00	897	873	850		849	826	804						818	796		
37.00	1009	983	958		952	928	908						928	904		
39.00	1118	1092	1066		1068	1041	1016						1036	1016		
42.00	1298	1269	1250		1238	1216	1196						1220	1192		
45.00	1500	1470	1440		1432	1407	1378						1412	1384		
48.00	1714	1681	1650		1644	1611	1580						804	788		
51.00	1939	1903	1868		1864	1837	1804						1834	1804		
54.00	2173	2135	2098		2098	2062	2026						2072	2036		
60.00	2692	2651	2612		2600	2560	2520						2584	2544		

Table 7-13
Continued

Tube Count for 1-in. OD Tubes on 1¼-in. △ Pitch

Shell	TEMA Type L or M Fixed Tubesheet No. of Passes				TEMA Type P Outside Packed Floating Head No. of Passes				TEMA Type S Inside Floating Head No. of Passes				TEMA Type U U-Tube No. of Passes			
ID in.	1	2	4	6	1	2	4	6	1	2	4	6	2	4	6	8
5.047	8	6	4		7	4	4						0	0		
6.065	14	14	8		10	10	4						4	4		
7.981	26	26	16		22	18	16	14	18	14	8	12	14	8		
10.02	42	40	36		38	36	28	24	33	28	16	18	26	24		
12.00	64	61	56		56	52	48	46	51	48	42	44	44	36		
13.25	85	76	72		73	72	60	44	73	68	52	44	56	52		
15.25	110	106	100		100	98	88	80	93	90	78	76	86	76		
17.25	147	138	128		130	126	116	104	126	122	112	102	114	104		
19.25	184	175	168		170	162	148	140	159	152	132	136	152	136		
21.25	227	220	212		212	201	188	176	202	192	182	172	192	176		
23.25	280	265	252		258	250	232	220	249	238	216	212	232	220		
25.00	316	313	294		296	294	276	250	291	278	250	240	270	256		
27.00	371	370	358		355	346	328	300	345	330	298	288	322	304		
29.00	434	424	408		416	408	392	360	400	388	356	348	378	364		
31.00	503	489	468		475	466	446	420	459	450	414	400	444	424		
33.00	576	558	534		544	529	510	498	526	514	484	464	508	492		
35.00	643	634	604		619	604	582	566	596	584	548	536	578	560		
37.00	738	709	684		696	679	660	646	672	668	626	608	660	632		
39.00	804	787	772		768	753	730	723	756	736	704	692	740	712		
42.00	946	928	898		908	891	860	840	890	878	834	808	872	836		
45.00	1087	1069	1042		1041	1017	990	968	1035	1008	966	948	1010	980		
48.00	1240	1230	1198		1189	1182	1152	1132	1181	1162	1118	1092	1156	1124		
51.00	1397	1389	1354		1348	1337	1300	1280	1350	1327	1277	1254	1322	1284		
54.00	1592	1561	1530		1531	1503	1462	1440	1520	1492	1436	1416	1496	1452		
60.00	1969	1945	1904		1906	1879	1842	1802	1884	1858	1800	1764	1866	1828		

Tube Count for 1-in. OD Tubes on 1¼-in. ◇ Pitch

Shell	TEMA Type L or M Fixed Tubesheet No. of Passes				TEMA Type P Outside Packed Floating Head No. of Passes				TEMA Type S Inside Floating Head No. of Passes				TEMA Type U U-Tube No. of Passes			
ID in.	1	2	4	6	1	2	4	6	1	2	4	6	2	4	6	8
5.047	8	6	4		5	4	4						0	0		
6.065	12	10	8		12	10	8						4	4		
7.981	24	20	16		21	18	16						10	8		
10.02	37	32	28		32	32	28						24	10		
12.00	57	53	48		52	46	40						36	32		
13.25	70	70	64		61	58	56						50	44		
15.25	97	90	84		89	82	76						70	64		
17.25	129	120	112		113	112	104						96	88		
19.25	162	152	142		148	138	128						124	120		
21.25	205	193	184		180	174	168						156	152		
23.25	238	228	220		221	210	200						200	188		
25.00	275	264	256		261	248	236						232	220		
27.00	330	315	300		308	296	286						282	268		
29.00	379	363	360		359	345	336						330	320		
31.00	436	422	410		418	401	388						382	368		

Table 7-13
Continued

Tube Count for 1-in. OD Tubes on 1¼-in. ◇ Pitch

	TEMA Type L or M				TEMA Type P				TEMA Type S				TEMA Type U			
	Fixed Tubesheet				Outside Packed Floating Head				Inside Floating Head				U-Tube			
Shell	No. of Passes				No. of Passes				No. of Passes				No. of Passes			
ID in.	1	2	4	6	1	2	4	6	1	2	4	6	2	4	6	8
33.00	495	478	472		477	460	448						440	424		
35.00	556	552	538		540	526	508						498	484		
37.00	632	613	598		608	588	568						562	548		
39.00	705	685	672		674	654	640						630	620		
42.00	822	799	786		788	765	756						744	728		
45.00	946	922	912		910	885	866						872	852		
48.00	1079	1061	1052		1037	1018	1000						1002	980		
51.00	1220	1159	1176		1181	1160	1142						1138	1116		
54.00	1389	1359	1330		1337	1307	1292						1292	1264		
60.00	1714	1691	1664		1658	1626	1594						1604	1576		

Tube Count for 1-in. OD Tubes on 1¼-in. □ Pitch

	TEMA Type L or M				TEMA Type P				TEMA Type S				TEMA Type U			
	Fixed Tubesheet				Outside Packed Floating Head				Inside Floating Head				U-Tube			
Shell	No. of Passes				No. of Passes				No. of Passes				No. of Passes			
ID in.	1	2	4	6	1	2	4	6	1	2	4	6	2	4	6	8
5.047	9	6	4		5	4	4	–	5	4	4	–	0	0		
6.065	12	12	12		12	6	4	–	12	6	4	–	4	4		
7.981	22	20	16		21	16	16	12	17	12	8	12	12	8		
10.02	38	38	32		32	32	32	18	30	30	16	18	24	20		
12.00	56	56	52		52	52	44	24	52	48	42	24	38	36		
13.25	69	66	66		61	60	52	50	61	56	52	50	52	48		
15.25	97	90	88		89	84	80	64	85	78	62	64	72	68		
17.25	129	124	120		113	112	112	96	108	108	104	96	98	96		
19.25	164	158	148		148	144	140	114	144	136	130	114	128	124		
21.25	202	191	184		178	178	172	156	173	166	154	156	166	156		
23.25	234	234	222		216	216	208	192	217	208	194	192	200	196		
25.00	272	267	264		258	256	256	212	252	240	230	212	240	232		
27.00	328	317	310		302	300	296	260	296	280	270	260	284	276		
29.00	378	370	370		356	353	338	314	345	336	310	314	332	332		
31.00	434	428	428		414	406	392	368	402	390	366	368	290	384		
33.00	496	484	484		476	460	460	420	461	452	432	420	442	436		
35.00	554	553	532		542	530	518	484	520	514	494	484	254	248		
37.00	628	621	608		602	596	580	550	588	572	562	548	574	560		
39.00	708	682	682		676	649	648	625	661	640	624	620	644	628		
42.00	811	811	804		782	780	768	730	776	756	738	724	758	748		
45.00	940	931	918		904	894	874	850	900	882	862	844	872	868		
48.00	1076	1061	1040		1034	1027	1012	980	1029	1016	984	972	1002	988		
51.00	1218	1202	1192		1178	1155	1150	1125	1170	1156	1126	1114	1146	1140		
54.00	1370	1354	1350		1322	1307	1284	1262	1310	1296	1268	1256	1300	1288		
60.00	1701	1699	1684		1654	1640	1632	1585	1641	1624	1598	1576	1620	1604		

Table 7-13
Continued

Tube Count for $1\frac{1}{4}$-in. OD Tubes on $1\frac{9}{16}$-in. △ Pitch

	TEMA Type L or M				TEMA Type P				TEMA Type S				TEMA Type U			
	Fixed Tubesheet				Outside Packed Floating Head				Inside Floating Head				U-Tube			
Shell	No. of Passes				No. of Passes				No. of Passes				No. of Passes			
ID in.	1	2	4	6	1	2	4	6	1	2	4	6	2	4	6	8
5.047	7	4	4		0	0	0	–					0	0		
6.065	8	6	4		7	6	4	–					0	0		
7.981	19	14	12		14	14	8	–					6	4		
10.02	29	26	20		22	20	16	–					14	12		
12.00	42	38	34		37	36	28	22					22	20		
13.25	52	48	44		44	44	36	28					32	28		
15.25	69	68	60		64	62	48	45					48	44		
17.25	92	84	78		85	78	72	69					64	60		
19.25	121	110	104		109	102	96	86					86	80		
21.25	147	138	128		130	130	116	112					114	104		
23.25	174	165	156		163	152	144	130					138	132		
25.00	196	196	184		184	184	172	164					162	152		
27.00	237	226	224		221	216	208	196					196	184		
29.00	280	269	256		262	252	242	228					232	220		
31.00	313	313	294		302	302	280	270					268	256		
33.00	357	346	332		345	332	318	305					310	296		
35.00	416	401	386		392	383	364	357					356	344		
37.00	461	453	432		442	429	412	407					404	388		
39.00	511	493	478		493	479	460	449					452	440		
42.00	596	579	570		576	557	544	512					534	522		
45.00	687	673	662		657	640	628	596					626	612		
48.00	790	782	758		756	745	728	696					720	700		
51.00	896	871	860		859	839	832	820					822	800		
54.00	1008	994	968		964	959	940	892					930	908		
60.00	1243	1243	1210		1199	1195	1170	1116					1160	1140		

Tube Count for $1\frac{1}{4}$-in. OD Tubes on $1\frac{1}{16}$-in. □ Pitch

	TEMA Type L or M				TEMA Type P				TEMA Type S				TEMA Type U			
	Fixed Tubesheet				Outside Packed Floating Head				Inside Floating Head				U-Tube			
Shell	No. of Passes				No. of Passes				No. of Passes				No. of Passes			
ID in.	1	2	4	6	1	2	4	6	1	2	4	6	2	4	6	8
5.047	4	4	4		0	0	0	0	0	0	0	0	0	0		
6.065	6	6	4		6	6	4	0	6	6	4	0	0	0		
7.981	12	12	12		12	12	12	0	12	12	4	0	6	4		
10.02	24	22	16		21	16	16	12	21	12	8	12	12	8		
12.00	37	34	32		32	32	32	18	29	28	16	18	20	20		
13.25	45	42	42		38	38	32	24	38	34	34	24	28	28		
15.25	61	60	52		52	52	52	48	52	48	44	48	42	36		
17.25	80	76	76		70	70	68	50	70	66	56	50	56	56		
19.25	97	95	88		89	88	88	80	85	84	70	80	74	68		
21.25	124	124	120		112	112	112	96	108	108	100	96	98	96		
23.25	145	145	144		138	138	130	114	136	128	128	114	124	120		
25.00	172	168	164		164	164	156	136	154	154	142	136	140	136		

Table 7-13
Continued

Tube Count for 1 1/4-in. OD Tubes on 1/16-in. □ Pitch

	TEMA Type L or M				TEMA Type P				TEMA Type S				TEMA Type U			
	Fixed Tubesheet				Outside Packed Floating Head				Inside Floating Head				U-Tube			
Shell	No. of Passes				No. of Passes				No. of Passes				No. of Passes			
ID in.	1	2	4	6	1	2	4	6	1	2	4	6	2	4	6	8
27.00	210	202	202		193	184	184	172	184	180	158	172	176	176		
29.00	241	234	230		224	224	216	198	217	212	204	198	200	196		
31.00	272	268	268		258	256	256	236	252	248	234	236	232	232		
33.00	310	306	302		296	296	282	264	289	276	270	264	272	268		
35.00	356	353	338		336	332	332	304	329	316	310	304	312	296		
37.00	396	387	384		378	370	370	358	372	368	354	340	348	348		
39.00	442	438	434		428	426	414	408	420	402	402	392	396	392		
42.00	518	518	502		492	492	484	464	485	476	468	464	472	456		
45.00	602	602	588		570	566	556	544	565	554	546	544	552	536		
48.00	682	681	676		658	648	648	620	653	636	628	620	628	620		
51.00	770	760	756		742	729	722	712	738	726	720	705	712	708		
54.00	862	860	856		838	823	810	804	837	820	812	804	808	804		
60.00	1084	1070	1054		1042	1034	1026	1008	1036	1028	1012	1008	1012	992		

Tube Count for 1 1/4-in. OD Tubes on 1 9/16-in. ◇ Pitch

	TEMA Type L or M				TEMA Type P				TEMA Type S				TEMA Type U			
	Fixed Tubesheet				Outside Packed Floating Head				Inside Floating Head				U-Tube			
Shell	No. of Passes				No. of Passes				No. of Passes				No. of Passes			
ID in.	1	2	4	6	1	2	4	6	1	2	4	6	2	4	6	8
5.047	5	4	4		0	0	0						0	0		
6.065	6	6	4		5	4	4						0	0		
7.981	13	10	8		12	10	8						4	4		
10.02	24	20	16		21	18	16						12	12		
12.00	37	32	28		32	28	28						20	20		
13.25	45	40	40		37	34	32						26	24		
15.25	60	56	56		52	52	48						40	36		
17.25	79	76	76		70	70	64						56	52		
19.25	97	94	94		90	90	84						74	68		
21.25	124	116	112		112	108	104						96	88		
23.25	148	142	136		140	138	128						120	112		
25.00	174	166	160		162	162	156						142	136		
27.00	209	202	192		191	188	184						170	164		
29.00	238	232	232		442	430	416						200	192		
31.00	275	264	264		261	249	244						228	220		
33.00	314	307	300		300	286	280						268	256		
35.00	359	345	334		341	330	320						306	296		
37.00	401	387	380		384	372	360						346	336		
39.00	442	427	424		428	412	404						390	380		
42.00	522	506	500		497	484	472						456	448		
45.00	603	583	572		575	562	552						542	528		
48.00	682	669	660		660	648	640						618	604		
51.00	777	762	756		743	728	716						708	692		
54.00	875	857	850		843	822	812						802	784		
60.00	1088	1080	1058		1049	1029	1016						1010	984		

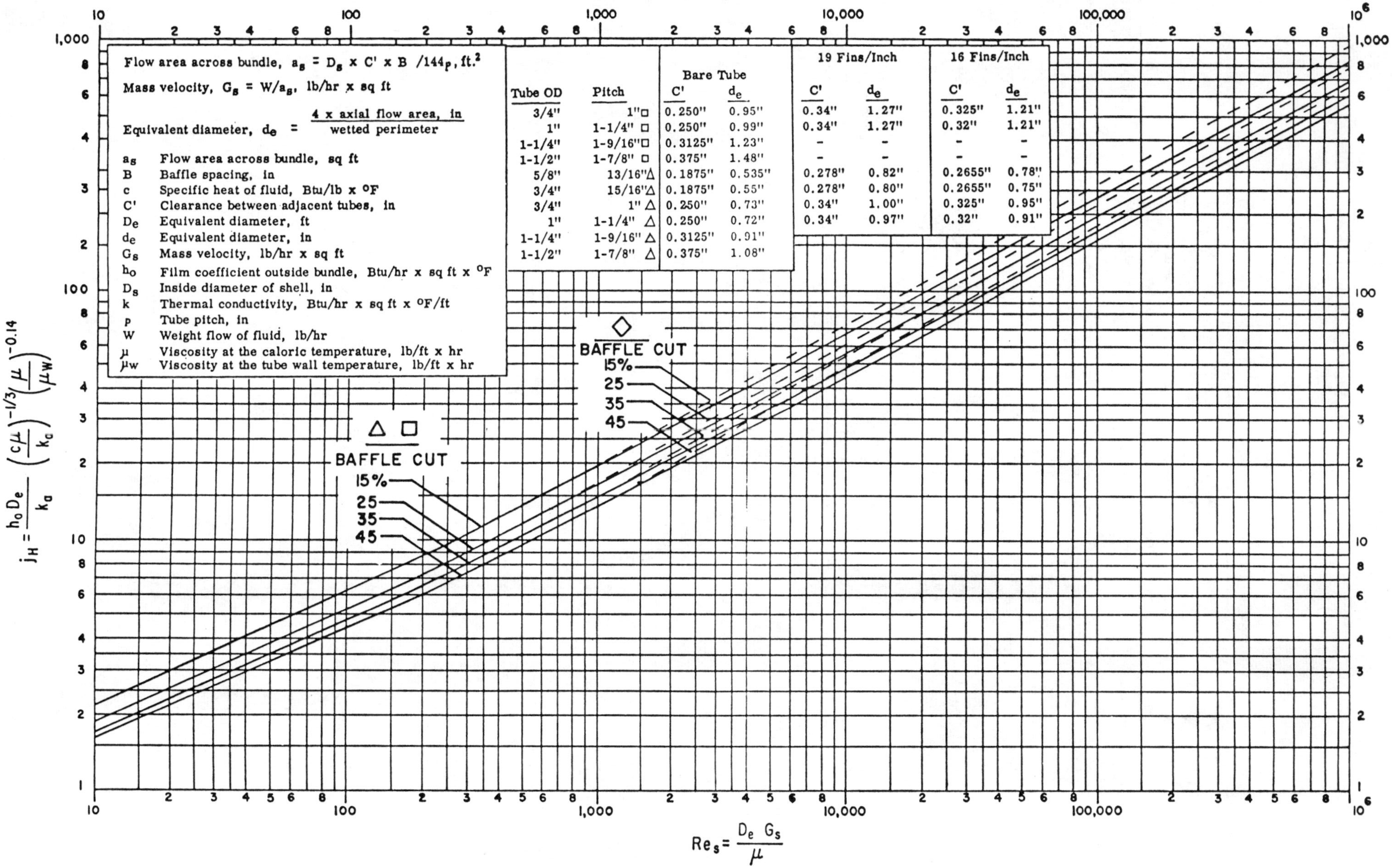

Figure 7-22A. Shell-side heat transfer parameter J_H versus N_{Re} for various baffle cuts. (By permission, *Engineering Data Book*, 1959, Wolverine, one of the Signal Companies.)

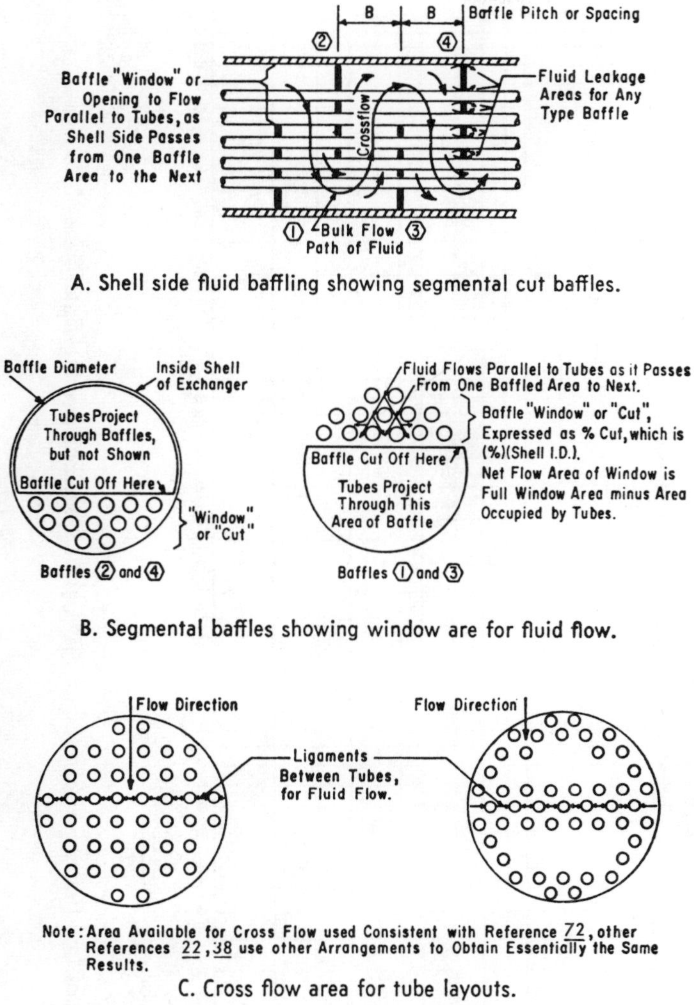

Figure 7-22B. Various baffle window schemes [4].

Baffles never, except for unusual designs such as orifice baffles, extend a full 360° around the shell. The baffle plate is cut such that the shell-side fluid can flow around its edge. The open area between the baffle edge and the shell wall is known as a baffle "window." Baffle windows are commonly referred to in terms of percentages of the entire circular shell area. Figure 7-22b illustrates various baffle window schemes.

Shell-Side Pressure Drop. There are several methods to calculate the frictional pressure drop across tube bundles, and the reader is referred to Ludwig [4] or Kern [9] who give comprehensive discussions of the various techniques. The method we will use is the one developed by Kern [9], where the expression for the shell-side pressure drop is given as follows:

$$\Delta P = \frac{f_s G_s^2 D_s (NB + 1)}{(5.22)(10)^{10} D_e \gamma \dfrac{\mu}{\mu_w}} \qquad (7\text{-}31)$$

where (μ/μ_w), G_s, D_s, D_e = are previously defined
NB = number of baffles
γ = specific gravity of shell-side fluid
f_s = combined friction factor determined from Figure 7-23

TUBE VIBRATIONS

Chapters 2 and 4 described how fluids moving around objects can produce vibrations. The same thing happens in shell and tube heat exchangers, but it creates a different problem. Chapters 2 and 4 were primarily concerned with vortex shedding. This chapter covers vortex shedding and several other types of vibration phenomena. Also, the problem is different from that in Chapter 4 because the boundary conditions of the system have changed. Chapter 4 used a cantilever beam to show how a tower or stack is restrained several different ways at the "ends."

There have been many research studies made in the field of tube vibrations. Probably the most numerous stem from the nuclear industry. The problem is complex and no one method proposed is a full and complete analysis of tube vibrations. Consequently, research is still being done to better understand the causes and prevention of tube vibrations. Here we will outline the causes of the phenomena and present some quantitative approaches to the problems.

Presented first is a simple and quick approach to predict tube vibrations caused by shell-side flow. This approach was originally developed by John T. Thorngren [14] in 1970 and is called the "maximum velocity method." We will present a modified version of the method proposed by Thorngren to encompass a wider range of applications and to specifically define all the variables in the equations. This method addresses the tube vibration caused by vortex shedding when the shell-side fluid alters direction at the baffle plate and strikes the tubes. The arrows in Figure 7-1 show how flow direction of a fluid turns at the baffle plate and strikes the tubes midway between the baffle plates. This causes the tubes to deflect and the hole in the baffle plate acts as a fulcrum for the tubes to deflect against. Two types of problems can result—a fatiguing of the tubes at the baffle hole and eventual tube rupture, or the tubes colliding

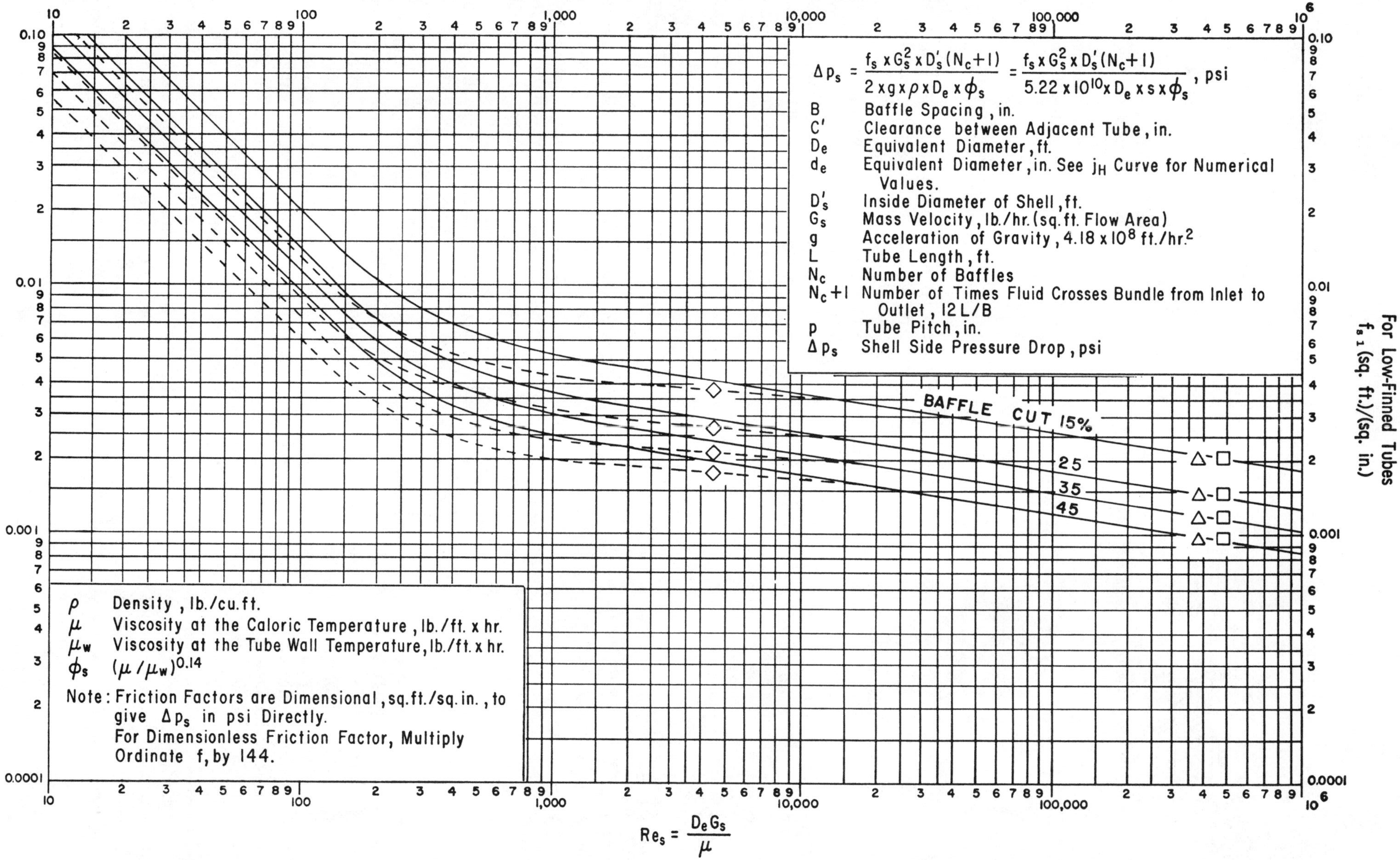

Figure 7-23. Friction factors for shell-side surfaces of low-finned and bare tubes. (By permission, *Engineering Data Book*, 1959, Wolverine, one of the Signal Companies.)

with each other resulting in abrasion and rupture. This phenomenon can be offset somewhat by anticipating shell-side velocities. The velocity of the shell-side fluid can be kept reasonable if the entry and exit nozzles are properly sized.

To determine the force exerted on the tubes by the fluid vortices we refer to Equation 4-80, the von Karman equation as follows:

$$F = \frac{C_F \rho V^2 \ell^2}{2g_c} \tag{4-80}$$

In the case of tubes the characteristic area, ℓ^2, can be replaced by the outside tube diameter, d_o (ft), and the value for the force becomes:

$$F_L = \frac{C_F \rho V^2 d_o}{2g_c} = \text{force per unit length of tubes} \tag{4-80a}$$

Chapter 2 introduced the concept of the Strouhal number as the numeric constant between the resonant frequency of vortex shedding, f_s, and the cylinder diameter, d_o, divided by the free stream velocity, V. This is written as:

$$s = \frac{f_s d_o}{V} \tag{2-37}$$

As observed in Figure 2-32 the vortices break up when N_{Re} approaches 3.0×10^5. Referring to Figures 4-21 and 4-29 we see that the force or drag coefficient is constant in this range. However, in Chapter 4 we are dealing with air as the flow medium. In our application here we are dealing with a wide range of fluids—gases to heavy, viscous liquids. Now, observing Figure 7-24, we notice how the drag coefficient varies over a wide range of Reynolds numbers. For viscous liquids moving at low velocities, it is very possible for the Reynolds number to be low, making the drag coefficient high. Thus, we keep C_F as a variable in Equation 4-80.

In Figure 2-32 we see that at $N_{Re} \geqslant 3.5 \times 10^6$ the vortex street is again developed. At this Reynolds number the fluid flow velocity would be too high to consider wind design as discussed in Chapters 2 and 4. Also, those two chapters were only concerned with a single cylinder, whereas here we are studying the effects of vortices on an array of tubes. At values of $N_{Re} \geqslant 3.5 \times 10^6$ for tube arrays we would be more concerned with flow-induced vibration, which is discussed in the next section. Also, Reynolds numbers in this range rarely, if ever, occur in heat exchangers.

In an array of tubes the maximum velocity of the shell-side fluid occurs at the minimum cross section between the tubes. Thus, V increases such that as the tubes get closer together the value of N_{Re} approaches 3.0×10^5 and the vortex street is broken up as previously described. Thus, the effect of vortex shedding in tube arrays is only significant for tubes spaced such that the Reynolds number falls within the appropriate range. The effects of tubes being spaced closer are discussed later in this chapter.

For considering vortex shedding, Equation 4-80 represents the maximum force exerted on the tubes. To analyze the tubes, one must consider them as continuous beams supported by baffle plates. The baffle plates act like beam supports as the fluid exerts a force in the tubes forcing them against the baffles. The general equations for the shear force and deflections of a continuous beam take the following general forms:

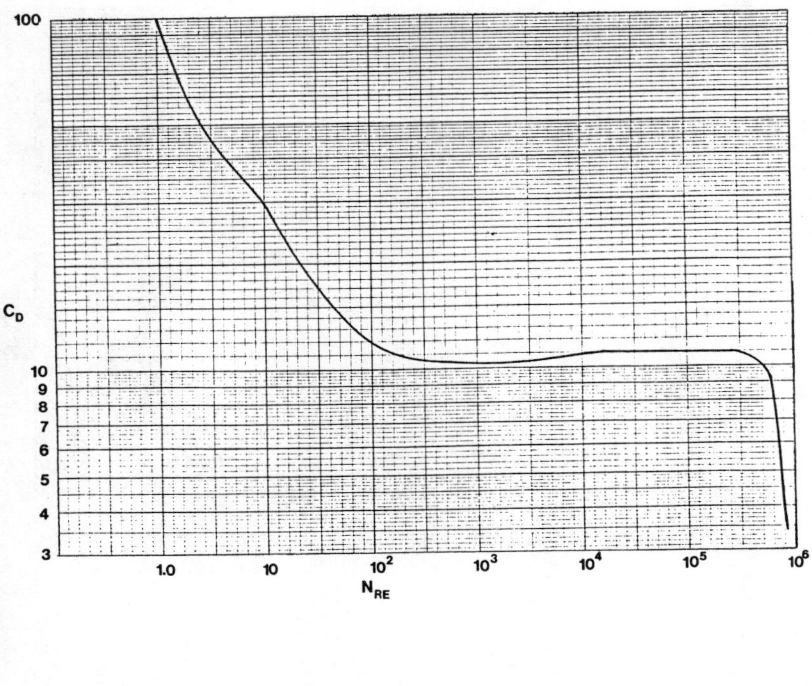

Figure 7-24. Drag coefficients versus Reynolds number for long circular cylinders in crossflow.

$$F_s = aF_LL \tag{7-32}$$

$$S = \frac{bF_LL^4}{EI} \tag{7-33}$$

The values for a and b are dependent upon the boundary conditions of the continuous beam. Typical values are presented in Figure 7-25 and are fairly comprehensive for most shell and tube exchangers. For cases not covered in Figure 7-25, the specific values must be solved for using the analysis for a continuous beam.

As the shell fluid exerts pressure on the tube, the tube deflects at mid-span forcing the tube at the baffle against the baffle hole. The stresses induced in the tube are a result of localized forces at the tube-baffle contact points. At these points the tube behaves similarly to a horizontal vessel such that only a portion of the tube wall offers effective resistance against collapse. Thus, Equation 4-2 predicts the amount of tube wall that effectively resists the baffle wall reaction, and is written as

$$\Delta = \frac{\pi}{180}\left(\frac{5\theta}{12} + 30\right) \tag{4-2}$$

Continuous Beams

Case	a_{max}	b_{max}
1	1.200	0.0059
2	0.550	0.0099
3	1.100	0.0069
4	1.223	0.0094
5	0.572	0.0097
6	1.143	0.0065

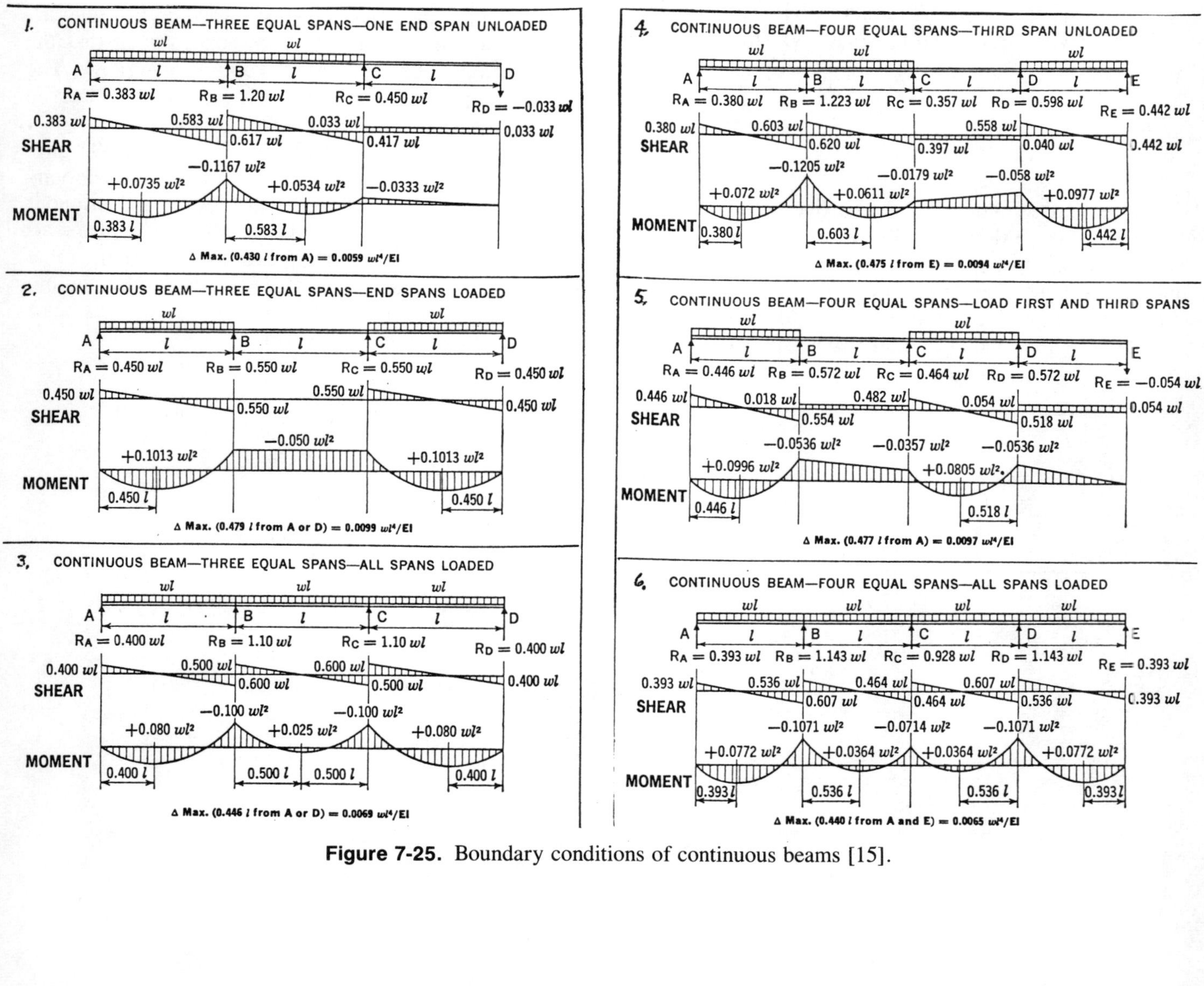

Figure 7-25. Boundary conditions of continuous beams [15].

where θ = angle of contact
Δ = radians

Thorngren [14] proposed that about 40% of the tube metal is effective in resisting wall membrane stresses. In Equation 4-2 this would make the value of $\theta = 144°$, greater than most saddle-shell connections for horizontal vessels. To take the problem further we consider the tube wall as a ring shown in Figure 7-26. The assumption is that neither the tube nor baffle hole will deform to reduce stresses, which is the worst condition. For determining contact stresses between the two bodies, Timoshenko [16] has shown that for the case in Figure 7-26 the diameter of the circle of contact is

$$d = 1.76\left[\frac{P(E_t + E_B)d_t d_B}{2E_t E_B(d_t + d_B)}\right]^{1/3} \tag{7-34}$$

From Equation 7-34 one can deduce that the tube-baffle interface should be analyzed as point loadings. For such loadings as shown in Figure 7-26 the contact force representing the shear of the tube against the baffle plate is

$$F_s + \left(\frac{d_t d_B}{d_t + d_B}\right) c \left(\frac{\sigma_{all}}{0.798}\right)^2 \tag{7-35}$$

Evaluating the relationship for shear in Equation 7-32 we have

$$F_s = \alpha a F_L L$$

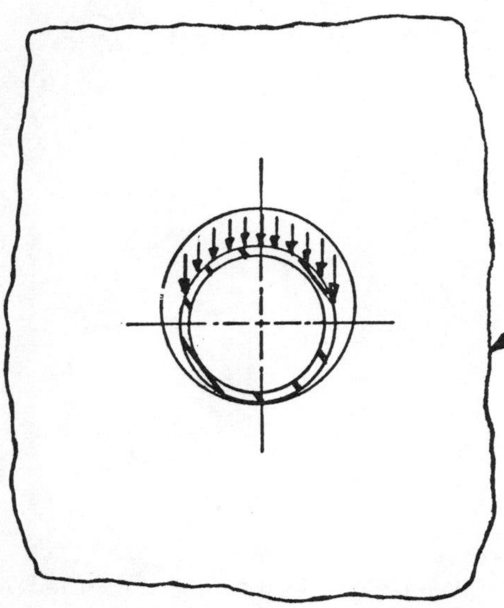

Figure 7-26. Fluid force causing tube to impinge on baffle plate.

where α = constant that represents the amount of effective resisting tube wall area

Now combining Equations 7-32 and 7-35 we have the following:

$$\alpha = \left(\frac{c}{aF_L L}\right)\left(\frac{d_t d_B}{d_t + d_B}\right)\left(\frac{\sigma_{all}}{0.798}\right)^2 \tag{7-36}$$

$$\text{where} \quad c = \frac{1 - \nu_t^2}{E_t}\,\frac{1 - \nu_B^2}{E_B} \tag{7-37}$$

ν_t = Poisson ratio for the tube material, dimensionless
ν_B = Poisson ratio for the baffle plate material, dimensionless
E_t = modulus of elasticity for tube material, psi
E_B = modulus of elasticity for baffle plate material, psi
c = constant, in./lb_f
α = constant, dimensionless

To arrive at the modified damage number for baffle damage we solve for F_L in Equation 7-32:

$$F_L = \frac{F_s}{\alpha a L}$$

Now dividing this relation into Equation 4-80 we obtain

$$\frac{C_D d_t \rho V^2 a L \alpha}{2g_c F_s} = 1.0$$

Letting the baffle damage number be represented by N_{BE}, a dimensionless parameter, we have:

$$N_{BE} = \frac{C_D d_t \rho V^2 a L \alpha}{2g_c F_s} \tag{7-38}$$

where $N_{BE} \leqslant 1.0$

If $N_{BE} > 1.0$, then tube damage at the baffle is very probable and a thicker tube should be selected and the analysis repeated.

The analysis of determining the dimensionless parameter, N_{CD}, which governs tube damage induced by excessive displacements in tube movements, is similar to that for the baffle damage parameter. Solving for F_L in Equation 7-33 we determine F_L as follows:

$$F_L = \frac{\delta EI}{bL^4}$$

Dividing this expression into Equation 4-80 we have

$$\frac{D_D d_t \rho V^2 b L^4}{2 g_c \delta E_t I} = 1.0$$

We define N_{CD} as

$$N_{CD} = \frac{C_D d_t \rho V^2 b L^4}{2 g_c \delta E_t I} \qquad (7\text{-}39)$$

where $N_{CD} \leqslant 1.0$

Once again, if $N_{CD} > 1.0$, then thicker tubes should be selected and the analysis repeated.

Equation 7-39 is similar to that obtained by Thorngren [14] and Coit [17]. The dimensionless parameters, N_{BE} and N_{CD}, in Equations 7-38 and 7-39 should be regarded as mere rules of thumb. Even though they are dimensionless, they do not have the same firm basis as do dimensionless parameters used in fluid mechanics and transport phenomena.

One can approximate the tube behavior by using the principles in Chapter 2, Example 2-6. Using the baffles as supports and spacing them (either equally or unequally), one can simulate the tube displacements. However, since we are not dealing with a single tube, vortex shedding around tube bundles can presently only be accounted for in design by being conservative.

Flow-induced vibration of exchanger tubes is another mode different and distinct from vortex shedding. In vortex shedding a component of the flow, the vortex, is the contributing cause to the tube vibration. In flow-induced vibration, forces are exerted on the tubes that are caused by flow field interactions around the tubes. Fluid that flows normal to the tubes is forced into a smaller area between the tubes resulting in a Venturi effect known as "jetting" or "jet switching." This phenomenon is shown in Figure 7-27 where a control volume of fluid is shown being compressed between two tubes. The result of this "jetting" effect is the fluid exiting the narrow area between the tubes diverges into a diffused mass that whips or whirls around remaining tubes. This "whirling" effect is another mode of vibration.

Vibration induced by turbulence is the most common mode. This phenomenon is commonly confused with the other modes because the term turbulence is viewed synonymously with fluid flow and vibration resulting from such flow. However, vortex shedding, jetting, and whirling are different from turbulence because even though they exist in turbulent flow, they can all be final causes of failure and each must be controlled. Turbulence can be best viewed as a pressure field around a tube shown in Figure 7-28. Here we see a pressure dis-

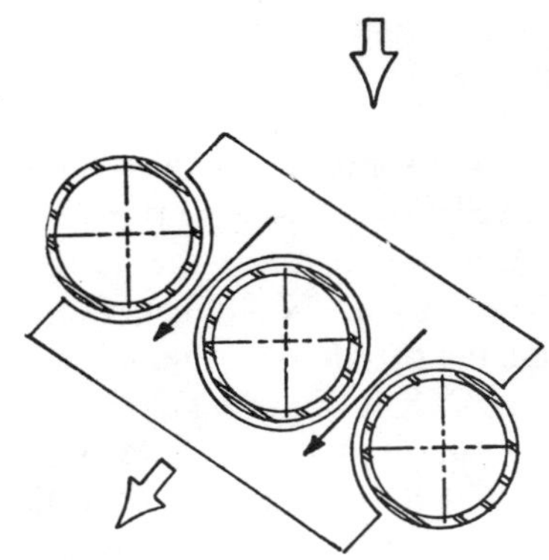

Figure 7-27. Jet switching in tube arrays.

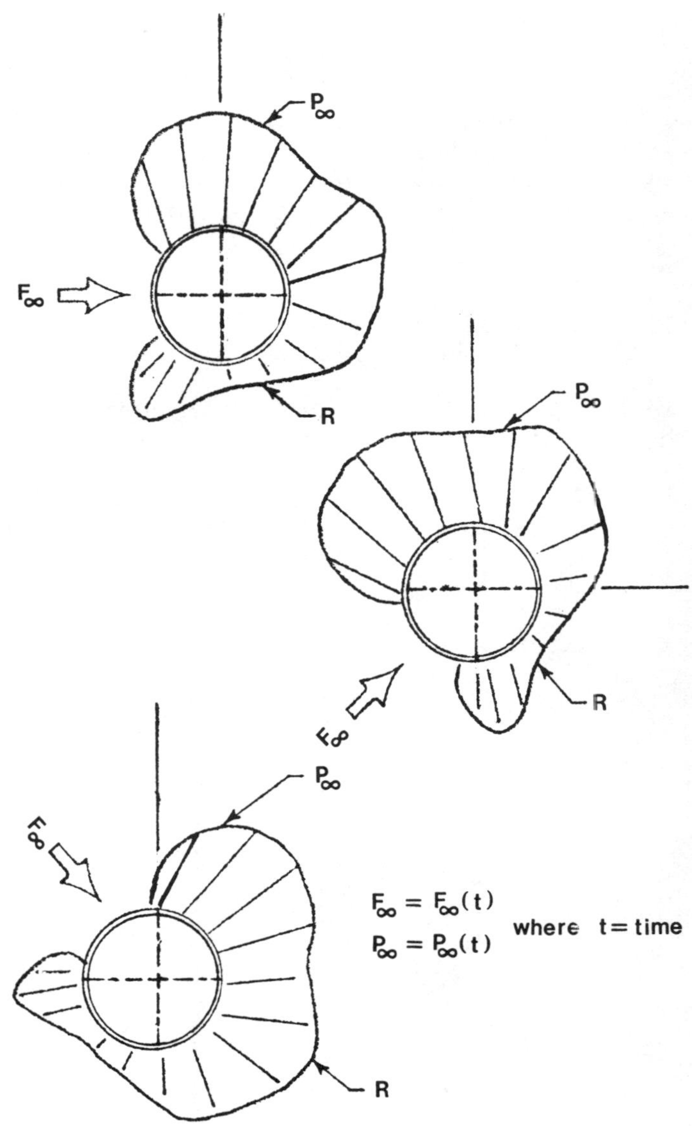

Figure 7-28. The magnitude of the direction of the fluid striking the cylinder can be thought of mathematically as a forcing function, F_∞, mapping a pressure distribution around the cylinder over region R.

tribution around a cylinder in much the same way as an aircraft airfoil. This pressure field, or distribution, varies all through the flow regime and is also a function of time. As this pressure distribution fluctuates and changes, the tube is subject to vibration as the fluctuation frequency approaches that of the natural frequency of the tube. It is this fluctuation frequency that is referred to as the turbulence response spectra. As the pressure distribution changes over random time intervals, vibration is generated.

We will describe briefly the methods to analyze these modes of vibration. The subject is exhaustive and is far from being a perfected science. One could spend many volumes the size of this one and not adequately cover the subject. For this reason the reader is referred to Blevins [18] for more details.

As previously stated, turbulent flow induces continuously varying pressure distributions all around the tube surface(s). The efficiency as to how the pressure forces excite the tubes in each mode (first, second, third, etc.) of vibration is termed "joint efficiency," denoted by $J_j(\omega)$. This parameter represents the model efficiency of the pressure forces at a given frequency ω, where ω_j is the natural frequency of the jth mode. Figure 7-29 shows the joint acceptance of a single tube plotted against $\omega L/V_c$ (U in Figure 7-29), where L is the tube length between spans and V is the fluid velocity. The ratio of the fluid velocity to the turbulence frequency, ω, is known as the correlation length and is given by

$$L_c = V/\omega \tag{7-40}$$

If $L_c >> 1$, then the varying pressure forces that act along the tube oscillate in phase. Also, if the value of L_c is much smaller than the length of the tube, such that

$$\frac{\omega L}{V} >> 1$$

then the joint acceptance is independent of the mode shape and is proportional to the value of L_c/L.

The mean response of the tube is the average tube displacement induced by the mean flow over the specified time interval. The greatest percentage of the response exists in the fundamental mode. For a tube that spans several baffle supports the following expression is used for the mean response:

$$\delta_{rms} = 0.036CV^2d_t\left(\frac{\rho}{m}\right)\left(\frac{d_o}{f_N}\right)^{1.5}\left(\frac{1}{L\zeta}\right)^{0.5}\sin\left(\frac{\pi x}{L_T}\right) \tag{7-41}$$

where $C = 9.7 \times 10^{-4}(\text{sec})^{0.5}/(\text{ft})^{1.5}$
ρ = density of fluid external to tubes, slugs/ft^3 (1 slug = 32.2 lb_m)

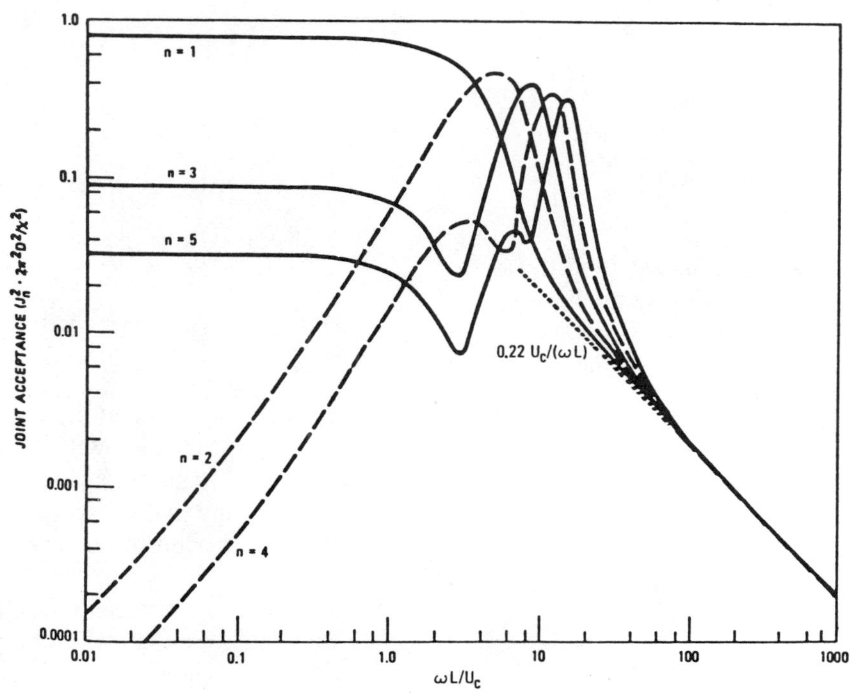

Figure 7-29. The joint acceptance of a simply supported tube subjected to turbulent flow. (From *Flow-Induced Vibration* by R. Blevins ©1977 by Van Nostrand Reinhold Company, Inc. Reprinted by permission.)

V = fluid velocity of fluid external to tubes, ft/sec
m = mass density of fluid external to tubes, slug/ft
d_t = tube OD, ft
L = tube length between baffles, ft
L_T = total length of tube between tubesheets, ft
f_N = fundamental natural frequency of tube portion between baffles, Hz
ζ = sum of structural damping and the fluid dynamic damping
x = distance along tube, ft
d_o = $4R_H$ = 4(hydraulic radius) =

$$\frac{4\text{ (flow area between tubes)}}{\text{wetted perimeter}}$$

$$= \frac{4d_t}{\pi}\left[\left(\frac{3P^2}{4d_t^2}\right) - \frac{\pi}{4}\right] \rightarrow \text{for tubes on an equilateral}$$

triangular pitch of P

$$= \frac{4d_t}{\pi}\left[\left(\frac{P}{d_t}\right) - \frac{\pi}{4}\right] \rightarrow \text{for tubes on a square pitch of P}$$

Using Figure 7-29 the value of the joint acceptance for the appropriate mode and the first mode are obtained. The ratio of the joint acceptance of the mode being considered to that of the first mode is multiplied by the value of δ_{rms} obtained from Equation 7-41. The relationship in Equation 7-41 is based on the theory of tube turbulence developed by Wambsganss and Chen [19], which yields the following maximum stress value:

$$\sigma_{max} = E\frac{\pi^2}{L^2}C_{D(max)} \tag{7-42}$$

where C_D = drag coefficient of tube surfaces

$$\delta_{max} = 2.58\delta_{rms} \text{ (for } x = L/2) \tag{7-43}$$

Equation 7-42 represents the maximum tube deflection to be incurred. The factor 2.58 represents the amplitude of the highest one percent of cycles.

The value for the natural frequency at the tube in Equation 7-41 takes on several forms. The easiest to use is the formulation developed by Blevins [18]:

$$f_N = \frac{\pi}{8L^2}\frac{E_t(d_t^2 + d_{ti}^2)^{0.5}}{m_t} \tag{7-44}$$

where E_t = modulus of elasticity of tube metal, psi
m_t = mass density of tube metal, slugs/ft^3
d_t = tube OD, in.
d_{ti} = tube ID, in.

TEMA gives a listing of expressions for the natural frequencies of the tubes based on several types of support end conditions, and tubes that have equal spans and unequal spans. These expressions were presented earlier in this chapter and in Table 7-6. Equation 7-44 is simplest to use because it requires less input. However, when the information is available and time permits, the expressions recommended by TEMA should be used.

The phenomena of "jetting" and "whirling" are not as well founded as vortex shedding and turbulence. This does not say that vortex shedding and turbulence are solidly based, but relatively speaking, they are compared to the other vibration modes, such as jetting and whirling.

From Figure 7-28 one can predict that when the tubes are inclined to the fluid flow, the results are force components about the x and y axes. Equation 4-80 illustrates how one can determine the force induced per unit length of a circular cylinder. In the case of whirling and jetting the term C_D is a variable. This term is called the force coefficient and is used in Equation 4-80 to evolve the following expressions:

$$F_y = \frac{\rho V^2 d_t}{2g_c}K_y\left(\frac{\delta_y}{d_t}\right) \tag{7-45}$$

$$F_x = \frac{\rho V^2 d_t}{2g_c}C_x\left(\frac{\delta_x}{d_t}\right)$$

$$\text{where } K_y = -\frac{2T}{D} + 5 - \frac{3D}{T} + \left(\frac{D}{T}\right)^2 \text{ for } \frac{T}{D} < 1.5$$

$$K_y = \left(\frac{D}{T}\right)^2 \text{ for } \frac{T}{D} \geqslant 1.5 \tag{7-46}$$

$$K_y = C_y(D/T)$$

$$C_x = -\left(\frac{D}{T}\right)^2 + 2\left(\frac{D}{T}\right)^3$$

where D and T are parameters defined in Figure 7-30 and Figure 7-31.

Values for K_y have been plotted against the parameter T/D. These values are shown in Figures 7-30 and 7-31 to represent the whirling parameter $2(2\pi)^{0.5}/(C_xK_y)^{0.25}$. Experiments indicate that the lower the whirling parameter the greater the probability that whirling (and jetting) will occur.

To determine if the tube deflections are within a safe range one must estimate the components F_y and F_x at their maximum values using Equation 4-80. From the tube spacing determine the force coefficients K_y and C_x from Equation 7-46. Then solve for δ_y and δ_x and determine if those deflections are acceptable. After determining that the deflections are in a safe range, use Figures 7-30 and 7-31 to determine the whirling parameter. If the parameter is on the low side, then the tube spacing should be increased to raise the whirling parameter. Unfortunately, at the current state of technology, there are no critical values to decide whether the whirling parameter is critical. One manner in which to avoid problems with whirling is to use Table 7-14 in determining the maximum shell-side fluid velocity flow. This table and the previous discussion will eliminate any problems with jetting or whirling. If the velocities cannot be controlled, because of someone else's design or a client's requests, then this procedure can give one an idea of whether whirling can be anticipated. The main focus is to keep the tubes spaced such that the maximum velocity will be reasonable. It has been confirmed by experiment that the critical velocity for whirling increases rapidly with the minimum spacing between the tubes and that in-line tube arrangements have lower critical velocities than staggered tube arrangements (refer to Figure 7-19 for the various illustrations of arrangements).

PLATE-FIN HEAT EXCHANGERS

These units use have been on the increase the past several years because of an increasing number of liquified gas and cryogenic plants. The plate fin heat exchanger is more efficient than the shell and tube exchanger because the comparable shell and tube exchanger required to replace a plate fin would be eight times the volume and twenty-four times the weight of the plate fin if constructed of aluminum. The reason for this is that if the plate-fin is made of brazed aluminum, the aluminum conducts heat better than most materials and can be used down to absolute zero (−460°F). Since the ductility of carbon steel is lost at −20°F, one must revert to expensive nickel alloys or stainless steels in the shell and tube design. Thus, for cold services, the plate-fin offers some advantages.

It is here that the advantages of the brazed plate-fin exchanger end. For the plate-fin to be applied, a very clean service is required. Even in clean services, these units can accommodate certain thermal shock and fatigue. It is quite possible after continued and repeated thermal loading in excess of differential temperatures of 50°F that internal components can fail. In addition, because these units are aluminum, external nozzle loadings induced by the piping can cause pipe stress problems. One must be extremely careful how much loading is induced to the nozzles, because even if failures do not occur, leaks are common if overloading exists. Thus, if the service is not clean, a shell and tube design must be used.

In gas processing and cryogenic services, the plate-fin exchanger suffices because in these applications the ser-

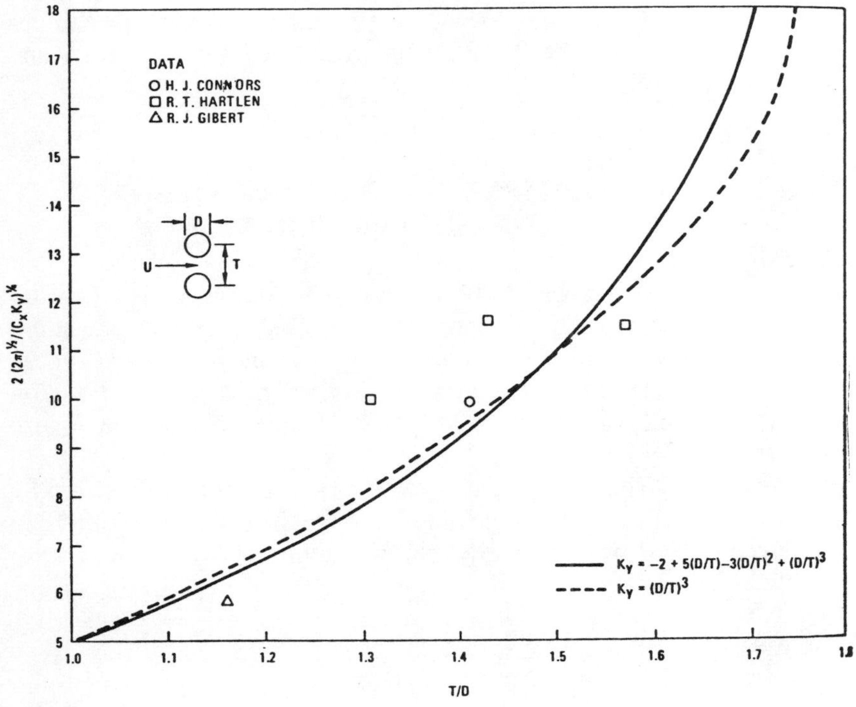

Figure 7-30. Whirling parameter of a tube row expressed as a function of transverse spacing. (From *Flow-Induced Vibration* by R. Blevins ©1977 by Van Nostrand Reinhold Company, Inc. Reprinted by permission.)

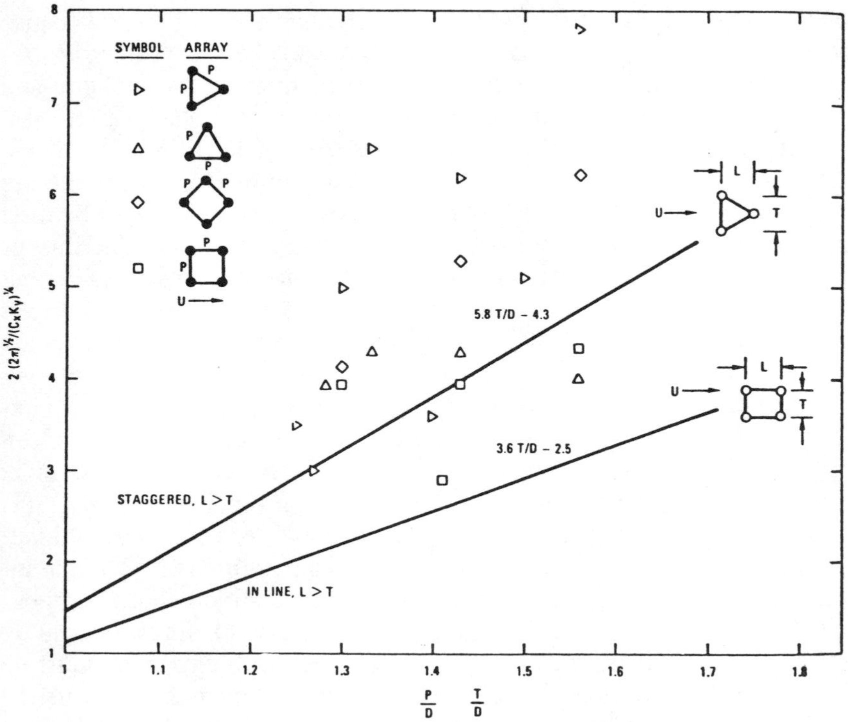

Figure 7-31. Whirling parameter for tube arrays. (From *Flow-Induced Vibration* by R. Blevins ©1977 by Van Nostrand Reinhold Company, Inc. Reprinted by permission.)

Table 7-14
Maximum Recommended Shell-Side Velocities

All liquids in 10 ft/sec
Gases and Vapors—in ft/sec

Pressure (psi)	Molecular Weight 18	30	50	100	150	200	400
27-in.(vac)	250	185	160	110	100	90	77
15-in.(vac)	130	100	85	65	60	52	45
0	100	80	70	50	45	40	35
50	65	55	45	35	30	25	20
100	55	45	35	25	20	18	16
200	50	40	30	23	19	17	—
500	40	30	20	—	—	—	—
1000	20	15	—	—	—	—	—

vices are relatively clean. However, it must be noted that shell and tube exchangers are more popular because of their flexibility of use. Certainly with moderate to heavy viscous fluids, the shell and tube exchanger is the only design to use.

Figure 7-32 shows a plate-fin exchanger with rectangular boxes containing an assortment of plates and fins resembling honeycomb structures. Fluids flow in tubular channels formed by fin attachments between plates (Figure 7-33). The plates that separate the two services vary from approximately 0.006 in. to 0.023 in. in thickness, depending on the pressure of the service. This design is commercially available at a temperature and pressure of approximately $-452°F$ at 1,400 psig.

With newly developed techniques in vacuum brazing, stronger bonds have been achieved that reduce failures of internal components subjected to thermal shock and fatigue.

The aluminum flanges used on these units are designed per ASME Section VIII Division I and, quite commonly, are identical to ANSI B16.5 flanges.

For further discussion on the thermal analysis and design of plate-fin units, the reader is referred to Kays and London [20].

EXAMPLE 7-1: REGENERATED GAS EXCHANGER DESIGN

A gas-gas shell and tube heat exchanger is to be designed. The exchanger is to be used to exchange heat between a hydrocarbon process gas and a gas used for regeneration. The unit is to be designed per specification sheet in Figure 7-34. The exchanger is shown in Figure 7-35.

The process gas is to be cooled from 965°F to 705°F. The regeneration gas is to be heated from 200°F to 661°F in a parallel configuration. Thus,

$$\begin{array}{rr} & 975°F \\ & \underline{200°F} \\ GTTD = & 775°F \end{array} \qquad \begin{array}{rr} & 750°F \\ & \underline{625°F} \\ LTTD = & 125°F \end{array}$$

Figure 7-32. The plate-fin exchanger. (Courtesy of Albraze International, Inc.)

$$\text{LMTD} = \frac{775 - 125}{\ln\left(\frac{775}{125}\right)} = 356°\text{F}$$

now,

$$q = \dot{m}C_p(\text{LMTD})$$

The shell-side mass flow rate = 22,050 lb_m/hr for the shell-side gas, C_p = 1.10 Btu/lb_m-°F. The required heat duty of the unit is

$$q = (22{,}050)\frac{lb_m}{hr}(1.10)\frac{Btu}{lb_m\text{-}°F}(356)°F$$

$$q = 8{,}634{,}780 \frac{Btu}{hr}$$

The available tube area in the exchanger is determined as follows: From Table 7-3, we determine that for a 1¼-in. tube the square feet of external surface per foot of tube is 0.3272 ft^2. Thus,

$$\text{Available area} = (0.3272)\frac{ft^2}{ft}(283)\text{ tubes }(15)\text{ ft}$$
$$= 1{,}388.95\ ft^2$$

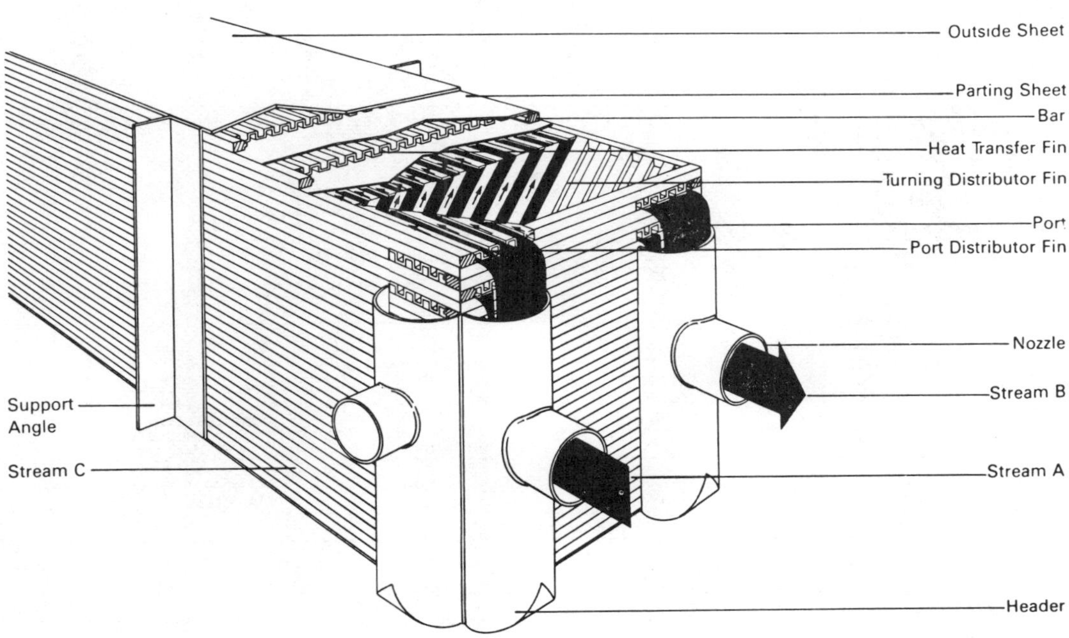

Figure 7-33. Tubular channels in plate surfaces result in excellent heat transfer in plate-fin heat exchangers. (Courtesy of Albraze International, Inc.)

HEAT EXCHANGER SPECIFICATION SHEET

1			Job No.	
2	Customer		Reference No.	
3	Address		Proposal No.	
4	Plant Location		Date	Rev.
5	Service of Unit		Item No.	
6	Size	Type (Hor/Vert)	Connected In	Parallel Series
7	Surf/Unit (Gross/Eff.)	Sq Ft; Shells/Unit	Surf/Shell (Gross/Eff.)	Sq Ft

	PERFORMANCE OF ONE UNIT		Shell Side		Tube Side	
8						
9	Fluid Allocation		Shell Side		Tube Side	
10	Fluid Name					
11	Fluid Quantity, Total	Lb/Hr				
12	Vapor (In/Out)					
13	Liquid					
14	Steam					
15	Water					
16	Noncondensable		22,050	22,050	41,884	41,884
17	Temperature (In/Out)	°F	200	625	975	750
18	Specific Gravity					
19	Viscosity, Liquid	Cp				
20	Molecular Weight, Vapor					
21	Molecular Weight, Noncondensable					
22	Specific Heat	Btu/Lb °F	1.10			
23	Thermal Conductivity	Btu Ft/Hr Sq Ft °F				
24	Latent Heat	Btu/Lb @ °F				
25	Inlet Pressure	Psig				
26	Velocity	Ft/S	48.75		48.48	
27	Pressure Drop, Allow./Calc.	Psi	10/10		10/10	
28	Fouling Resistance (Min.)					
29	Heat Exchanged		Btu/Hr: MTD (Corrected)			°F
30	Transfer Rate, Service		Clean			Btu/Hr Sq Ft °F

	CONSTRUCTION OF ONE SHELL		Shell Side	Tube Side	Sketch (Bundle/Nozzle Orientation)
32			Shell Side	Tube Side	
33	Design/Test Pressure	Psig	50/75	50/75	
34	Design Temperature	°F	700	1000	
35	No. Passes per Shell				
36	Corrosion Allowance	In.			
37	Connections Size & Rating	In	16 IN	40 IN	
38		Out	16 IN	40 IN	
39		Intermediate			

40	Tube No. 283 OD 1.25 In.; Thk (Min/Avg) .120 In.; Length 15 Ft; Pitch 1.75 In. 30 (60 circled) 90 45	
41	Tube Type B163 NCI	Material
42	Shell B168 ID 40 OD In.	Shell Cover (Integ.) (Remov.)
43	Channel or Bonnet	Channel Cover
44	Tubesheet-Stationary B168	Tubesheet-Floating
45	Floating Head Cover	Impingement Protection
46	Baffles-Cross Type 25	% Cut (Diam/Area) Spacing: c/c Inlet In.
47	Baffles-Long	Seal Type
48	Supports-Tube U-Bend	Type
49	Bypass Seal Arrangement	Tube-Tubesheet Joint
50	Expansion Joint	Type
51	ρv^2-Inlet Nozzle Bundle Entrance	Bundle Exit
52	Gaskets-Shell Side	Tube Side
53	-Floating Head	
54	Code Requirements	TEMA Class
55	Weight/Shell Filled with Water	Bundle Lb
56	Remarks	
57		
58		
59		
60		
61		

Figure 7-34. Heat exchanger specification sheet. (©1978 Tubular Exchanger Manufacturers Association.)

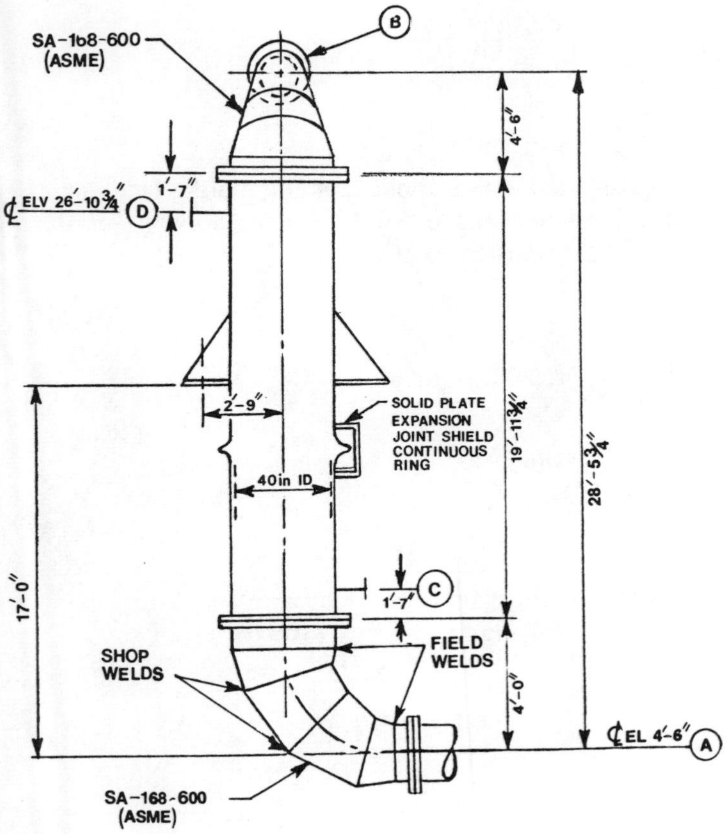

Figure 7-35. Vertical gas-gas exchanger.

Shell-side nozzles C and D are 16 in. in diameter, which makes the flow area

$$A_F = \frac{\pi(16)^2}{4} = 201.06 \text{ in.}^2 = 1.396 \text{ ft}^2$$

Shell-side mass density $= \rho_s = 0.09 \text{ lb}_m/\text{ft}^3$

$$V = \frac{22{,}050 \dfrac{\text{lb}_m}{\text{hr}}\left(\dfrac{1 \text{ hr}}{3600 \text{ sec}}\right)}{0.09 \dfrac{\text{lb}_m}{\text{ft}^3}(1.396) \text{ ft}^2} = 48.75 \text{ ft/sec}$$

From Table 7-14 we observe that this is a reasonable velocity.

Tube-Side Film Coefficient

For turbulent flow inside tubes we use Equation 7-19, the Sieder-Tate correlation,

$$N_{Nu} = 0.027(N_{Re})^{0.8}(N_{Pr})^{1/3}\left(\frac{\mu}{\mu_w}\right)^{0.14}$$

For the tube-side gas, 1¼-in.—11 gauge tubes

$N_{Pr} = 0.7$, obtained from process data
$k = 0.03 \text{ Btu/hr-ft}^2\text{-°F}$
$\mu = 0.01 \text{ C}_p = 0.024 \text{ lb/ft-hr}$

Tube-side mass flow rate $= 41{,}884 \text{ lb}_m/\text{hr}$

For each tube,

$$\dot{m} = \frac{41{,}884 \text{ lb}_m/\text{hr}}{283 \text{ tubes}} = 148 \text{ lb}_m/\text{hr}$$

$\rho = 0.1524 \text{ lb}_m/\text{ft}^3$,
$d_o = 1.25$ in.,
$d_i = 1.010$ in.;
$A_T = 0.8012 \text{ in.}^2$

$$V = \frac{(148)\dfrac{\text{lb}_m}{\text{hr}}\left(\dfrac{1 \text{ hr}}{3600 \text{ sec}}\right)}{(0.1524)\dfrac{\text{lb}_m}{\text{ft}^3}\left(\dfrac{0.8012 \text{ in.}^2}{144 \text{ in.}^2}\right)} = 48.48 \text{ ft/sec}$$

From Table 7-14 this velocity is reasonable

$$N_{Re} = \frac{(48.48)\dfrac{\text{ft}}{\text{sec}}(1.010) \text{ in.}\left(\dfrac{1 \text{ ft}}{12 \text{ in.}}\right)(0.1524)\dfrac{\text{lb}_m}{\text{ft}^3}}{0.024\dfrac{\text{lb}_m}{\text{ft-hr}}\left(\dfrac{1 \text{ hr}}{3{,}600 \text{ sec}}\right)}$$

$N_{Re} = 93{,}278 > 10{,}000$ and Equation 7-19 applies

$$N_{Nu} = 0.027(93{,}278)^{0.8}(0.7)^{1/3}(1.0) = 226.78$$

$$N_{Nu} = \frac{h_i d_i}{k}$$

From which,

$$h_i = \frac{(226.78)(0.03)\dfrac{\text{Btu}}{\text{hr-ft-°F}}}{\left(\dfrac{1.010}{12}\right)\text{ft}}$$

$$= 80.83 \frac{\text{Btu}}{\text{hr-ft}^2\text{-°F}}$$

Shell-Side Film Coefficient

$$N_{Nu} = \frac{h_o D_e}{k} = 0.36\left(\frac{D_e G_s}{\mu}\right)^{0.55}(N_{Pr})^{1/3}\left(\frac{\mu}{\mu_w}\right)^{0.14} \quad (7\text{-}26)$$

For 60°-Δ arrangement, p = 1.75 in.

$$D_e = \frac{8[0.43P^2 - 0.5\pi d_o^2/4]}{\pi d_o} \quad (7\text{-}29)$$

$$D_e = \frac{8[0.43(1.75)^2 - 0.5\pi(1.25)^2/4]}{\pi(1.25)} = 1.433 \text{ in.}$$

or

$$D_e = 0.119 \text{ ft}$$

c = tube clearance = 1.75 − 1.25 = 0.50 in.

B = distance between baffles

$$B = \frac{(15)\text{ft }(12)\ \frac{\text{in.}}{\text{ft}}}{8 \text{ baffles}} = 22.50 \text{ in.}$$

Computing the flow area of tube bundle = a_s,

$$a_s = \frac{D_s(c)(B)}{p(144)} \text{ ft}^2 \quad (7\text{-}30)$$

$$a_s = \frac{(40)\text{ in. }(0.50)\text{ in. }(22.50)}{(1.75)(144)\ \frac{\text{in.}^2}{\text{ft}^2}} = 1.79 \text{ ft}^2$$

$$G_s = \frac{\dot{m}}{a_s}$$

For the shell-side gas, ρ = 0.09 lb_m/ft^3 average for temperatures specified, and μ = 0.05 lb_m/ft-hr

$$G_s = \frac{22{,}050\ \frac{lb_m}{hr}}{1.79\ ft^2} = 12{,}348.00\ \frac{lb_m}{hr\text{-}ft^2}$$

$$N_{Re} = \frac{D_eG_s}{\mu} = \frac{(0.119)\text{ ft }(12{,}348.00)\ lb_m/hr\text{-}ft^2}{0.05\ lb_m/ft\text{-}hr}$$

$$N_{Re} = 29{,}388.24$$

The exchanger has baffles with 25% cut, thus from Figure 7-22,

$$j_H = 100$$

$$h_o = \frac{j_H k}{D_e}(N_{Pr})^{1/3}\left(\frac{\mu}{\mu_m}\right)^{0.14};$$

N_{Pr} = 0.8 from process data

$$h_o = (100)\left(\frac{0.03}{0.119}\right)(0.8)^{1/3}(1)$$

$$= 23.40\ \frac{Btu}{hr\text{-}ft^2\text{-}°F}$$

For gases used in this application the fouling factors are 0.001 shell-side and 0.001 tube-side. Solving for the overall heat transfer coefficient,

$$U = \frac{1}{\frac{1}{80.83} + 0.001 + 0.001 + \frac{1}{23.40}}$$

$$U = 17.51\ \frac{Btu}{hr\text{-}ft^2\text{-}°F}$$

$$\text{Area required} = \frac{8{,}634{,}780\ \frac{Btu}{hr}}{(17.51)\ \frac{Btu}{hr\text{-}ft^2\text{-}°F}\ (356)°F}$$

$$= 1{,}384.91 \text{ ft}^2$$

From previous calculation,

Available area = 1,388.95 ft^2

In most applications the available area should not be 10% greater than the required area, such material is not wasted.

Shell-Side Pressure Drop

$$\Delta P_s = \frac{f_s G_s^2 D_s (N_B + 1)}{(5.22)(10)^{10} D_e \gamma \phi} \quad (7\text{-}31)$$

N_B = 8 baffles
D_s = shell ID = 40 in. = 3.333 ft
G_s = 12,348.00 $lb_m/hr\text{-}ft^2$

For N_{Re} = 29,388.24, f = 0.0022 from Figure 7-23.

For plain and bare tubes,

$$f_s = \frac{f}{1.2} = 0.0018$$

$$D_e = 0.119$$

γ = specific gravity of shell-side gas = 0.9

$$\phi = 1.0 = \left(\frac{\mu}{\mu_w}\right)^{0.14}$$

$$\Delta P_s = \frac{(0.0018)(12{,}348.00)^2(3.333)(8 + 1)}{(5.22)(10)^{10}(0.119)(0.9)(1.0)}$$

$\Delta P_s = 0.0015$ psi $<<$ 10 psi allowed on data sheet, Figure 7-34

EXAMPLE 7-2: VIBRATION CHECK FOR REGENERATED GAS EXCHANGER

The exchanger in Example 7-1 is to be checked for possible vibration problems. To accomplish this we compute the damage numbers of equations 7-38 and 7-39.

B = tube span between baffles = 22.50 in.

Shell-side gas density = 0.09 lb_m/ft^3

d_t = 1.25 in.

From ASME Section VIII Div. I (see Chapter 4) for the tube material at design temperature,

σ_{all_t} = 18,052 psi at shell-side conditions

d_B = 1.25 in. + 1/64 in.

= 1.156, where 1/64 in. is the baffle hole clearance (see Figure 7-34)

From Figure 7-20 we compute the shell-side gas velocity between tubes.

D_s = 40.0 in., D_{to} = 37.125, P = 1.75

$$A_F = \frac{B}{144}\left[D_s - D_{to} + \frac{D_{to} - d_t}{P}(P - d_t)\right] ft^2$$

$$A_F = \frac{22.50}{144}\left[40 - 37.125 + \frac{37.125 - 1.25}{1.75}(1.75 - 1.25)\right]$$

$$A_F = 2.051 \text{ ft}^2$$

$$G_s = 18{,}522.0 \frac{lb_m}{ft^2\text{-}hr} = 5.145 \frac{lb_m}{ft^2\text{-}sec}$$

$$V = \frac{(5.145)\dfrac{lb_m}{ft^2\text{-}sec}(2.051)\ ft^2}{0.09\dfrac{lb_m}{ft^3}}$$

$$= 117.236 \frac{ft}{sec} \text{ between tubes}$$

From Equation 7-35 we compute the shear force induced on the tube at the baffle hole,

$$F_s = \left(\frac{d_t d_B}{d_t + d_B}\right) C \left(\frac{\sigma_{all}}{0.798}\right)^2 \tag{7-35}$$

$$c = \frac{1 - \nu_t}{E_t} + \frac{1 - \nu_B}{E_B},\ E_t = E_B = 27.0 \times 10^6 \text{ psi}$$

$$c = \frac{2(1.00 - 0.333)}{27 \times 10^6} = 4.941 \times 10^{-8}$$

$$F_s = \left[\frac{(1.25)(1.156)}{2.281}\right](4.94 \times 10^{-8})\left(\frac{18{,}052}{0.798}\right)^2$$

$$= 16.015 \ lb_F/ft$$

From Figure 7-25, a = 1.10 and b = 0.0069

$$F_L = \frac{c_F \rho V^2 d_t}{2g_c} = \frac{(0.6)(0.09)\dfrac{lb_M}{ft^3}(117.236)^2\dfrac{ft^2}{sec^2}\left(\dfrac{1.25}{12}\right) ft}{2(32.2)\dfrac{lb_M - ft}{lb_f - sec^2}}$$

$$F_L = 1.200 \ lb_f/ft$$

$$F_s = \alpha a F_L L \Rightarrow \alpha = \frac{F_s}{aF_L L}$$

$$\alpha = \frac{16.015 \ lb_f/ft}{(1.10)(1.200)\dfrac{lb_f}{ft}\left(\dfrac{22.50 \text{ in.}}{12 \text{ in.}}\right) ft} = 6.471$$

$$N_{BE} = \frac{C_D d_t \rho V^2 a L \alpha}{2g_c F_s} \tag{7-38}$$

$$N_{BE} = \left[\frac{(0.6)\left(\dfrac{1.25}{12}\right) ft\ (0.09)\dfrac{lb_m}{ft^3}(117.236)^2\dfrac{ft^2}{sec^2}}{2(32.2)\dfrac{lb_m - ft}{lb_f - sec^2}(16.015)\dfrac{lb_f}{ft}}\right] \times (1.1)\left(\frac{22.50}{12}\right) ft\ (6.471)$$

$$N_{BE} = 1.00$$

$$N_{CD} = \frac{C_D d_t \rho V^2 b L^4}{2g_c \delta E_t I} \tag{7-39}$$

$$\delta = \frac{bF_LL^4}{E_tI}$$

$$= \frac{(0.0069)(1.20)\ \frac{lb_f}{ft}\left(\frac{1\ ft}{12\ in.}\right)(22.50)^4\ ft}{(27 \times 10^6)\ \frac{lb_f}{ft^2}\ (0.0688)\ in.^4}$$

$$\delta = 9.520 \times 10^{-5}\ in.$$

$$N_{CD} = \left[\frac{(0.6)\left(\frac{1.25}{12}\right) ft\ (0.09)\ \frac{lb_m}{ft^3}}{2(32.2)\ \frac{lb_m - ft}{lb_f - sec^2}\left(\frac{9.520 \times 10^{-5}}{12\ in.}\right)}\right] \times \left[\frac{(117.236)^2\ \frac{ft^2}{sec^2}\ (0.0069)\left(\frac{22.50}{12}\right)^4 ft^4}{(27 \times 10^6)\ \frac{lb_f}{in.^2}\left(\frac{144\ in.^2}{ft^2}\right)(0.0688\ in.^4)\ \frac{1\ ft^4}{(12)^4 in.^4}}\right]$$

$$N_{CD} = 1.000$$

With N_{BE} and N_{CD} not exceeding 1.0, we do not expect vibration trouble. To be certain we compute the maximum tube deflections as follows:

$$\delta_{rms} = 0.036CV^2d_t\left(\frac{\rho}{m}\right)\left(\frac{d_o}{f_n}\right)^{1.5}\left(\frac{1}{L\zeta}\right)^{0.5}\sin\left(\frac{\pi x}{L}\right) \quad (7\text{-}41)$$

$$f_n = \frac{\pi}{8L^2}\left[\frac{E_t(d_t^2 d_{ti}^2)}{M_t}\right] \quad (7\text{-}44)$$

$$d_t = 1.25\ in. = 0.104\ ft$$

$$d_{ti} = 1.084\ in. = 0.090\ ft$$

$$f_n = \frac{\pi}{8(22.50)^2\ in.^2} \times \left[\frac{(27 \times 10^6)\ \frac{lb_f}{in.^2}\ [(1.25)^2\ in.^2 + (1.084)^2\ in.^2}{0.2833\ \frac{lb_m}{in.^3}\left(\frac{1\ slug}{32.2\ lb_m}\right)\left(\frac{1728\ in.^3}{ft^3}\right)}\right]^{0.5}$$

$$f_n = 1.710\ Hz$$

For shell-side fluid,

$$\rho = 0.09\ \frac{lb_m}{ft^3}\left(\frac{1\ slug}{32.2\ lb_m}\right) = 0.003\ \frac{slug}{ft^3}$$

For the tube-side,

$$M_t = 1.448\ \frac{lb_m}{ft} = 0.045\ \frac{slug}{ft}$$

$$\delta_{rms} = 0.036\ (4.941 \times 10^{-8})(117.236)^2\left(\frac{1.25}{12}\right)\left(\frac{0.09}{0.045}\right) \times \left(\frac{1.25}{(12)(1.710)}\right)\left[\frac{1}{\frac{(22.50)}{12}(0.09)}\right]^{0.5}\sin\left(\frac{12\pi x}{22.5}\right)$$

with $\sin\left(\frac{12\pi x}{22.5}\right) = 1.0$ for worst case

$$\delta_{rms} = 7.553 \times 10^{-7}\ ft = 9.063 \times 10^{-6}\ in.$$

With this magnitude of tube displacement and N_{BE} and N_{CD} being in the safe zone, we conclude that the exchanger will not have vibration problems.

EXAMPLE 7-3: CHLORINE SUPERHEATER DESIGN

A plant wishes to use hot oil to heat chlorine gas. The exchanger unit, a chlorine superheater, is to be a TEMA 18-150 AEL. The chlorine gas is to be heated from 77°F to 158°F and the hot oil is cooled from 250°F to 176°F. The exchanger is to be rated and analyzed for tube—tubesheet loading. The exchanger specification is shown in Figure 7-36. The thermal duty is 600,000 Btu/hr. The exchanger is a parallel flowing unit.

Shell-side (hot oil)	250°F in.	176°F out
Tube-side (chlorine gas)	77°F in.	158°F out
GTTD =	173°F	LTTD = 18°F

$$LMTD = \frac{GTTD - LTTD}{\ln\left(\frac{GTTD}{LTTD}\right)} = \frac{173 - 18}{\ln\left(\frac{173}{18}\right)} = 68.496°F$$

(with a parallel exchanger no correction is needed for LMTD)

Tube-Side Film Coefficient

For chlorine gas, $C_p = 0.116$ Btu/lb_m-°F; $\rho = 1.667$ lb_m/ft^3

$$q = \dot{m}C_p(LMTD)$$

HEAT EXCHANGER SPECIFICATION SHEET

#					
1				Job No.	
2	Customer			Reference No.	
3	Address			Proposal No.	
4	Plant Location			Date	Rev.
5	Service of Unit			Item No.	
6	Size	Type	(Hor/Vert)	Connected In	Parallel Series
7	Surf/Unit (Gross/Eff.)	Sq Ft; Shells/Unit		Surf/Shell (Gross/Eff.)	Sq Ft
8	**PERFORMANCE OF ONE UNIT**				
9	Fluid Allocation		Shell Side		Tube Side
10	Fluid Name		HOT OIL		CHLORINE GAS
11	Fluid Quantity, Total	Lb/Hr			
12	Vapor (In/Out)			88,100	
13	Liquid				
14	Steam				
15	Water				
16	Noncondensable				
17	Temperature (In/Out)	°F	250°/176°		77°/158°
18	Specific Gravity	lbm/ft3	1.00		1.667
19	Viscosity, Liquid	Cp			
20	Molecular Weight, Vapor				
21	Molecular Weight, Noncondensable				
22	Specific Heat	Btu/Lb °F	0.426		0.116
23	Thermal Conductivity	Btu Ft/Hr Sq Ft °F			
24	Latent Heat	Btu/Lb @ °F			
25	Inlet Pressure	Psig			
26	Velocity	Ft/S			
27	Pressure Drop, Allow./Calc.	Psi	/		/
28	Fouling Resistance (Min.)				
29	Heat Exchanged 600,000		Btu/Hr: MTD (Corrected)	68.5	°F
30	Transfer Rate, Service		Clean		Btu/Hr Sq Ft °F
31	**CONSTRUCTION OF ONE SHELL**				Sketch (Bundle/Nozzle Orientation)
32			Shell Side	Tube Side	
33	Design/Test Pressure	Psig	150 / 175	/	
34	Design Temperature	°F			
35	No. Passes per Shell				
36	Corrosion Allowance	In.			
37	Connections	In			
38	Size &	Out			
39	Rating	Intermediate			
40	Tube No. 150 OD 1 In.;Thk (Min/Avg)		In.; Length 15ft. Ft; Pitch 1 1/4 In.	◁ 30 △ 60 □ 90 ◇ 45	
41	Tube Type		Material		
42	Shell 18" ID OD In.		Shell Cover	(Integ.) (Remov.)	
43	Channel or Bonnet		Channel Cover		
44	Tubesheet-Stationary		Tubesheet-Floating		
45	Floating Head Cover		Impingement Protection		
46	Baffles-Cross 6	Type 48	% Cut (Diam/Area) 1 1/4" Spacing: c/c	Inlet In.	
47	Baffles-Long		Seal Type		
48	Supports-Tube	U-Bend		Type	
49	Bypass Seal Arrangement		Tube-Tubesheet Joint		
50	Expansion Joint		Type		
51	ρv^2-Inlet Nozzle	Bundle Entrance		Bundle Exit	
52	Gaskets-Shell Side		Tube Side		
53	-Floating Head				
54	Code Requirements			TEMA Class	
55	Weight/Shell	Filled with Water		Bundle	Lb
56	Remarks				
57					
58					
59					
60					
61					

Figure 7-36. Chlorine superheater heat exchanger specification sheet.

$$\dot{m} = \frac{700{,}000 \dfrac{\text{Btu}}{\text{hr}}}{0.116 \dfrac{\text{Btu}}{\text{lb}_m - {}^\circ\text{F}} (68.496)^\circ\text{F}} = 88{,}099.783 \frac{\text{lb}_m}{\text{hr}}$$

$$Q = \frac{88{,}099.783 \dfrac{\text{lb}_m}{\text{hr}} \left(\dfrac{1 \text{ hr}}{3600 \text{ sec}}\right)}{1.667 \dfrac{\text{lb}_m}{\text{ft}^3}} = 14.680 \frac{\text{ft}^3}{\text{sec}}$$

For each of the 150–1-in.–14 BWG tubes,

$$V = \frac{14{,}680 \dfrac{\text{ft}^3}{\text{sec}}}{(0.003794)\text{ft}^2(150)\text{tubes}} = 25.796 \text{ ft/sec} \rightarrow \text{Reasonable}$$

$$N_{Re} = \frac{VD\rho}{\mu}; \ \mu = 0.0148 \text{ Cp} = 0.036 \text{ lb}_m/\text{ft-hr}$$

$$N_{Re} = \frac{(25.796) \dfrac{\text{ft}}{\text{sec}} \left(\dfrac{0.834}{12}\right) \text{ft } (1.667) \dfrac{\text{lb}_m}{\text{ft}^3} \left(\dfrac{3600 \text{ sec}}{1 \text{ hr}}\right)}{(0.036) \dfrac{\text{lb}_m}{\text{ft-hr}}}$$

$$= 298{,}860.527$$

$$N_{Pr} = \frac{\mu C_p}{k}; \ k = 5.0 \times 10^{-3} \frac{\text{Btu}}{\text{hr-ft-}^\circ\text{F}}$$

$$N_{Pr} = \frac{(0.036) \dfrac{\text{lb}_m}{\text{ft-hr}} (0.116) \dfrac{\text{Btu}}{\text{lb}_m\text{-}^\circ\text{F}}}{(5.0 \times 10^{-3}) \dfrac{\text{Btu}}{\text{hr-ft-}^\circ\text{F}}} = 0.835$$

$$N_{Nu} = 0.027(N_{Re})^{0.8}(N_{Pr})^{1/3} \left(\frac{\mu}{\mu_w}\right)^{0.14}, \ 0.7 < N_{Pr} < 17{,}000$$

$$N_{Nu} = 0.027(298{,}860.527)^{0.8}(0.835)^{1/3}(1.0) = 610.464$$

$$N_{Nu} = \frac{h_i d_i}{k} \Rightarrow h_i = \frac{(610.464)(5.0 \times 10^{-3}) \dfrac{\text{Btu}}{\text{hr-ft-}^\circ\text{F}}}{\left(\dfrac{0.835}{12}\right) \text{ft}}$$

$$h_i = 43.866 \frac{\text{Btu}}{\text{hr-ft}^2\text{-}^\circ\text{F}}$$

Shell-Side Film Coefficient

$$q = \dot{m}C_p(\text{LMTD})$$

$$C_p = 0.426 \text{ Btu/lb}_m\text{-}^\circ\text{F};$$

$$\rho = 62.46 \frac{\text{lb}_m}{\text{ft}^3}; \ k = 0.077 \text{ Btu/hr-ft-}^\circ\text{F}$$

$$\dot{m} = \frac{700{,}000 \dfrac{\text{Btu}}{\text{hr}} \left(\dfrac{1 \text{ hr}}{3600 \text{ sec}}\right)}{(.426) \dfrac{\text{Btu}}{\text{lb}_m\text{-}^\circ\text{F}} (68.496)^\circ\text{F}}$$

$$= 6.664 \frac{\text{lb}_m}{\text{sec}}$$

$$Q = \frac{6.664 \dfrac{\text{lb}_m}{\text{sec}}}{62.46 \dfrac{\text{lb}_m}{\text{ft}^3}} = 0.107 \text{ ft}^3/\text{sec}$$

The smallest shell-side nozzle is the 3-in. outlet, where

$$A_i = 7.393 \text{ in.}^2 = 0.051 \text{ ft}^2$$

$$V = \frac{0.107 \dfrac{\text{ft}^3}{\text{sec}}}{0.051 \text{ ft}^2} = 2.092 \frac{\text{ft}}{\text{sec}} \rightarrow \text{Very reasonable}$$

$$N_{Pr} = \frac{\mu C_p}{k}; \ \mu = 2.544 \text{ lb}_m/\text{ft-hr}$$

$$N_{Pr} = \frac{(2.544) \dfrac{\text{lb}_m}{\text{ft-hr}} (0.426) \dfrac{\text{Btu}}{\text{lb}_m\text{-}^\circ\text{F}}}{0.077 \dfrac{\text{Btu}}{\text{hr-ft-}^\circ\text{F}}} = 14.075$$

For 60° Δ arrangement,

$$D_e = \frac{8[0.43 \text{ p}^2 - 0.5\pi d_o^2/4]}{\pi d_o}$$

$$= \frac{8[(0.43)(1.25)^2 - 0.5\pi(1.0)^2/4]}{\pi(1.0)}$$

$$D_e = 0.711 \text{ or } D_e = 0.059 \text{ ft}$$

$$c = 1.25 - 1.00 = 0.25 \text{ in.}$$

$$B = 30 \text{ in. for 6 baffles}$$

$$D_s = 18.00 \text{ in.} = \text{shell ID}$$

$$a_s = \frac{D_s cB}{p(144)} = \frac{(18.00)(0.25)(30)}{(1.25)(144)} = 0.75 \text{ ft}^2$$

$$G_s = \frac{\dot{m}}{a_s}$$

$$G_s = \frac{(6.664) \dfrac{\text{lb}_m}{\text{sec}} (3{,}600) \dfrac{\text{sec}}{\text{hr}}}{0.75 \text{ ft}^2} = 31{,}987.20 \frac{\text{lb}_m}{\text{hr-ft}^2}$$

$$N_{Re} = \frac{D_e G_s}{\mu} = \frac{(0.059)\text{ft}(31{,}987.20)\frac{\text{lb}_m}{\text{hr-ft}^2}}{2.544\ \frac{\text{lb}_m}{\text{ft-hr}}} = 741.842$$

The exchanger has baffles with 45% cut, so from Figure 7-22,

$$j_H = 12$$

$$h_o = \frac{j_H k}{D_e}(N_{Pr})^{1/3}\left(\frac{\mu}{\mu_w}\right)^{0.14}$$

$$h_o = \frac{(12)(0.077)}{0.059}(14.075)^{1/3}(1) = 37.779\ \frac{\text{Btu}}{\text{hr-ft}^2\text{-°F}}$$

Since both gases are relatively clean, the fouling factors for both sides are 0.0001.

$$U = \frac{1}{\frac{1}{37.779} + 0.001 + 0.0008 + \frac{1}{43.866}}$$

$$U = 19.582\frac{\text{Btu}}{\text{hr-ft}^2\text{-°F}}$$

$$\text{Area required} = \frac{700{,}000\ \frac{\text{Btu}}{\text{hr}}}{(19.582)\frac{\text{Btu}}{\text{hr-ft}^2\text{-°F}}(68.496)} = 521.875\ \text{ft}^2$$

$$\text{Available area} = (0.2618)(150)(15) = 589.050\ \text{ft}^2$$

This implies a 12.87% excess, which is acceptable.

Shell-Side Pressure Drop

$$\Delta P_s = \frac{f_s G_s^2 D_s (N_c + 1)}{(5.22)(10)^{10} D_e \gamma \phi}$$

$N_c = 6$ baffles

D_s = shell ID = 18 in. = 1.5 ft

$$G_s = 31{,}987.20\frac{\text{lb}_m}{\text{hr-ft}^2}$$

From Figure 7-24 for $N_{Re} = 741.842$, $f = 0.0027$

$$f_s = \frac{f}{1.2} = 0.00225$$

$D_e = 0.059$ ft

$\gamma = 1.001$

$\phi = 1.0$

A_R = ratio of OD to ID of tube

$A_R = 1.199$

$$h_{io} = \frac{h_i}{A_R} = \frac{73.629}{1.199} = 61.409\ \frac{\text{Btu}}{\text{hr-ft}^2\text{-°F}}$$

$$t_w = 196.720 - \frac{61.409}{61.409 + 107.480}(196.720 - 99.680)$$

$t_w = 161.436°\text{F}$

Maximum allowable tube joint load = L_{max}

$$L_{max} = A_t \sigma_{all} f_r$$

For SB-161-200 at 162°F, $\sigma_{all} = 10{,}000$ psi

$$L_{max} = (0.239)\text{in.}^2(10{,}000)\ \frac{\text{lb}}{\text{in.}^2}(1.0) = 2{,}390.00\ \text{lb}_f$$

The tube wall temperature is used in a method developed by Miller [21], which is a more exact approach than most and consequently results in a more economical design.

P_s = shell-side pressure = 100 psi

$\Delta t = t_w - t_a$; t_a = ambient air temperature = 70°F

$\Delta t = 161.436°\text{F} - 70°\text{F} = 91.436°\text{F}$

D_s = shell ID = D_s − CA

CA = corrosion allowance = 0 for pure helium (inert)(erosion is negligible)

$D_s = 18.0 - 0.0 = 18.0$ in.

$$t_{shell} = \frac{PR}{\sigma E - 0.6P} = \frac{(100)(9.0)}{(16{,}200)(1.0) - 0.6(100)} = 0.0558\ \text{in.}$$

Use t_{shell} = 3/16-in. = 0.1875 in.

For the shell, $E_s = 27.546 \times 10^6$ psi at 161.436°F

$$\Delta P_S = \frac{(0.00225)(31{,}987.20)^2(1.5)(6 + 1)}{(5.22)(10)^{10}(0.059)(1.001)(1.0)}$$

= 0.008 psi which is acceptable

Tube Metal Temperature

For parallel flow,

$$\Delta t_h = 250 - 77 = 173°F$$

$$\Delta t_c = 176 - 158 = 18°F$$

$$\frac{\Delta t_c}{\Delta t_h} = \frac{18}{173} = 0.104$$

For hot end,

$$U_h = \frac{600{,}000}{(510.510)(173)} = 6.794$$

For cold end,

$$U_c = \frac{600{,}000}{(510.510)(18)} = 65.294$$

$$c = \left|\frac{U_h - U_c}{U_c}\right| = \left|\frac{6.794 - 65.294}{65.294}\right| = 0.896$$

From Figure 7-17, $F_c = 0.28$

$$t_{ch} = t_{ho} + F_c(t_{hi} - t_{ho}) \tag{7-11}$$

$$t_{ch} = 176 + (0.28)(250 - 176) = 196.720°F$$

$$t_{cc} = t_{ci} + F_c(t_{co} - t_{ci}) \tag{7-12}$$

$$t_{cc} = 77 + (0.28)(158 - 77) = 99.680°F$$

$$t_w = t_{ch} - \frac{h_{io}}{h_{io} + h_o}(t_{ch} - t_{cc}) \tag{7-13}$$

α = coefficient of thermal expansion, in./in.-°F

For the shell material,

$\alpha_s = 6.090 \times 10^{-6}$ in./in.-°F at 161.436°F

P_{t_i} = tube-side (channel-side) pressure = 100 psi

$\Delta t = 161.436°F - 70°F = 91.436°F$

$d_t = 0.834$

$$\left.\begin{aligned} E_t &= 13.799 \times 10^6 \text{ psi} \\ \alpha_t &= 7.010 \times 10^{-6} \text{ in./in.-°F} \\ E_{Ts} &= 27.546 \times 10^6 \text{ psi} \end{aligned}\right\} \text{at } 161.436°F$$

where E_{Ts} = modulus of elasticity of tubesheet metal
T = tubesheet thickness = 1.1875 in.
a_t = cross-sectional area of tube (see Table 7-3) = 0.239 in.^{2}
n = number of tubes = 150
$na_t = (150)(0.239) = 35{,}850$ in.^{2}

$$A = \frac{\pi(18.0)^2}{4} A = 254.469 \text{ in.}^2 = \text{shell cross-sectional area}$$

$$c = \frac{\pi(1.0)^2}{4}(110) = 86.394 \text{ in.}^2 = \text{total cross-sectional area of tube holes}$$

$$B = \frac{\pi}{4}(D_{S_o}^2 - D_{S_i}^2) = \frac{\pi}{4}[(18.375)^2 - (18.125)^2]$$

$$= 7.1668 \text{ in.}^2$$

$$A - C = 254.469 \text{ in.}^2 - 86.394 \text{ in.}^2 = 168.075 \text{ in.}^2$$

The ratio of the inside shell bore area to the net tubesheet area minus the tubes is the net area that resists the tube and shell reaction forces and moments. This ratio is referred to as the ligament or deflexion efficiency and is expressed as

$$\eta = \frac{(A - C)}{A}$$

$$\text{Let } \psi = \frac{E_t na_t}{E_s B} = \frac{(13.799 \times 10^6)(35.850)}{(27.546 \times 10^6)(7.1668)} = 2.5058$$

Let ΔP_E = equivalent pressure difference, psi

$$\Delta P_E = P_s - P_{t_i} - \frac{P_{t_i}\, n\, a_t}{A - C} = 100 - 100 - \frac{(100)(35.850)}{168.075}$$

$$= -21.3298 \text{ psi}$$

Computing the differential thermal expansion = $\Delta\alpha$

$$\Delta\alpha = \alpha_t \Delta_t - \alpha_s \Delta_s$$

$$\Delta\alpha = (7.010 \times 10^{-6})(91.436) - (6.090 \times 10^{-6})(91.436)$$

$$= 0.000084$$

P_E = the effective pressure differential induced by the equivalent pressure difference, ΔP_E, and thermal expansion, $\Delta\alpha$

$$P_E = P + (\Delta\alpha)\frac{E_t na_t}{A - C} \tag{7-47}$$

$$P_E = -21.3298 + (0.000084)\frac{(13.799 \times 10^6)(35.85)}{168.075}$$

$$= 226.263 \text{ psi}$$

Assume the normal tube projection beyond the tubesheet to be 1/8 in.,

$$L = (13)(12) - 2(1.1875) - 2(0.125) = 153.375 \text{ in.}$$

Defining the dimensionless parameter, λ, as

$$\lambda = 1.08\left[\frac{E_t n a_t}{LT^3 E_{Ts}(A - C)}\right]^{0.25} D_s \qquad (7\text{-}48)$$

$$\lambda = 1.08 \times \left[\frac{(13.799 \times 10^6)(35.850)}{(153.375)(1.125)^3(27.546 \times 10^6)(168.075)}\right]^{0.25} (18.125)$$

$$\lambda = 2.696$$

$\sigma_{r(max)} = -415.968$ psi for $\Gamma = \Gamma_4 = -0.046$

$\sigma_{r(max)} = -415.968$ psi is well below the maximum allowable stress, which means that the tubesheet is of sufficient thickness.

One could repeat the process if it was desired to use a thinner tubesheet. Had $\sigma_{r(max)}$ exceeded the maximum allowable stress for the tubesheet material, then a greater tubesheet thickness would have to be selected and the process repeated.

From Figures 7-37, 7-38, 7-39, and 7-40:

$$\Gamma_1 = 1.55;\ \Gamma_2 = 3.12;\ \Gamma_3 = -0.046;\ \Gamma_4 = 1.970$$

The maximum radial stress in the tubesheet is expressed as

$$\sigma_{r(max)} = \frac{\left[P_E - P_{ti}\left[\frac{A\psi}{(A - C)}\right]\right]\left(\frac{D_s}{T}\right)^2}{4(\psi\Gamma_1 + \Gamma_2)} \qquad (7\text{-}49)$$

$$\sigma_{r(max)} = \frac{\left[226.263 - \frac{(100)(254.469)(2.5058)}{(168.075)}\right]\left(\frac{18.125}{1.125}\right)^2}{4[(2.5058)(1.55) + 3.12]}$$

$\sigma_{r(max)} = -1{,}418.659$ compression $< 16{,}200$ psi allowable for the tubesheet material

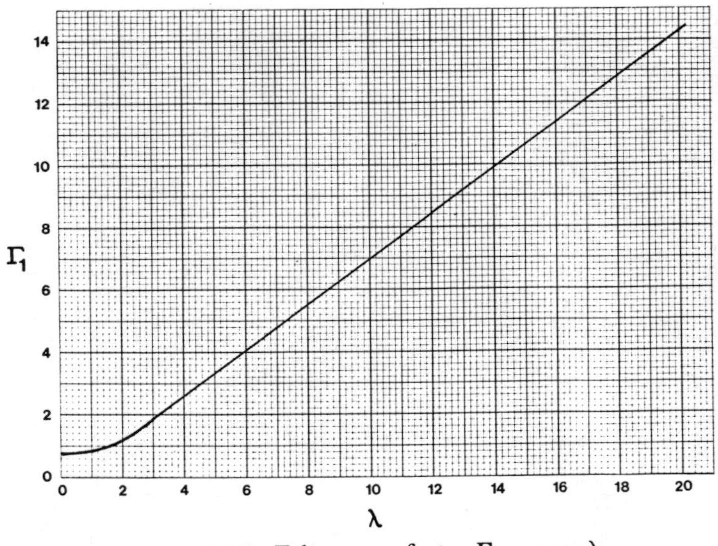

Figure 7-37. Tube stress factor Γ_1 versus λ.

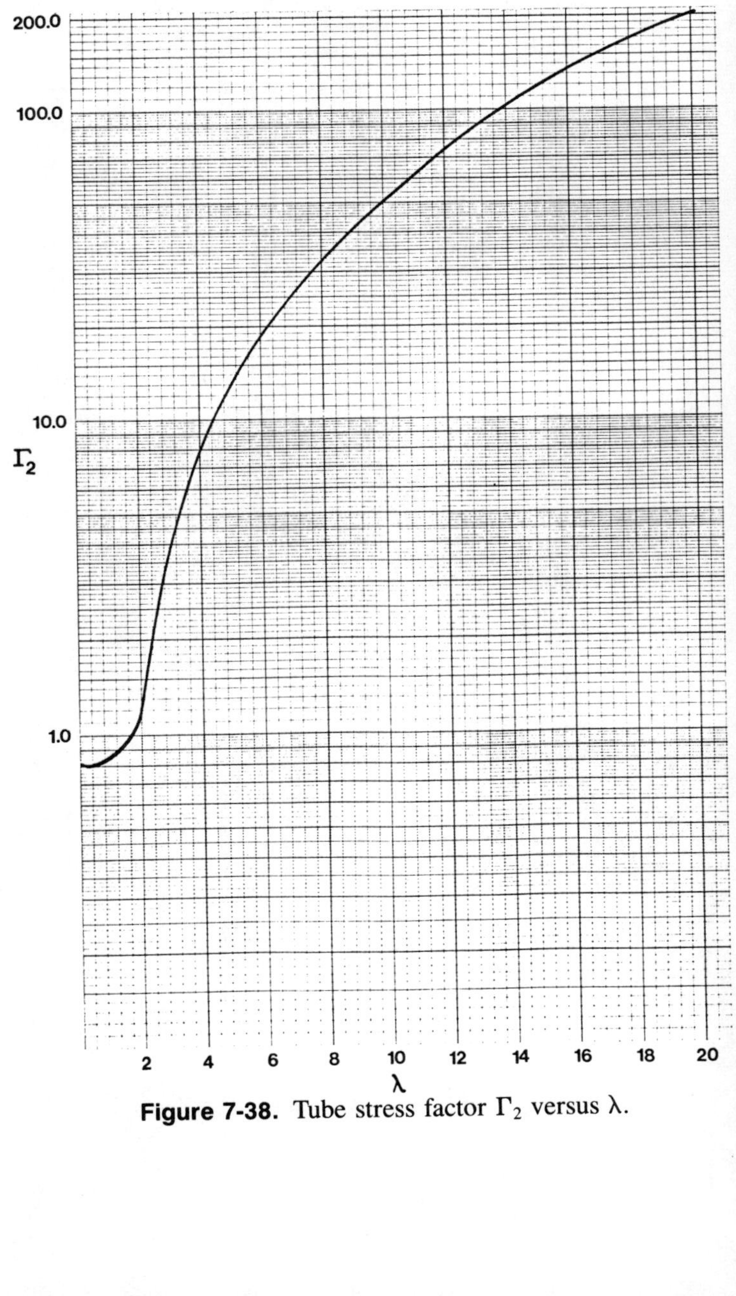

Figure 7-38. Tube stress factor Γ_2 versus λ.

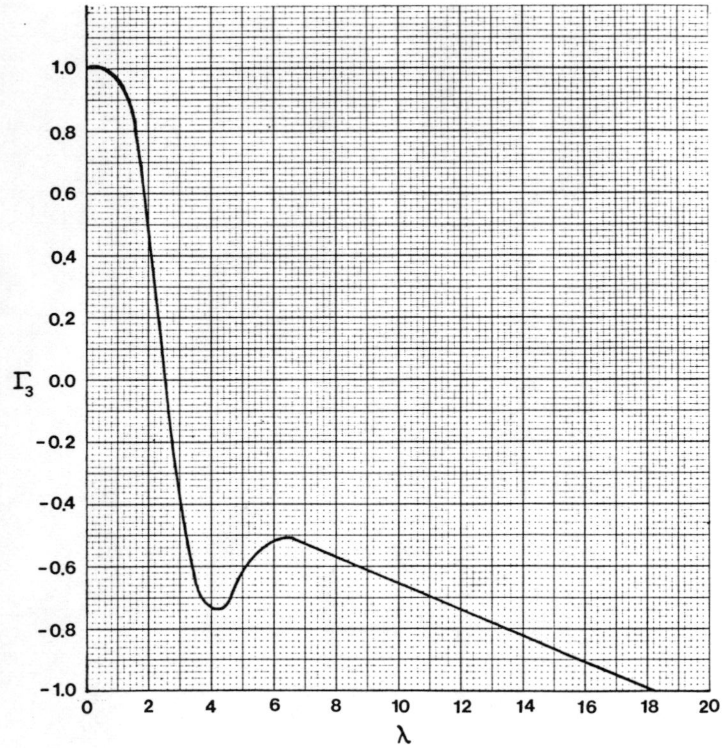

Figure 7-39. Tube stress factor Γ_3 versus λ.

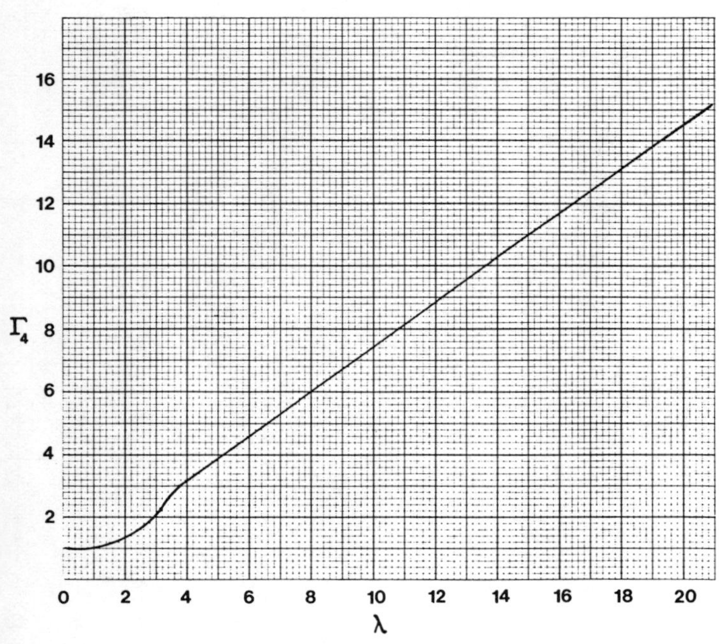

Figure 7-40. Tube stress factor Γ_4 versus λ.

The maximum stress in the tubes is the greater of the following:

$$\sigma_{r(max)} = \frac{A - C}{na_t}\left[\Delta P_E - \frac{\Gamma_3\left(P_E - \dfrac{AP_{ti}\psi}{A - C}\right)}{(\psi + \Gamma_4)}\right] \tag{7-50}$$

or

$$\sigma_{r(max)} = \frac{A - C}{na_t}\left[\Delta P_E - \frac{\Gamma_4\left(P_E - \dfrac{AP_{ti}\psi}{(A - C)}\right)}{(\psi + \Gamma_4)}\right] \tag{7-51}$$

in which,

$$\sigma_{r(max)} = \frac{(168.075)}{(35.850)}\left[-21.3298 - \frac{\Gamma\left[226.263 - \dfrac{(100)(254.469)(2.5058)}{168.075}\right]}{(2.5058 + 1.970)}\right]$$

$$\sigma_{r(max)} = -92.62 \text{ psi for Equation 7-51}$$

EXAMPLE 7-4: ASPHALT COATING MIX HEATER—A NON-NEWTONIAN FLUID APPLICATION

A roofing manufacturer needs a shell and tube heat exchanger to heat an asphalt coating mix from 425°F to 500°F to improve flow characteristics. The fluid to heat the asphalt coating mix is a leading manufacturer's hot oil heat transfer fluid. The asphalt coating mix is to be tube-side and the hot oil is to be shell-side. Determine the size of unit required with the design to be counterflow. The process is described in Example 3-6. The exchanger heat duty is to be 1,000,000 Btu/hr. See Figure 7-41 for complete exchanger specifications. First we compute the LMTD for a counterflow exchanger,

Shell-side (hot oil)	650°F in	550°F out
Tube-side (asphalt coating mix)	500°F out	425°F in
GTTD =	150°F	LTTD = 125°F

$$\text{LMTD} = \frac{\text{GTTD} - \text{LTTD}}{\ln\left(\dfrac{\text{GTTD}}{\text{LTTD}}\right)} = \frac{150 - 125}{\ln\left(\dfrac{150}{125}\right)} = 137.12°\text{F}$$

HEAT EXCHANGER SPECIFICATION SHEET

1				Job No.	
2	Customer			Reference No.	
3	Address			Proposal No.	
4	Plant Location			Date	Rev.
5	Service of Unit ASPHALT HEATER			Item No.	
6	Size	Type	(Hor/Vert)	Connected In	Parallel Series
7	Surf/Unit (Gross/Eff.)	Sq Ft; Shells/Unit		Surf/Shell (Gross/Eff.)	Sq Ft
8	**PERFORMANCE OF ONE UNIT**				
9	Fluid Allocation	Shell Side		Tube Side	
10	Fluid Name	HOT OIL		ASPHALT COATING MIX	
11	Fluid Quantity, Total Lb/Hr				
12	Vapor (In/Out)				
13	Liquid				
14	Steam				
15	Water				
16	Noncondensable				
17	Temperature (In/Out) °F	650	550	425	500
18	Specific Gravity @ 550°F	1.00	0.997	1.43	1.43
19	Viscosity, Liquid Cp	0.30	0.30	9.38	9.38
20	Molecular Weight, Vapor				
21	Molecular Weight, Noncondensable				
22	Specific Heat Btu/Lb °F	0.526	0.526	0.368	0.368
23	Thermal Conductivity Btu Ft/Hr Sq Ft °F				
24	Latent Heat Btu/Lb @ °F				
25	Inlet Pressure Psig				
26	Velocity Ft/S				
27	Pressure Drop, Allow./Calc. Psi	10 / 10		10 / 10	
28	Fouling Resistance (Min.)				
29	Heat Exchanged		Btu/Hr; MTD (Corrected)		°F
30	Transfer Rate, Service 1,000,000		Clean		Btu/Hr Sq Ft °F
31	**CONSTRUCTION OF ONE SHELL**			Sketch (Bundle/Nozzle Orientation)	
32		Shell Side	Tube Side		
33	Design/Test Pressure Psig	150 / 225	150 / 225		
34	Design Temperature °F				
35	No. Passes per Shell				
36	Corrosion Allowance In.				
37	Connections In				
38	Size & Out				
39	Rating Intermediate				
40	Tube No. 594 OD 3/4 In.; Thk (Min/Avg) 14 BWG In.; Length 20 Ft; Pitch 1.0 In.	30	60 (circled)	90	45
41	Tube Type		Material		
42	Shell 27 IN ID OD In.		Shell Cover		(Integ.) (Remov.)
43	Channel or Bonnet		Channel Cover		
44	Tubesheet-Stationary		Tubesheet-Floating		
45	Floating Head Cover		Impingement Protection		
46	Baffles-Cross	Type	% Cut (Diam/Area)	Spacing: c/c	Inlet In.
47	Baffles-Long		Seal Type		
48	Supports-Tube	U-Bend		Type	
49	Bypass Seal Arrangement		Tube-Tubesheet Joint		
50	Expansion Joint		Type		
51	ρv^2-Inlet Nozzle	Bundle Entrance		Bundle Exit	
52	Gaskets-Shell Side		Tube Side		
53	-Floating Head				
54	Code Requirements			TEMA Class	
55	Weight/Shell	Filled with Water		Bundle	Lb
56	Remarks				
57					
58					
59					
60					
61					

Figure 7-41. Asphalt heater heat exchanger specification sheet. (©1978 Tubular Exchanger Manufacturers Association.)

In a counterflow exchanger we must correct the LMTD. Using Figure 7-16 we have for a one-shell-pass, two-tube-pass,

$$P = \frac{500 - 425}{650 - 425} = 0.333; \; R = \frac{650 - 550}{500 - 425} = 1.333$$

From Figure 7-16, F = 0.93. Thus, the corrected LMTD becomes

$$LMTD = (0.93)(137.12) = 127.522°F$$

Tube-Side Film Coefficient

For asphalt coating mix at 450°F we have the following properties:

$$C_p = 0.368 \text{ Btu/lb}_m\text{-°F}; \; \rho = 89.232 \text{ lb}_m/\text{ft}^3; \; \mu = 938 \text{ cp} = 2{,}251.20 \text{ lb}_m/\text{ft-hr}$$

$$q = \dot{m}C_p\,(LMTD) = 1{,}000{,}000 \text{ Btu/hr}$$

$$\dot{m} = \frac{1{,}000{,}000 \dfrac{\text{Btu}}{\text{hr}}}{(0.368)\dfrac{\text{Btu}}{\text{lb}_m\text{-°F}}(127.522)°F} = 21{,}309.196 \text{ lb}_m/\text{hr}$$

$$Q = \frac{21{,}309.196 \dfrac{\text{lb}_m}{\text{hr}}\left(\dfrac{1 \text{ hr}}{3600 \text{ sec}}\right)}{89.232 \dfrac{\text{lb}_m}{\text{ft}^3}} = 0.066 \frac{\text{ft}^3}{\text{sec}}$$

We will try 594-3/4-in. tubes—14 BWG. Checking the tube wall thickness for internal pressure, 150 psig

$$t_{min} = \frac{P(ID)}{\sigma_{all}E - 0.6P}$$

where σ_{all} = maximum allowable stress for tube material, psi
E = tube weld joint efficiency = 1.0
P = internal pressure, psig
ID = tube ID, in.

$$t_{min} = \frac{(150)(0.584)}{(17{,}500)(1.0) - 0.6(150)} = 0.005 \text{ in.}$$

$$t_{actual} = 0.083 \text{ in.}$$

Flow velocity through each tube is

$$V = \frac{0.066 \dfrac{\text{ft}^3}{\text{sec}}}{(0.0019)\text{ft}^2(594)\text{tubes}} = 0.059 \text{ ft/sec}$$

To obtain the tube-side film coefficient we must obtain the Reynolds number. The asphalt base coating mix is a non-Newtonian fluid (see Chapter 1), so Equation 1-6 is not valid. So, to compute the Reynolds number, we must use Equation 1-7,

$$N_{Re} = \frac{D^nV^{2-n}\rho}{\gamma} \tag{1-7}$$

When working with non-Newtonian fluids, rheological data are necessary. The reader is encouraged to refer to Govier [22], but will often find that rheological data are not available in literature. In this situation a sample of the fluid must be sent to a testing lab. Do not attempt to approximate a non-Newtonian fluid with Newtonian equations and assumptions—the results can be a catastrophe. At the current state-of-the-art there are no simple answers for such complicated subjects such as non-Newtonian fluids.

Samples of our fluid were sent to a testing lab to have the properties evaluated. Some of these properties have already been given. The fluid is determined by the lab to be a Bingham fluid, in which the shear stress and velocity gradient of the fluid particles are linearly related. For a Bingham plastic, n in Equation 1-7 is

$$n = \frac{1 - 4x/3 + x^4/3}{1 - x^4}$$

where x = ratio of the fluid particle yield stress to the shear stress in the fluid particles at the tube wall

Lab tests reveal that x = 0.5 and γ = 8.0 for which

$$n = \frac{1 - \dfrac{4}{3}(0.5) + \dfrac{(0.5)^4}{3}}{1 - (0.5)^4} = 0.378$$

Now,

$$N_{Re} = \frac{(0.584)^{0.378}(0.059)^{2-n}\dfrac{\text{ft}}{\text{sec}}(89.232)\dfrac{\text{lb}_m}{\text{ft}^3}}{8.0} = 0.092$$

The film coefficient is determined from Figure 7-42, which is the Metzner-Reed-Reynolds number (Equation 1-7) versus friction factor, f. From this figure we obtain

$$f = 180$$

Now, we must compute the pressure drop through each tube to determine if a 3/4 in.—14 BWG tube is adequate.

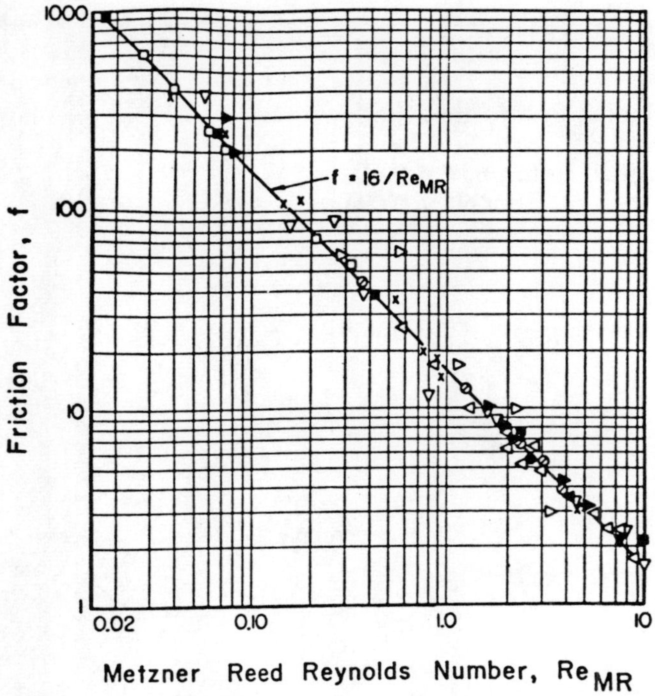

Figure 7-42. Friction factors for flow of non-Newtonian fluids [22].

For our velocity heads we use the entrance and exit loses and get

$$\sum K = 0.78 + 1.00 = 1.78 \text{ (see Figure 1-11)}$$

Using Equation 1-4 we compute the pressure drop over a 20-ft-long tube as

$$\Delta P_f = \left(\frac{fL}{d} + \sum K_i\right)\frac{\rho V^2}{2g_c} \tag{1-4}$$

$$\Delta P_f = \left[\frac{(180)(20)(12)\text{ in.}}{(0.584)\text{ in.}} + 1.78\right] \times \frac{(89.232)\,\dfrac{lb_m}{ft^3}\,(0.059)^2\,\dfrac{ft^2}{sec^2}\left(\dfrac{1\text{ ft}^2}{144\text{ in.}^2}\right)}{2(32.2)\,\dfrac{ft\text{-}lb_m}{sec^2\text{-}lb_f}}$$

$\Delta P_f = 2.478$ psi = Acceptable

Looking at this pressure drop one realizes that a flow velocity of 0.059 ft/sec is not so slow for a bulky fluid with a viscosity of almost 1,000 cp. The Prandtl number for our fluid is

$$N_{Pr} = \frac{\mu C_p}{k} = \frac{(2251.20)\,\dfrac{lb_m}{ft\text{-}hr}\,(0.368)\,\dfrac{Btu}{lb_m\text{-}°F}}{(0.10)\,\dfrac{Btu}{hr\text{-}ft\text{-}°F}}$$

$$N_{Pr} = 8284.416$$

For laminar flow, the Sieder-Tate correlation is

$$N_{Nu} = \frac{h_iD}{k} = 1.86\left[(N_{Re})(N_{Pr})\left(\frac{D}{L}\right)\right]^{1/3}\left(\frac{\mu}{\mu_w}\right)^{0.14}$$

$$N_{Nu} = 1.86\left[(0.092)(8284.416)\left(\frac{0.584}{(20)(12)}\right)\right]^{1/3}(1.0)$$

$$\frac{h_iD}{k} = 2.285$$

$$h_i = \frac{(2.285)(0.1)\,\dfrac{Btu}{hr\text{-}ft\text{-}°F}}{\left(\dfrac{0.584}{12}\right)ft}$$

$$= 4.695\,\frac{Btu}{hr\text{-}ft^2\text{-}°F}$$

Shell-Side Film Coefficient

$$q = \dot{m}C_p(LMTD)$$

For the hot oil at 600°F the following properties exist:

$$C_p = 0.526\text{ Btu/lb}_m\text{-°F};\ \rho = (0.997)(62.4) = 62.213\text{ lb}_m/\text{ft}^3$$

$$k = 0.076\text{ Btu/hr-ft-°F};\ \mu = 0.30\text{ cp} = 0.720\text{ lb}_m/\text{ft-hr}$$

$$\dot{m} = \frac{1{,}000{,}000\,\dfrac{Btu}{hr}\left(\dfrac{1\text{ hr}}{3{,}600\text{ sec}}\right)}{(0.526)\,\dfrac{Btu}{lb_m\text{-}°F}\,(127.522)°F} = 4.141\,\frac{lb_m}{sec}$$

$$Q = \frac{4.141\,\dfrac{lb_m}{sec}}{62.213\,\dfrac{lb_m}{ft^3}} = 0.067\,\frac{ft^3}{sec}$$

The smallest nozzle shell-side is a 3-in. nozzle, making the maximum shell-side velocity

$$V = \frac{0.067 \frac{ft^3}{sec}}{0.051\ ft^2}$$
$$= 1.305\ ft/sec \rightarrow \text{Very reasonable}$$

$$N_{Pr} = \frac{\mu C_p}{k} = \frac{(0.720)\frac{lb_m}{ft\text{-}hr}(0.526)\frac{Btu}{lb_m\text{-}°F}}{(0.076)\frac{Btu}{hr\text{-}ft\text{-}°F}} = 4.983$$

For a 60° Δ arrangement,

$$D_e = \frac{8[0.43\ p^2 - 0.5\pi d_o^2/4]}{\pi d_o} = \frac{8[0.43(1.00) - 0.5\pi(1.0)/4]}{\pi(0.75)}$$

$D_e = 0.127$ or $D_e = 0.011$ ft

c= 1.00 − 0.75 = 0.25 in.

B = 15.0 in. for 16 equally spaced baffles over 20 ft

$D_s = 27.00$ = shell ID

$$a_s = \frac{D_s(c)B}{144p} = \frac{(27.00)(0.25)(15.0)}{(1.00)(144)} = 0.703\ ft^2$$

$$G_s = \frac{\dot{m}}{a_s} = \frac{4.141 \frac{lb_m}{sec}(3{,}600)\frac{sec}{hr}}{0.703\ ft^2}$$
$$= 21{,}201.920 \frac{lb_m}{hr\text{-}ft^2}$$

$$N_{Re} = \frac{D_e G_s}{\mu} = \frac{(0.011)\ ft\ (21{,}201.920)\frac{lb_m}{hr\text{-}ft^2}}{(0.720)\frac{lb_m}{ft\text{-}hr}} = 323.918$$

From Figure 7-21 $j_H = 12$ for baffles with 15% cut

$$h_o = \frac{j_H K}{D_e}(N_{Pr})^{1/3}\left(\frac{\mu}{\mu_w}\right)^{0.14}$$

From laboratory tests it was determined that μ/μ_w = 2.0.

$$h_o = \frac{(12)(0.076)\frac{Btu\ ft}{hr\text{-}ft^2\text{-}°F}}{(0.011)\ ft}(4.983)^{1/3}(2.0)^{0.14}$$

$$h_o = 155.959 \frac{Btu}{hr\text{-}ft^2\text{-}°F}$$

Fouling factors are as follows:

Asphalt coating mix = 0.01
Hot oil = 0.004

$$U = \frac{1}{\frac{1}{155.959} + 0.004 + 0.01 + \frac{1}{4.695}}$$

$$U = 4.284 \frac{Btu}{hr\text{-}ft^2\text{-}°F}$$

$$\text{Area required} = \frac{1{,}000{,}000 \frac{Btu}{hr}}{(4.284)\frac{Btu}{hr\text{-}ft^2\text{-}°F}(127.522)°F}$$
$$= 1{,}830.308\ ft^2$$

$$\text{Available area} = (0.1963)\frac{ft^2}{ft}(20)\ ft\ (594) = 2{,}332.044\ ft^2$$

Twenty-seven percent of the excess area can be eliminated by reducing the number of tubes. This would increase the flow rate in each tube and thus the pressure drop, which already is at 2.5 psi. For non-Newtonian fluids, properties can vary from sample to sample and extra margin is needed, so 25% to 30% excess area is not unreasonable. For more heat exchange it would be better to consider a surge tank with interior and exterior heating elements, since we are at the limits of the shell and tube design and, with a more viscous fluid, a surge tank of the type in Examples 3-3 and 3-4 is more practical.

Shell-Side Pressure Drop

$$\Delta P_s = \frac{f_s G_s^2 D_s (N_B + 1)}{(5.22)(10)^{10} D_e \gamma \phi}$$

N_B = 16 baffles

D_s = shell ID = 27 in. = 2.25 ft

$G_s = 21{,}201.92\ lb_m/hr\text{-}ft^2$

N_{Re} = 324 and from Figure 7-24, f = 0.0075

For plain and bare tubes,

$$f_s = \frac{f}{1.2} = \frac{0.0075}{1.2} = 0.00625$$

$$D_e = 0.011 \text{ ft}$$

$$\gamma = \text{specific gravity} = 0.997$$

$$\phi = 1.0 = \left(\frac{\mu}{\mu_w}\right)^{0.14}$$

$$\Delta P_s = \frac{(0.00625)(21{,}201.92)^2(2.25)(17)}{(5.22)(10)^{10}(0.011)(0.997)(1.0)}$$

$$= 0.188 \text{ psi, which is acceptable}$$

EXAMPLE 7-5: ZERO LMTD EXCHANGER

A candy manufacturer wishes to cool hot molasses to 140°F for the food processing of various confectionaries. The molasses is coming from a heating-blend kettle at 200°F. Spring water is to be used and it never varies (± 1/4°F) from 60°F. The water is to be heated to 120°F, and held at that temperature to heat honey. Determine the LMTD. The exchanger is a counterflow design.

Tube-side	200°F in	140°F out
Shell-side	120°F out	60°F in
	$\Delta t =$ 80°F	$\Delta t =$ 80°F

Now using Equation 3-23 we have

$$\text{LMTD} = \frac{80 - 80}{\ln\left(\frac{80}{80}\right)} = \frac{0}{0} = \text{undefined}$$

This problem is somewhat similar to that of Example 3-4 in its formulation. We must define the LMTD as the ratio of two functions T(t) and L(t) for which

$$\text{LMTD} = \frac{T(t)}{L(t)} = \frac{(200 - t) - (140 - 60)}{\ln\left(\frac{200 - t}{80}\right)}$$

As temperature t approaches a certain value such that T(t) and L(t) become zero being divided by zero. The derivatives of T(t) and L(t) exist when t approaches this value of t, so we can apply l'Hospital's rule that if T(t) = L(t) = 0 when t = a, and if the ratio of dT(t)/dt to dL(t)/dt exists, then

$$T(t) = (200 - t) - (140 - 60)$$

$$\frac{dT(t)}{dt} = -1$$

$$L(t) = \ln\left(\frac{200 - t}{80}\right)$$

$$\frac{dL(t)}{dt} = \left(\frac{80}{200 - t}\right)\left(\frac{-1}{80}\right) = \frac{-1}{200 - t}$$

l'Hospital's rule states that

$$\lim_{t \to a} \frac{T(t)}{L(t)} = \lim_{t \to a} \frac{dT(t)/dt}{dL(t)/dt}$$

or, with a = 120°F,

$$\lim_{t \to a} \left[\frac{-1}{\frac{-1}{200 - t}}\right] = \lim_{t \to 120} (200 - t) = 80°\text{F}$$

Therefore, LMTD = 80°F

With this value of LMTD, the exchanger can be designed, using the correction factor in the case of a counterflow unit.

NOTATION

- A = tube surface area, ft^2
- A_t = cross-sectional area of tube, in.^{2}
- a = constant for a continuous beam shear, dimensionless
- b = constant for a continuous beam deflection
- c = tube clearance, in.
- c = constant, in.^{2}/lb$_f$ (Equation 7-37)
- C = constant = $9.7 \times 10^{-4}(\text{sec})^{0.5}/(\text{ft})^{1.5}$ (Equation 7-41)
- C_c = $(2\pi^2 E_t/\sigma_y)^{0.5}$ (Equation 7-2)
- C_D = drag or force coefficient for a body immersed in a fluid, dimensionless
- C_p = specific heat at constant pressure, Btu/lb$_m$-°F
- D = 4 × hydraulic radius, in.
- D = tube diameter, in.
- D = parameter (Equation 7-27)
- d_B = diameter of baffle hole, in.
- d_i = inside tube diameter, in.

d_o = outside tube diameter, in.
d_t = outside tube diameter, ft
E_B = modulus of elasticity of baffle material, psi
E_t = modulus of elasticity of tube material, psi
F_c = correction factor, dimensionless (Figure 7-16)
F_{cr} = critical buckling strength for tubes, lb_r
F_L = force induced by fluid flowing around immersed body, lb_f
F_s = shear force against tube at baffle, lb_f
f_e, f_r, f_y = constants used in determining tube joint force, lb_f (Equations 7-3 and 7-4)
f_n = fundamental natural frequency of tube, Hz
g_c = gravitational constant = 32.2 lb_m-ft/lb_f-sec^2
GTTD = greatest temperature difference between the shell and tube side fluids, °F
h = film coefficient, Btu/hr-ft^2-°F
h_i = film coefficient inside tube, Btu/hr-ft^2-°F
h_o = film coefficient outside tube, Btu/hr-ft^2-°F
h_{io} = outside film coefficient of tube, using outside tube surfaces temperature, Btu/hr-ft^2-°F
I = moment of inertia, in.^{4}
I_t = moment of inertia of tube cross section, in.^{4}
$k\ell$ = structural constant, dimensionless (Equation 7-2)
k = equivalent effective unsupported length of the tube, in.
k_w = coefficient of thermal conductivity of tube wall, Btu/hr-ft-°F
k_f = thermal conductivity of fluid, Btu
k_{fi} = thermal conductivity of foreign deposits inside of tube, Btu/hr-ft-°F
k_{fo} = thermal conductivity of deposits on outside of tube, Btu/hr-ft-°F
L = tube length or span length of tube, ft
LMTD = logarithmic mean temperature difference, °F
LTTD = lesser temperature difference between shell and tube-side fluids, °F
ℓ = typical dimension of body immersed in fluid, ft
$\dot{m}$ = mass flow rate, lb_m/sec
m_t = mass density of tube metal, slugs/ft^3
NB = number of baffles
N_{BE} = baffle damage number, dimensionless
N_{CD} = critical damage number, dimensionless (Equation 7-39)
N_{Nu} = Nusselt number, dimensionless
N_{Pr} = Prandtl number, dimensionless
N_{Re} = Reynolds number, dimensionless
P = axial force, lb_f
p = tube pitch, in.
q = rate of heat transfer, Btu/hr
r = radius of gyration of tube, in. (Equation 7-2)
T = parameter (Figures 7-30 and 7-31)
T_{fi} = thickness of inside tube deposits, ft
T_{fo} = thickness of outside tube deposits, ft
T_w = tube wall thickness, ft
t_{cc} = caloric temperature of cold fluid, °F
t_{ch} = caloric temperature of hot fluid, °F
t_{ci} = inlet cold fluid temperature, °F
t_{co} = caloric temperature of cold fluid, °F
t_{hi} = inlet hot fluid temperature, °F
t_{ho} = outlet hot fluid temperature, °F
t_t = tube wall thickness, in.
t_w = outside tube wall temperature, °F
Δt = temperature differential ($t_1 - t_2$), °F
U = overall heat transfer coefficient for exchanger, Btu/hr-ft^2-°F
U_x = the value of the overall heat transfer coefficient at the caloric temperature, Btu/hr-ft^2-°F
V = flow velocity, ft/sec

Greek Terms

α = factor of effective tube resistant area, dimensionless
δ = deflection or displacements, in.
μ = dynamic viscosity of the fluid inside tube, lb_m/ft-hr
μ_w = dynamic viscosity of fluid at tube wall, lb_m/ft-hr
υ_B = Poisson ratio for baffle material
υ_t = Poisson ratio for tube material
ω = frequency of a given mode, Hz
ρ = density, lb_m/ft^3
σ_{all} = allowable stress for tube, psi
σ_c = allowable tube compressive stress, psi, for the tubes at the outer periphery of tube bundle (Equations 7-1 and 7-2)
σ_y = minimum yield stress of tube material at design temperature, psi
ζ = sum of structural damping and the fluid damping, dimensionless

REFERENCES

1. *Heat Exchangers,* Howell Training Company, Houston, Texas, 1975.
2. *Standards of the Tubular Exchanger Manufacturers Association* (TEMA), 6th Edition, Tarrytown, New York, 1978.
3. Rubin, F. L., "What's the Difference Between TEMA Exchanger Classes," *Hydrocarbon Processing,* 59, June p. 92, 1980.
4. Ludwig, E. E., *Applied Process Design for Chemical and Petrochemical Plants,* Volume 3, Second

Edition, Gulf Publishing Company, Houston, Texas, 1983.

5. Small, W. M. and R. K. Young, "The Rodbaffle Heat Exchanger," *Heat Trans. Eng.*, 1, no. 2, Oct.-Dec. (1979), p. 21.
6. Skrotzki, B. G. A., "Heat Exchangers," *Power,* June, 1954.
7. ASME Boiler and Pressure Vessel Code, Section VIII Division 1, American Society of Mechanical Engineers, New York.
8. Colburn, A. P., *Ind. Eng. Chem.*, 35, pp. 873–877, 1933.
9. Kern, Donald Q., *Process Heat Transfer,* McGraw-Hill Book Company, New York, 1950.
10. McAdams, W. H., *Heat Transmission,* Third Edition, McGraw-Hill Book Company, New York, 1954.
11. Jakob, M. *Heat Transfer,* Vol. 1, John Wiley & Sons, New York, 1959.
12. Grimson, E. D., "Correlation and Utilization of New Data on Flow Resistance and Heat Transfer for Crossflow Over Tube Banks," *"Transactions of the ASME,"* Vol. 59, pp. 583–584, 1937.
13. *Engineering Data Book,* Wolverine Division of UOP, Inc., A Signal Company, 1959.
14. Thorngren, John T., "Predict Exchanger Tube Damage," *Hydrocarbon Processing,* Vol. 49, no. 4, p. 129, 1970.
15. American Institute of Steel Construction, *Manual of Steel Construction,* Eighth Edition, AISC, Chicago, Illinois, 1980.
16. Timoshenko, S., and J. N. Goodier, *Theory of Elasticity,* Second Edition, Engineering Societies Monograph, McGraw-Hill Book Company, 1951.
17. Coit, R. L., C. C. Reak, and A. Lohmeier, "Design and Manufacture of Large Surface Condensers—Problems and Solutions," American Power Conference, April 1965.
18. Blevins, R. D., *Flow-Induced Vibration,* Van Nostrand Rheinhold Company, New York, 1977.
19. Wambsganss, M. W., and S. S. Chen, "Tentative Design Guide for Calculating the Vibration Response of Flexible Cylindrical Elements in Axial Flow," Argonne National Laboratory Report ANL-ETD-71-07, 1971.
20. Kays, William M. and A. L. London, *Compact Heat Exchangers,* Third Edition, McGraw-Hill Book Company, New York, 1984.
21. Miller, K. A. G., "The Design of Tube Plates in Heat Exchangers," *Proceedings of the Institution of Mechanical Engineers,* Vol. 1B, pp. 215–231.
22. Govier, G. W. and K. Aziz, *The Flow of Complex Mixtures in Pipes,* Robert E. Krieger Publishing Company, New York, 1977.
23. Metzner, A. B. and J. C. Reed, *AIChE Journal,* 1, p. 434, 1955.

8

External Loadings on Shell Structures

In a book about the mechanical design of process systems it is impossible to ignore the phenomenon of external loadings on shell structures. Such loadings occur when piping is flanged to pressure vessels and the vessel nozzle is exposed to loads induced by the piping, and when vessels are erected and the force of gravity induces loads at the lifting lugs.

We have already discussed external loadings in the design of piping supports in Chapter 2. Vessels require a similar analysis, but the phenomenon is different because in a vessel the loadings are more localized, particularly in a large vessel. In the case of external loadings on vessel nozzles one must consider primary stresses induced by internal pressure and secondary stresses induced by the external loadings. In the design of the lifting lugs only secondary stresses need to be considered, since vessels being lifted almost never have internal pressure.

The two "standards" that are most widely accepted for external loadings on pressure vessel nozzles are the WRC (Welding Research Council) Bulletin 107 [1] and the WRC 297 [2]. The latter is an expanded version with more curves to cover more cases, but it is only for cylindrical shells. Neither WRC 107 nor WRC 297 are considered standards per se. Therefore, one must take the results of the methods outlined here and add the primary stress, which is the internal pressure stress.

The reader is cautioned that the WRC 297 Bulletin is under evaluation at the time of this writing. Shell theory was used to develop the WRC 297, and the results are being compared to finite element studies currently being made. The reader is especially cautioned to use the Bulletin when the ratio of the diameter of the branch to the diameter of the header is between 0.5 and 1.0, expressed mathematically as

$$0.5 < d_b/D_H < 1.0$$

where d_b = diameter of the branch
D_H = diameter of the header

Also, WRC 297 and WRC 107 do not consider the case of external loading combined with internal pressure. Current studies are being made to accomplish this task.

Stress induced by internal pressure at the nozzle-shell intersection are extremely complex, so an analytical solution is impractical. Discontinuity stresses at the nozzle-shell juncture are caused by the change in geometry from the nozzle shell into the vessel shell. Consequently, a stress concentration factor, k_p, must be applied when using the following expression for internal pressure stress:

$$\sigma_p = \frac{P(ID)k_p}{2t} \tag{8-1}$$

where P = internal pressure, psi
ID = inside diameter of shell, in.
t = shell thickness, in.
k_p = internal pressure stress concentration factor, dimensionless

Values of k_p are far too exhaustive to be listed here, but are available in a work by Forman [3].

For many years reinforcing pads have been used for external loadings and it has been accepted practice to assume that such pads remove discontinuity stresses at the nozzle-shell juncture. While this is true, one must realize that the reinforcement decreases the flexibility of the nozzle-shell attachment. As shown in Figure 8-1a, the nozzle with the reinforcement will have maximum membrane stresses occurring at the nozzle-shell juncture (assuming the circumferential bending stresses are negligible compared with the membrane stresses). As Figure 8-1b shows that as the reinforcement thickness increases,

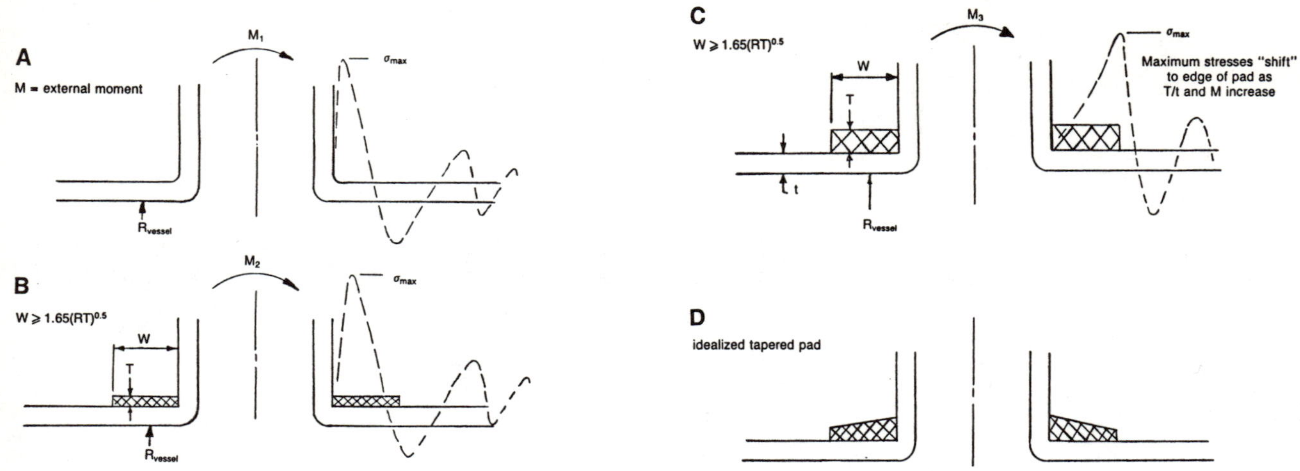

Figure 8-1. Simple schematic of maximum combined stress distribution, as supported by field tests and finite element studies.

the maximum stress shifts towards the edge of the pad, and as the ratio of the reinforcement pad to the shell thickness approaches a "critical value," the maximum stress induced by external loading occurs at the reinforcement edge-shell juncture point, shown in Figure 8-1c. Considering this it would intuitively appear that a tapered pad would ideally be the best in application, especially for thick pads (pad thickness relative to shell thickness), as shown in Figure 8-1d. The disadvantage of such a pad would be the increased difficulty and expense to fabricate such a pad. Analytical, finite element studies, and field experience bear the previous facts out. The width of a pad, from the nozzle edge to the pad edge, should not exceed $1.65\sqrt{RT}$. Beyond this range a pad has been shown to be ineffective.

Pads can be even dangerous on thin-walled shells. In many instances, adding a 1/2-in. pad to a nozzle on a thin-walled pipe, such as Schedule 55 (0.083 in. on a 4-in. pipe), is prohibitive. Such a pad could very easily transfer the maximum loading to the pad edge as shown in Figure 8-1c, resulting in crack propagation or even rupture. Caution should be taken in working with thin-walled shells, where the flexibility of the shell is often sufficient to decrease induced stresses from external loadings.

LIFTING LUG DESIGN

The design of lifting lugs can become an arduous task if one is not familiar with the erection of equipment. Lifting lugs must be designed to withstand the stresses induced from all the loading conditions; allow lifting and setting the equipment in one operation without readjusting or re-rigging the crane or other equipment, and protect equipment and personnel. The lugs must not interfere with vessel components, such as platforms, ladders, or piping.

The advantage to lifting lug design is that only secondary stresses must be considered—primary stress, such as internal pressure stress, can be ignored. We can assume that the vessels are not lifted while they are pressurized. Consequently, the AISC *Manual of Steel Construction* [4] can be used in which the factor of safety is 2:1 (unlike ASME's 4:1).

The vessel is to be considered as a simply supported horizontal beam. All non-shell components, head, ladders, etc. are considered as concentrated loads. The total erection weight is the sum of the concentrated loads and the distributed loads of the shell weight and internals.

Various types of lifting lugs are shown in Figure 8-2. Lifting and erection procedures are shown in Figure 8-3. Techniques for designing the lugs are given in the following examples.

EXAMPLE 8-1: LIFTING LUG DESIGN AND LOCATION

A 96-in. ID shell and tube heat exchanger is to be lifted from a dock onto an offshore structure. The exchanger weighs 158,750 lbs, which is the total erection weight. The objective is to locate and design the lifting lugs, and determine the minimum chocker length and maximum chocker angle.

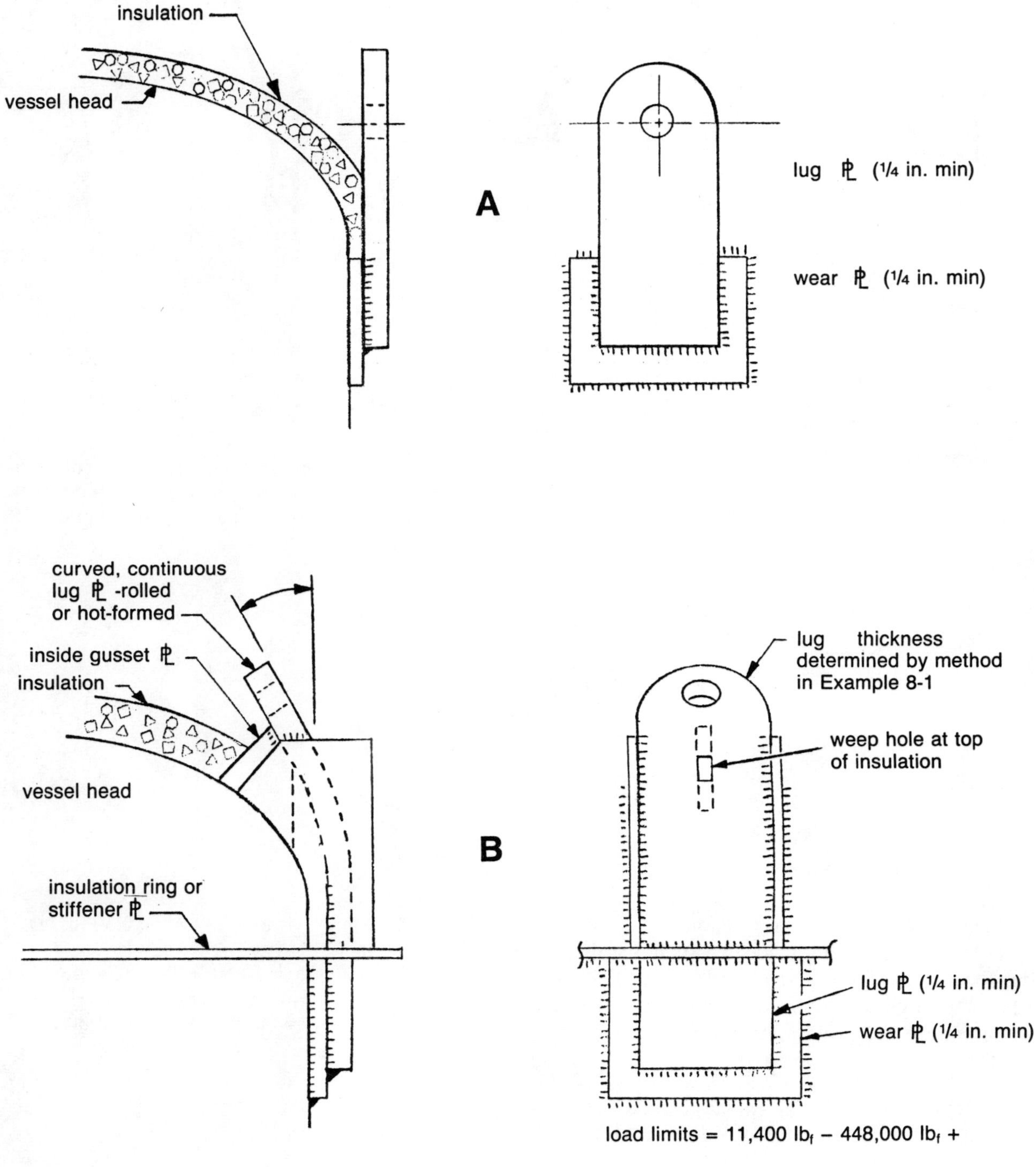

Figure 8-2. Standard lifting lug designs: (A) This design has the advantage of simplicity, but can only handle vessels less than 200,000 lb_f and 100 ft tall and may require a spreader bar if lift moments are excessive; (B) This design seldom requires a spreader bar and can lift any size vessel—however, it's a more complex design.

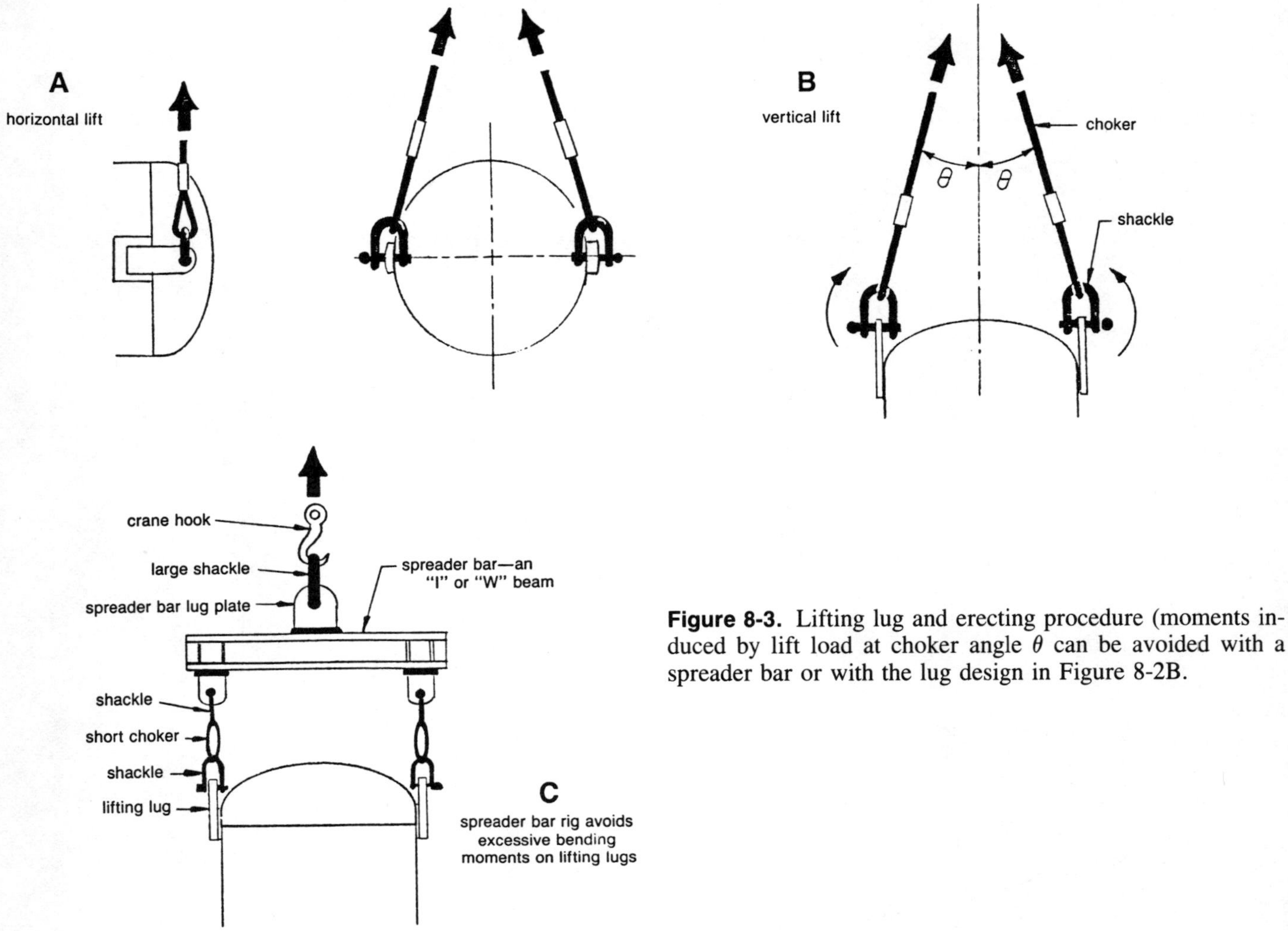

Figure 8-3. Lifting lug and erecting procedure (moments induced by lift load at choker angle θ can be avoided with a spreader bar or with the lug design in Figure 8-2B.

First, we construct a free body diagram, as shown in Figure 8-4. Each lifting lug is located such that the point of lift is located on a hypothetical vertical line that passes close to or through the centroid of the ellipsoidal head, shown in Figure 8-5. Summing moments to zero and solving for the reactions we have

$$\overset{+}{\circlearrowleft}\sum M_L = 0 = -R_R(46.542) + (346)(44.000) + (2{,}283)(40.75) + (346)(2.542) - (2{,}094)(46.75) + (151{,}587)(23.271)$$

$$R_R = 75{,}888.874 \text{ lb}$$

and

$$R_L = 75{,}698.126 \text{ lb}$$

For lug supporting the full vessel weighing 158,750 lb, referring to Table 8-1 we write

A = 16.50 in., B = 6.50 in., C = 4.50 in., D = 4 in., E = 6.50 in.

Hole diameter = a = (4.50 + 0.125) = 4.625 in. = minimum

Lug width = W_L = 3a = 3(4.50) = 13.50 in. = minimum

Lug Thickness, t_L

$$t_L = \text{Larger of} \begin{cases} \dfrac{W_L}{8} = \dfrac{13.50}{8} = 1.688 \text{ in. ✓ use 1.75 in.} \\[2ex] \dfrac{W}{1.6a\sigma_y} = \dfrac{158{,}750}{(1.6)(4.625)(38{,}000)} = 0.565 \text{ in.} \end{cases}$$

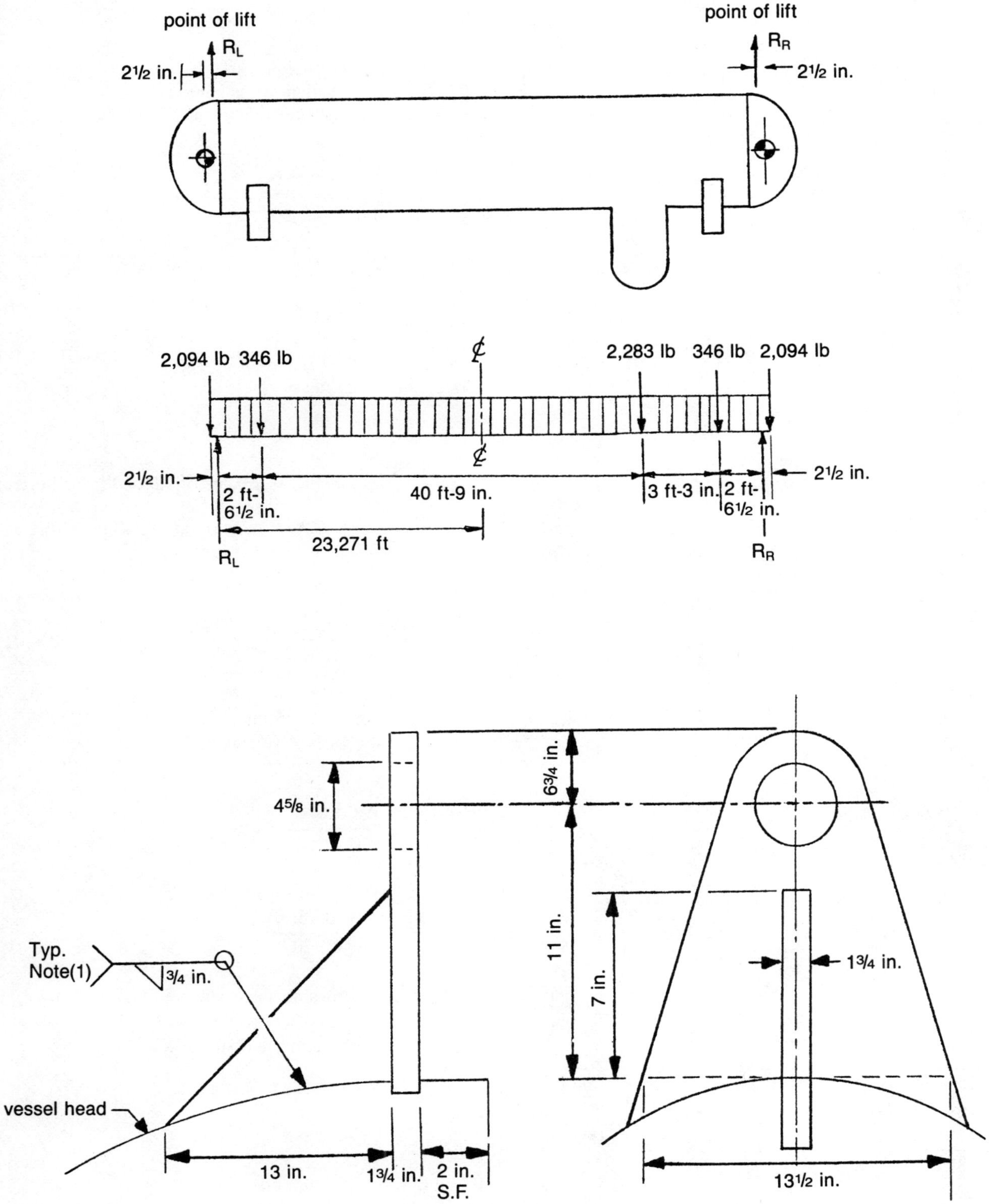

Figure 8-5. Detail of vessel lifting lug.

Table 8-1
Anchor Shackles

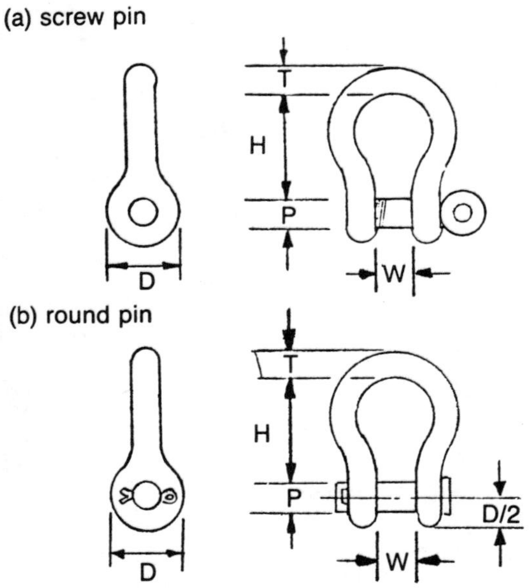

T (in.)	H (in.)	W (in.)	Pin Dia. P. (in.)	D (in.)	Safe Lifting Load lb_f
1/4	1 1/8	1/2	5/16	11/16	780
5/16	1 1/4	7/16	3/8	13/16	1,100
3/8	1 1/2	5/8	7/16	1	1,675
7/16	1 3/4	3/4	1/2	1 1/16	2,200
1/2	1 7/8	13/16	5/8	1 5/16	2,900
5/8	2 7/16	1 1/16	3/4	1 9/16	4,500
3/4	2 7/8	1 1/4	7/8	1 7/16	6,400
7/8	3 1/4	1 7/16	1	2 1/8	8,700
1	3 3/4	1 11/16	1 1/8	2 3/8	11,400
1 1/8	4 1/4	1 13/16	1 1/4	2 5/8	13,500
1 1/4	4 3/4	2 1/16	1 3/8	3	16,500
1 3/8	5 1/4	2 1/8	1 1/2	3 5/16	20,500
1 1/2	5 3/4	2 1/4	1 5/8	3 5/8	24,000
1 3/4	7	2 7/8	2	4 5/16	33,600
2	7 3/4	3 1/4	2 1/4	5	44,800
2 1/4	9 1/4	3 3/4	2 1/2	5 1/4	56,000
2 1/2	10 1/2	4 1/8	2 3/4	6	67,200
2 3/4	11 1/2	4 1/2	3	6 1/4	81,000
3	13	5	3 1/4	6 1/2	100,800
3 1/4	15	5 3/4	3 3/4	6 3/4	125,000
3 1/2	15 1/2	6	4	7	150,000
3 3/4	16 1/2	6 1/4	4 1/4	7 1/4	179,200
4	17 1/2	6 1/2	4 1/2	7 1/2	200,000
4 1/8	18 1/2	6 11/16	4 3/4	7 3/4	224,000
4 1/2	20	7	5	8	313,600
5 1/2	21	8	5 1/2	9	448,000

For lug material SA-516-GR 70, $\sigma_y = 38{,}000$ psi

Lug Height (assume 2 in. fireproofing)

$$H = \frac{E}{2} + \left(\frac{R_o}{2} + t_i\right)\left[1 - \left(\frac{R_o + t_L - B - D}{R + t_i}\right)^2\right]^{0.5}$$

where, H = lug height, in.
t_i = insulation thickness
R_o = shell outside radius, in.

$$H = \frac{6.50}{2} + \left(\frac{50}{2} + 2\right)$$

$$\times \left[1.0 - \left(\frac{50.0 + 1.75 - 6.50 - 4.00}{50.00 + 2.00}\right)^2\right]^{0.5}$$

$$H = 19.690 \text{ in.}$$

Check lug thickness

$t_L \geqslant \dfrac{5HR}{\sigma_y W^2}$, where R = greater of R_R or R_L for horizontal vessel
R = reaction at lug when lifting at skirt and lug end

$$t_L = \frac{5(19.690) \text{ in. } (75{,}888.874) \text{ lb}}{(38{,}000) \dfrac{\text{lb}}{\text{in.}^2} (13.50)^2 \text{ in.}}$$

$= 1.079$ in. < 1.75 in.
Lug thickness is sufficient

Minimum Weld Size

$$t_w = \frac{R[0.47 + 0.45(h/w)]}{\tau_A(w_L)(0.707)}$$

where τ_A = allowable shear stress in weld $= 0.3\sigma_{tw}$

σ_{tw} = weld minimum yield stress in tension

$$t_w = \frac{(75888.874)\left[0.47 + 0.45\left(\dfrac{19.690}{13.500}\right)\right]}{(0.30)(70{,}000)(13.50)(0.707)}$$

$t_w = 0.426$ in. minimum

Actual weld size $= t_{wa}$

$$t_{wa} = \text{Larger of} \begin{cases} t_v - 1/16 \text{ in.} \\ t_L - 1/16 \text{ in.} \end{cases}$$

and $t_{wa} > t_w$

where t_v = vessel thickness, in.

In this case, $t_L > t_v$, so that

$$t_{wa} = 1.75 \text{ in.} - 0.0625 = 1.688$$

For each side of weld

$$t_{wa} = \frac{1.688}{2} = 0.844$$

since $t_{wa} >> t_w$, A 3/4-in. weld is sufficient

Choker Angle (θ)

$$\theta = \arctan\left[\frac{\sigma_y W_1 t_L^2}{3W\left(H + A + \frac{c}{2}\right)}\right]$$

$$\theta = \arctan\left[\frac{(38,000)(13.50)(1.75)^2}{3(1,58750.00)\left(19.690 + 16.50 + \frac{4.50}{2}\right)}\right]$$

$$\theta = 4.905°$$

$$L_c = \frac{R_o}{12 \sin\theta} = \text{minimum choker length}$$

$$L_c = \frac{50.00}{12 \sin(4.905)} = 48.731 \text{ ft}$$

Because of height restrictions, the lug had to be lowered from 19.690 in. to 11.00 in. Thus, we now have the following:

$$\theta = \arctan\left[\frac{(38,000)(13.50)(1.75)^2}{3(158,750.00)\left(11.00 + 16.50 + \frac{4.50}{2}\right)}\right]$$

$$\theta = 6.327°$$

and

$$L_c = \frac{50}{12 \sin(6.327)} = 37.807 \text{ ft}$$

NOTATION

A,B,C,D,E = constants (Figure 8-6)
D_H = header diameter, in.
d_b = branch diameter, in.
H = constant (Table 8-1)
K_P = internal pressure stress concentration factor, dimensionless
L_c = minimum chocker length, ft
M_L = moment resolved about the left end (Figure 8-4), ft-lb
M_R = moment resolved about the right end (Figure 8-4), ft-lb

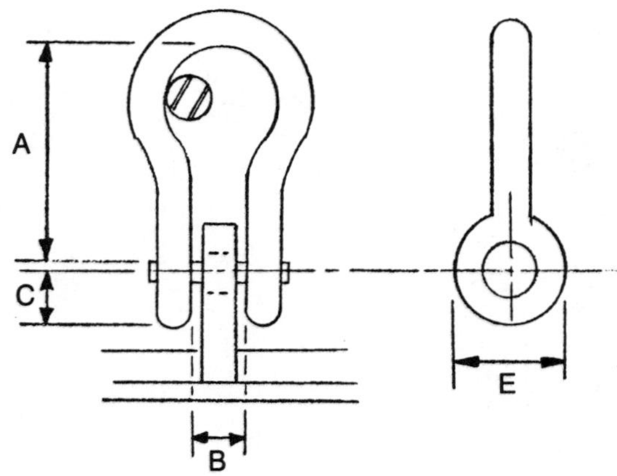

A = 16½ in., B = 6½ in., C = 4½ in., D = 4 in., E = 6½ in.

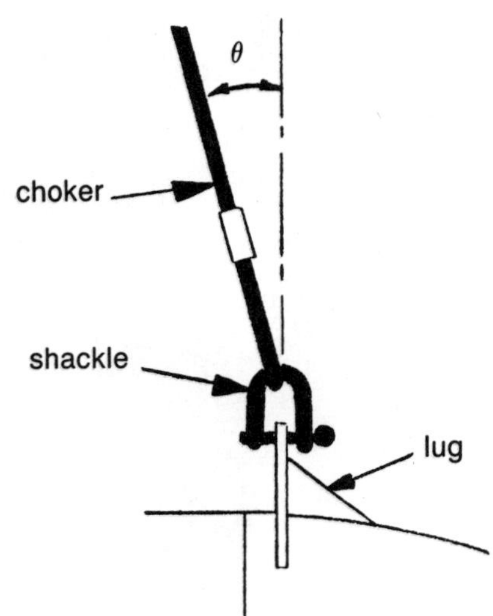

Figure 8-6. Detail of choker and shackle.

P = constant (Table 8-1)
R_L = reaction at left side (Figure 8-4), ft-lb
R_o = shell outside radius, in.
R_R = reaction at right side (Figure 8-4), ft-lb
t = shell thickness, in.
t_L = lug thickness, in.
t_w = weld size, in.
W_L = lug width, in.

Greek Symbols

σ_{tw} = minimum weld yield stress in tension, psi
τ_A = allowable shear stress in weld, psi
θ = chocker angle, degrees

REFERENCES

1. Welding Research Council, *Welding Research Council Bulletin WRC 107 Local Stresses in Spherical and Cylindrical Shells Due to External Loadings,* March, New York, 1979.
2. Welding Research Council, *Welding Research Council Bulletin WRC 297, Local Stresses in Cylindrical Due to External Loadings on Nozzles—Supplement to WRC Bulletin No. 107,* New York, August, 1984.
3. Forman, B. Fred, *Local Stresses in Pressure Vessels,* Second Edition, Pressure Vessel Handbook Publishing, Inc. Tulsa, OK., 1979.
4. American Institute of Steel Construction, *Manual of Steel Construction,* Eighth Edition, AISC, Chicago, Illinois, 1980.

Appendix A

Partial Volumes and Pressure Vessel Calculations

PARTIAL VOLUME OF A CYLINDER

$$V_P = \frac{R_i^2 L}{2}\left(\frac{\pi\alpha^\circ}{180^\circ} - \sin\alpha\right) = \text{partial volume shown in shaded region} \qquad \text{(A-1)}$$

L = length of cylinder

R = inside radius of cylinder

Example (Figure A-1)

For a cylinder with 144-in. ID find the partial volume of a fluid head of 60 in., if L = 100 ft:

$$\frac{\alpha}{2} = 80.41^\circ$$

$$V_P = \frac{(72)^2(1200)}{2}\left[\frac{\pi(160.81)}{180} - \sin\,(160.81^\circ)\right]$$

$$V_P = 7{,}707{,}650.2 \text{ in.}^3 = 33{,}366.5 \text{ gal}$$

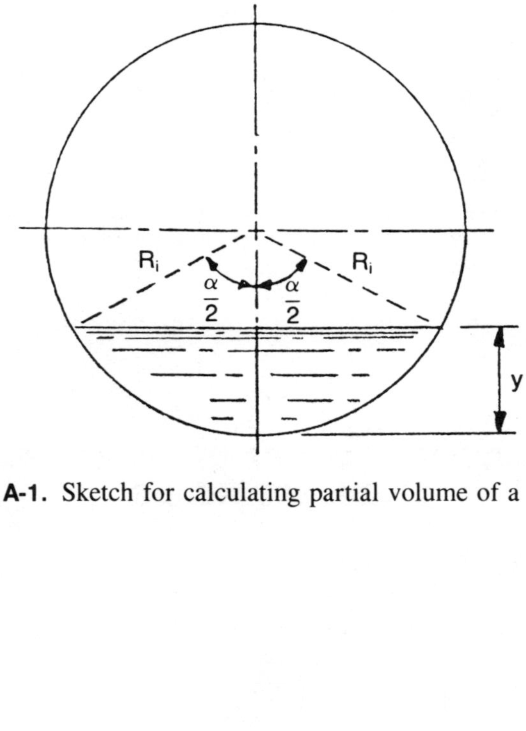

Figure A-1. Sketch for calculating partial volume of a cylinder.

PARTIAL VOLUME OF A HEMISPHERICAL HEAD

$$V_P = \frac{\pi y^2(3R_i - D)}{3} \qquad \text{(A-2)}$$

V_P = partial volume shown in shaded region

Example

For vertical volume in Figure A-2a find partial volume for a head with R_i = 50 in. and y = 35 in.:

$$V_P = \frac{\pi(35)^2[3(50) - 100]}{3} = 64{,}140.85 \text{ in.}^3 = 277.7 \text{ gal}$$

Example

For horizontal volume in Figure A-2b find partial volume for a head with R_i = 50 in. and y = 35 in.:

$$V_P = \frac{277.7}{2} = 138.85 \text{ gal}$$

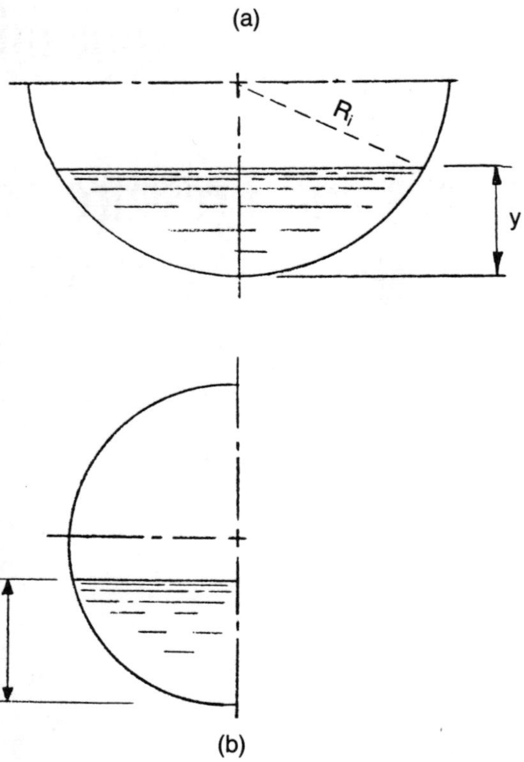

Figure A-2. Partial volume of vertical hemispherical head. (B) Partial volume of horizontal hemispherical head.

PARTIAL VOLUMES OF SPHERICALLY DISHED HEADS

Horizontal Head

The partial volume of a horizontal head (Figure A-3) is

$$V = \alpha \left| \frac{\sqrt{(\rho^2 - y^2)^3} - \sqrt{(\rho^2 - R_i^2)^3}}{3} - \frac{L(R_i^2 - y^2)}{2} \right| \qquad \text{(A-3)}$$

Vertical Head

The partial volume of a vertical head (Figure A-4) is

$$V = \frac{\pi y(3x^2 + y^2)}{6} \qquad \text{(A-4)}$$

or

$$V = \frac{\pi y^2(3\rho - y)}{3} \qquad \text{(A-5)}$$

Example—Spherically Dished Horizontal Head

A spherically dished head with a 114-in. ϕ OD is spun from 1-in. plate. Determine the partial volume of 10 in. of liquid. From vessel head manufacturer's catalog we determine the following:

IDD = 16.786 in. (Figure A-5), ρ = 108 in.

$$R_i = \frac{114 - 2(1.0)}{2} = 56.0 \text{ in.}$$

$$\alpha = 159.43° = 2.78$$

$$L = 108 - 16.786 = 91.21 \text{ in.}$$

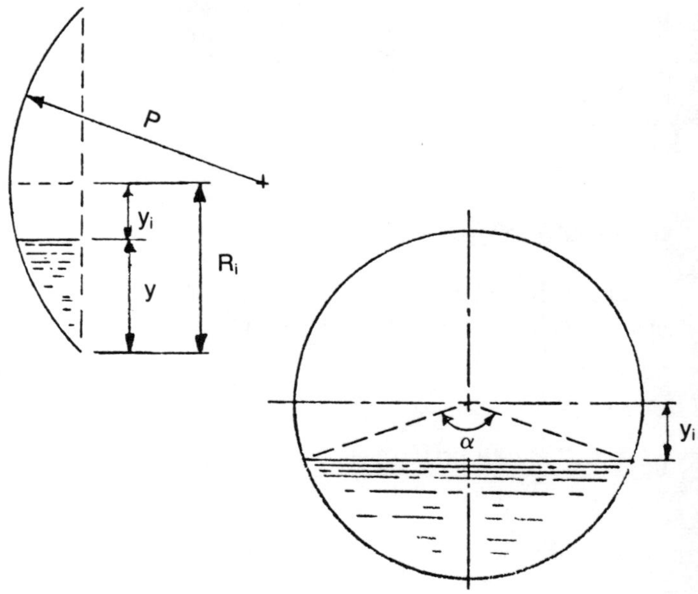

Figure A-3. Partial volume of spherically dished horizontal heads.

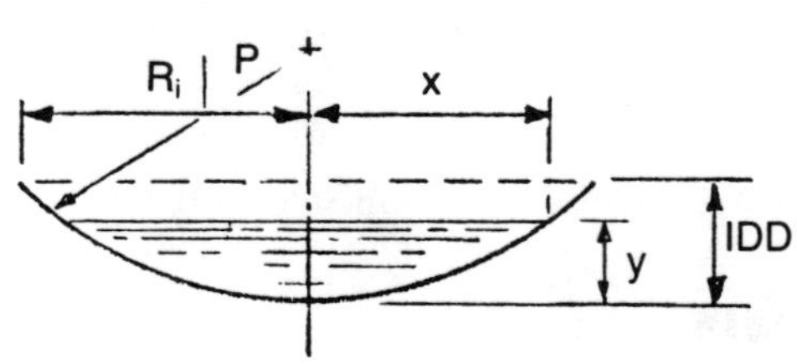

Figure A-4. Partial volume of spherically dished vertical heads.

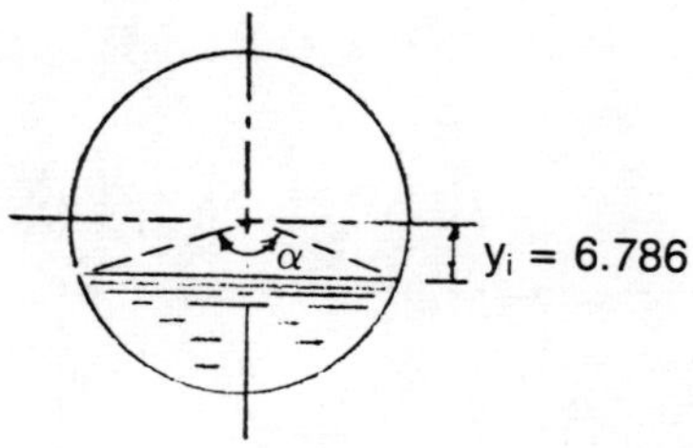

Figure A-5.

$$V = (2.78)\left|\frac{\sqrt{(108^2 - 6.786^2)^3} - \sqrt{(108^2 - 56^2)^3}}{3} - \frac{(91.21)(56^2 - 6.786^2)}{2}\right|$$

$V = 38{,}893.21 \text{ in.}^3 = 168.37 \text{ gal}$

Example— Spherically Dished Vertical Head

For the same head above, determine the partial volume of a head of liquid of 9 in.

$x = 55.456$ in.

$$V = \frac{\pi(9)[3(55.456)^2 + 9^2]}{6} = 14{,}874 \text{ in.}^3 = 64.4 \text{ gal}$$

PARTIAL VOLUMES OF ELLIPTICAL HEADS

The exact partial volume of a horizontal elliptical head (Figure A-6) is as follows:

$$V = \frac{(IDD)\alpha}{6R_i}\sqrt{(R_i^2 - y_i^2)^3} \qquad (A\text{-}6)$$

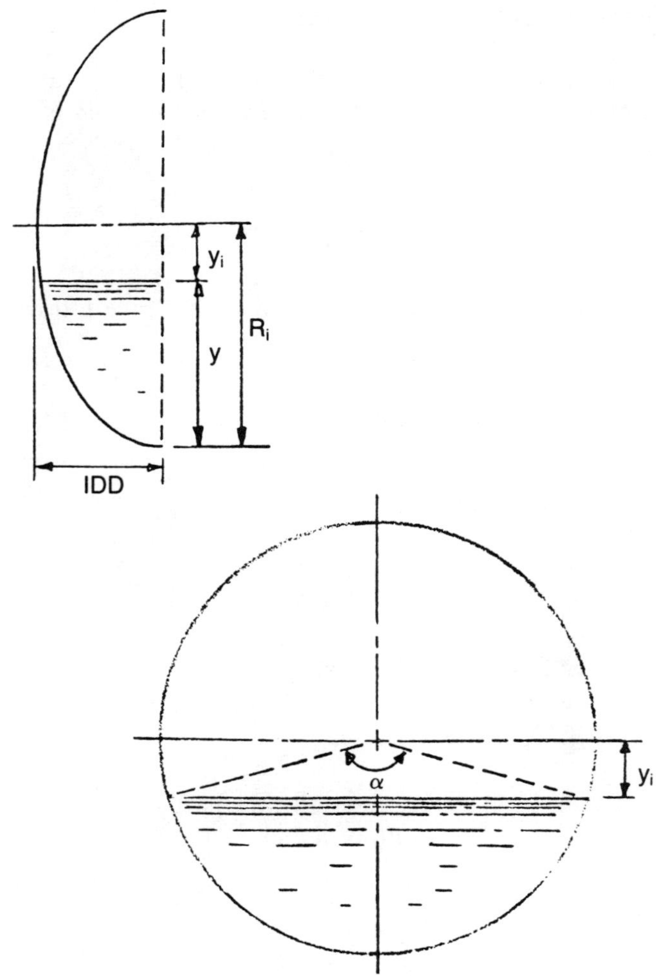

End View of Horizontal Head

Figure A-6. Partial volume of horizontal elliptical head.

Vertical Elliptical Heads

Volume of top portion ② of Figure A-7 is

$$V_{②} = \frac{\pi R_i^2}{2}\left[y - \frac{y^3}{3(IDD)^2}\right] \qquad (A\text{-}7)$$

Volume of bottom portion ① is

$$V_{①} = \frac{2\pi(IDD)R_i^2}{6} - \frac{\pi R_i^2}{2}\left[y - \frac{y^3}{3(IDD)^2}\right] \qquad (A\text{-}8)$$

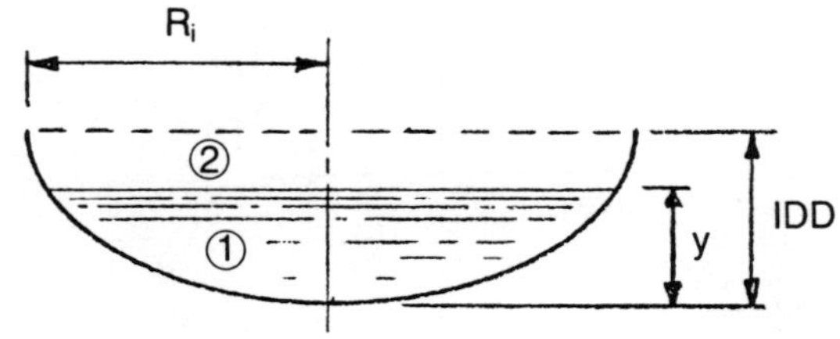

Figure A-7. Partial volume of vertical elliptical head.

Horizontal Head Example

Find the partial volume of a 2:1 (R_i/IDD = 2) elliptical head that is 108-in. OD. The level of the liquid is 35 in., and the head is spun from 1-in. plate.

$$\text{IDD} = \frac{108 - 2(1.0)}{4} = 26.50 \text{ in.}$$

From Equation A-6 and Figure A-8 we have the following:

$$V = \frac{(\text{IDD})\ \alpha}{6R_i} \sqrt{(R_i^2 - y_i^2)^3}$$

$$\alpha = 138.80° = 2.42$$

$$V = \frac{(19.0)(2.42)}{6(53)} \sqrt{((53)^2 - (19)^2)^3}$$

$$V = 17{,}512.94 \text{ in.}^3 = 75.81 \text{ gal}$$

Vertical Head Example

For some head above, determine the partial volume for a vertical head with 19 in. of liquid. Using Equation A-8 we have the following:

$$V = \frac{2\pi(\text{IDD})R_i^2}{6} - \frac{\pi R_i}{2}\left[y_i - \frac{y_i^3}{3(\text{IDD})^2}\right]$$

$$V = \frac{2\pi(26.50)(53.0)^2}{6} - \frac{\pi(53.0)}{2}\left[19.0 - \frac{(19.0)^3}{3(26.50)^2}\right]$$

$$V = 77{,}951.81 \text{ in.}^3 - 1310.75 \text{ in.}^3$$

$$V = 76{,}641.06 \text{ in.}^3 = 331.78 \text{ gal}$$

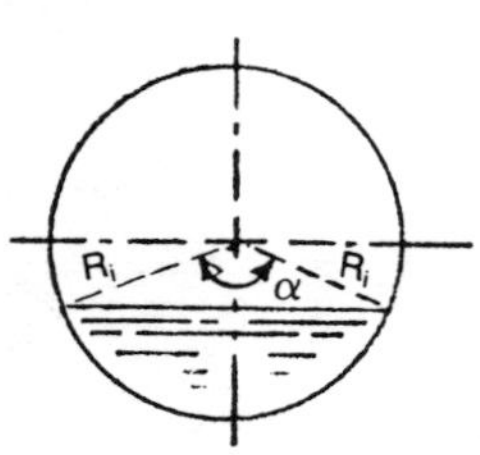

Figure A-8.

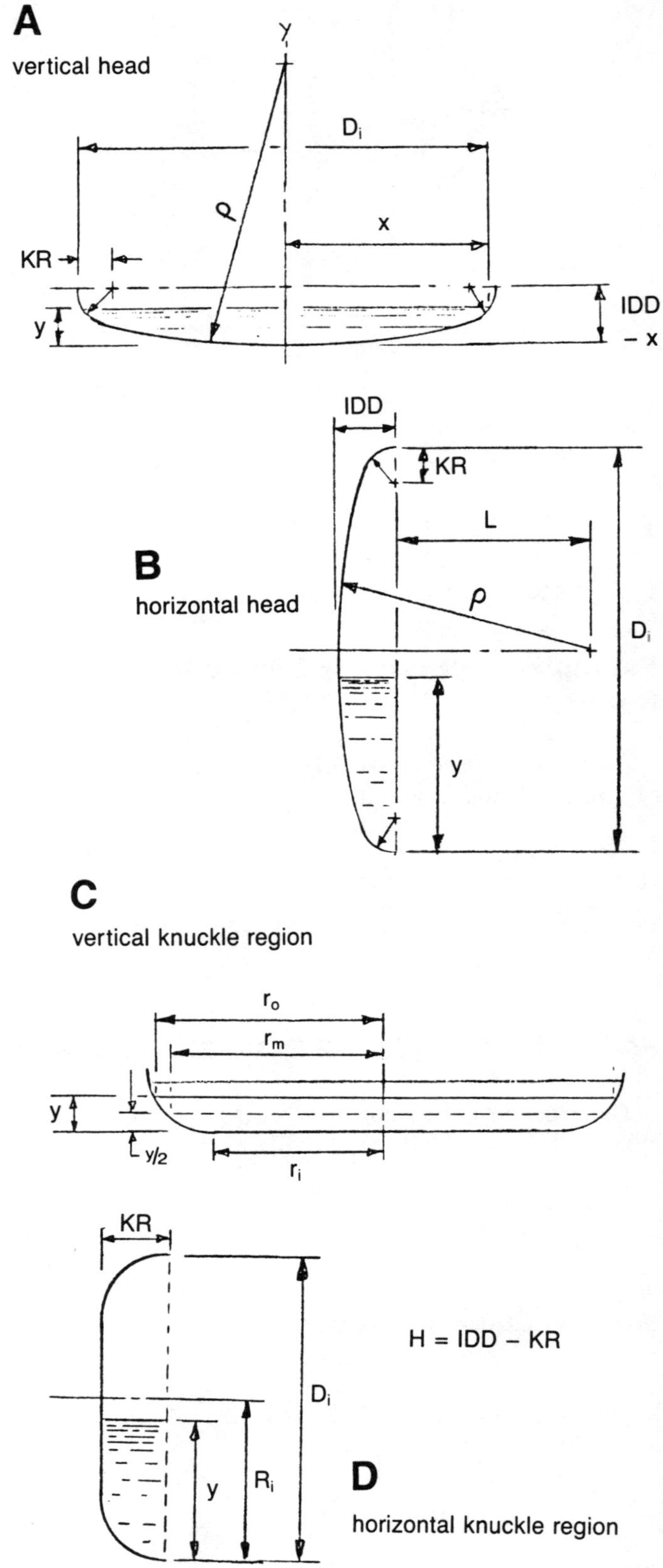

Figure A-9. Partial volumes of torispherical heads: (A) vertical, (B) horizontal, (C) vertical knuckle region, (D) horizontal knuckle region.

PARTIAL VOLUMES OF TORISPHERICAL HEADS

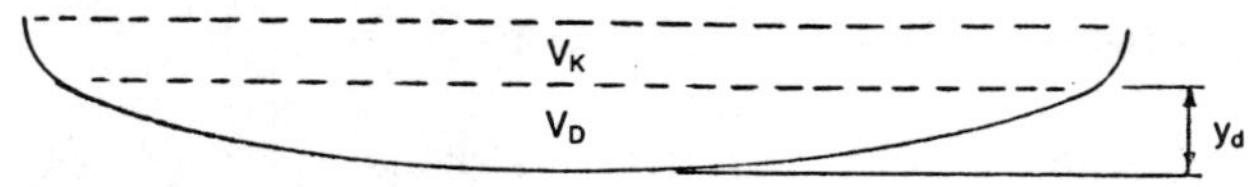

Figure A-10.

For Figures A-9 and A-10,

V_k = knuckle volume
V_D = dish volume
KR = knuckle radius
y = height of liquid
IDD = inside depth of dish
ρ = inside dish radius

For vertical heads (Figure A-9c) the knuckle-cylinder partial volume is

$$V_k = \frac{\pi y}{6}(r_o^2 + 4r_M^2 + r_i^2) \tag{A-9}$$

The partial volume of the dish region of a vertical head is

$$V_D = \frac{\pi y(3x^2 + y^2)}{6} \tag{A-10}$$

The total partial volume in a vertical head is

$$V_V = \frac{\pi H}{6}(r_o^2 + 4r_M^2 + r_i^2) + \frac{\pi y(3x^2 + y^2)}{6} \tag{A-11}$$

where y_d = IDD − KR

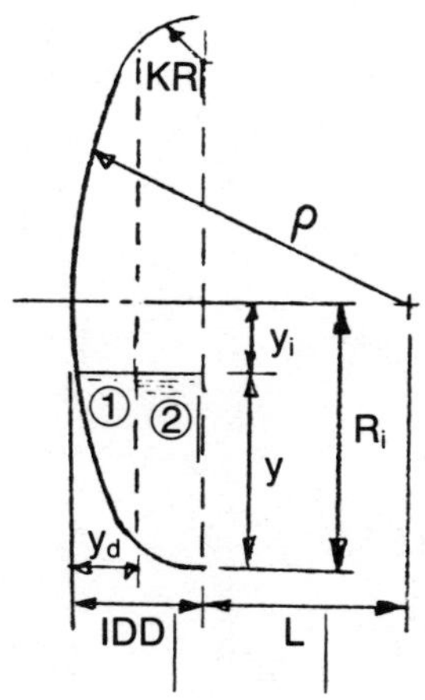

Horizontal Torispherical Heads

Partial Volume of Dish ① (Figure A-11)

$$V_{①} = \alpha\left|\frac{\sqrt{(\rho^2 - y_i^2)^3} - \sqrt{(\rho^2 - R_i^2)^3}}{3} - \frac{L(R_i^2 - y_i^2)}{2}\right| \tag{A-12}$$

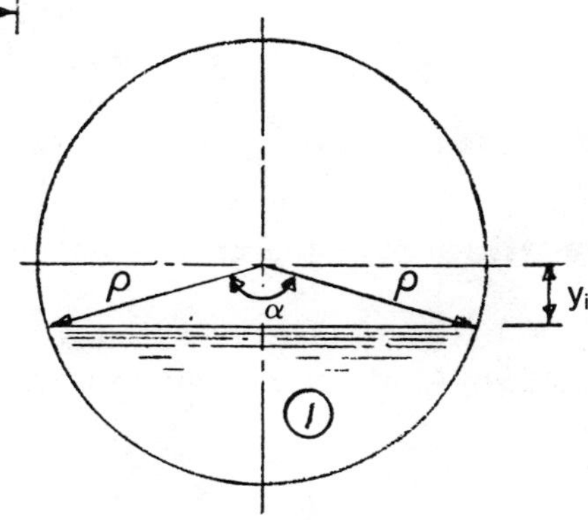

Figure A-11. Sketch for example partial volume calculation of horizontal torispherical head.

Volume of Knuck-Cylinder Region ② (Figure A-12)

$$V_{②} = \alpha\left[\frac{4(KR)}{3\pi} + (R_i - KR) + (R_i - KR)^2\right] \tag{A-13}$$

The total partial volume for a horizontal torispherical head is as follows:

$$V_T = V_{①} + V_{②}$$

$$V_T = \alpha\left|\frac{\sqrt{(\rho^2 - y_i^2)^3} - \sqrt{(\rho^2 - R_i^2)^3}}{3} - \frac{L(R_i^2 - y_i^2)}{2}\right|$$

$$+ \alpha\left[\frac{4(KR)}{3\pi} + (R_i - KR) + (R_i - KR)^2\right] \tag{A-14}$$

where L = ρ − IDD

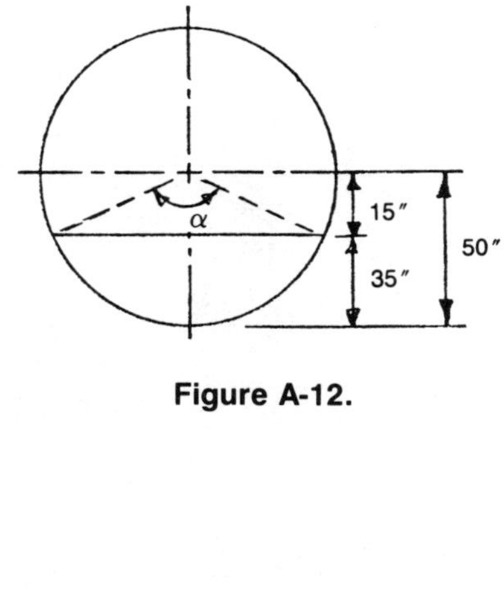

Figure A-12.

Horizontal Head Example

A 102-in. ϕ OD flanged and dished (torispherical) head made to ASME specifications (KR $\geqslant 0.60\rho$ and KR $> 3t_h$, t_h = head thickness) is spun from 1-in. plate. The head is horizontal and the liquid level is 35-in. determine the partial volume.

From the vessel head manufacturer's catalog and Figure A-12 we determine the following:

ρ = 96 in., KR = 6.125 in., IDD = 17.562 in.

$$R_i = \frac{100}{2} = 50 \text{ in., } L = 96.0 - 17.562 = 78.438 \text{ in.}$$

From Equation A-14 we have

$$V_T = (2.532)\left|\frac{\sqrt{(96^2 - 15^2)^3} - \sqrt{(96^2 - 50^2)^3}}{3} - \frac{(78.438)(50^2 - 15^2)}{2}\right| + (2.532)\left[\frac{4(6.125)}{3\pi} + (50.00 - 6.125) + (50.00 - 6.125)^2\right]$$

$$V_T = 34{,}093.44 \text{ in.}^3 = 147.59 \text{ gal}$$

Vertical Head Example

A 138-in. ϕ OD F&D (flanged and dished) head *not* made to ASME specifications is spun from 1½-in. plate. The head is vertical and the liquid level is 18-in. Determine the partial volume.

From the vessel head manufacturer's catalog we determine the following:

ρ = 132 in., KR = 3 in., IDD = 20.283 in.

$$R_i = \frac{138 - 2(1.5)}{2} = 67.50 \text{ in.;}$$

$$x = 67.50 - [(3)^2 - H^2]^{0.5} = 66.446 \text{ in.}$$

For knuckle-cylinder region,

$r_o = R_i = 67.50$; $r_i \cong R_i - KR = 67.50 - 3.00 = 64.50$ in.

$$r_m = \frac{67.50 + 64.50}{2} = 66.0;$$

$$h = |20.283 - (3.0 + 15.0)| = 2.283 \text{ in.}$$

$$V_V = \frac{\pi(2.283)}{6}[(67.50)^2 + 4(66.0)^2 + (64.50)^2] + \frac{\pi(17.283)[3(64.500)^2 + (17.283)^2]}{6}$$

$$V_V = 31{,}247.726 \text{ in.}^3 + 115{,}645.832 \text{ in.}^3$$

$$V_V = 146{,}893.558 \text{ in.}^3 = 635.903 \text{ gal}$$

INTERNAL PRESSURE ASME FORMULATIONS WITH OUTSIDE DIMENSIONS

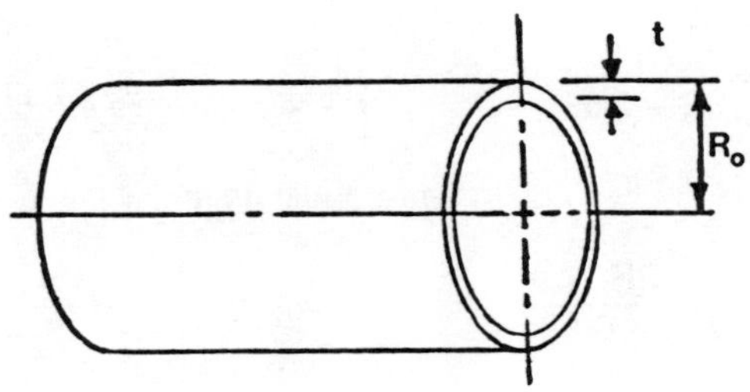

Cylindrical Shell
Longitudinal Joint

$$t = \frac{PR}{\sigma E + 0.4P} \qquad P = \frac{\sigma Et}{R - 0.4t}$$

Circumferential Joint

$$t = \frac{PR_o}{2\sigma E + 1.4P} \qquad P = \frac{2\sigma Et}{R_o - 1.4t}$$

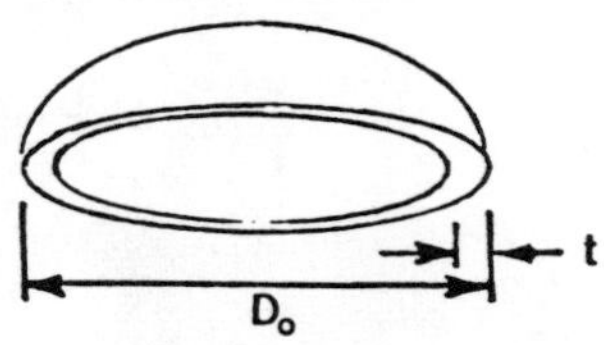

2:1 Ellipsoidal Head

$$t = \frac{PD_o}{2\sigma E + 1.8P} \qquad P = \frac{2\sigma Et}{D_o - 1.8t}$$

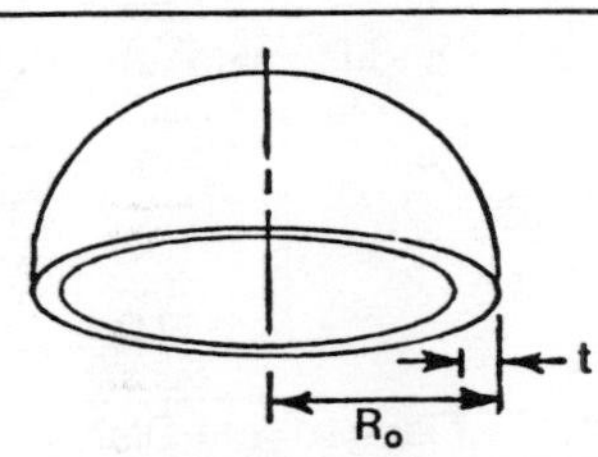

Sphere and Hemispherical Head

$$t = \frac{PR_o}{2\sigma E + 0.8P} \qquad P = \frac{2\sigma Et}{R_o - 0.8t}$$

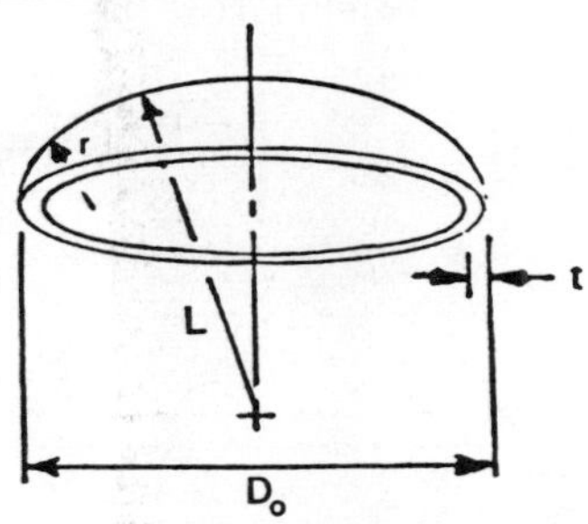

ASME Flanged and Dished Head
when L/R = 16⅔

$$t = \frac{0.885PL}{\sigma E + 0.8P} \qquad P = \frac{\sigma Et}{0.885L - 0.8t}$$

When L/R < 16⅔

$$t = \frac{PLM}{2\sigma E + P(M - 0.2)} \qquad P = \frac{2\sigma Et}{ML - t(M - 0.2)}$$

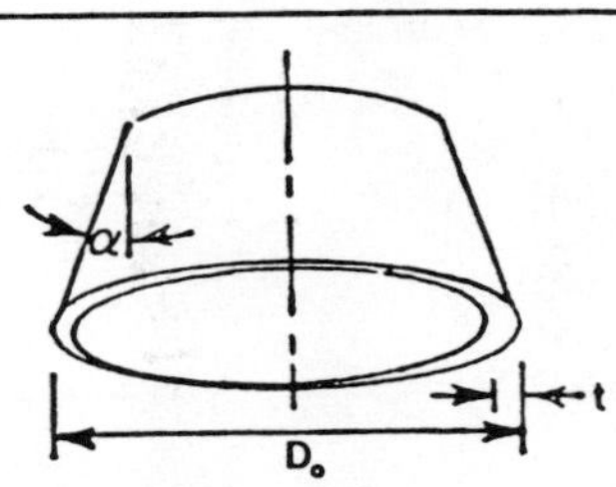

Conical Section

$$t = \frac{PD_o}{2 \cos \alpha(\sigma E + 0.4P)} \qquad P = \frac{2SEt \cos \alpha}{D_o - 0.8t \cos \alpha}$$

INTERNAL PRESSURE ASME FORMULATIONS WITH INSIDE DIMENSIONS

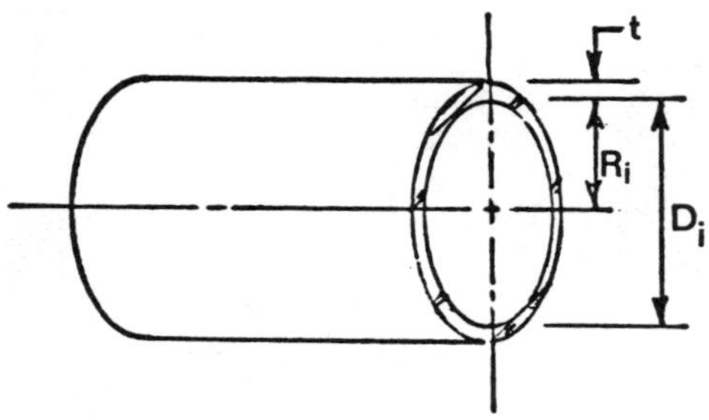

Cylindrical Shell
Longitudinal Joint

$$t = \frac{PR_i}{\sigma E - 0.6P} \qquad P = \frac{\sigma Et}{R_i + 0.6t}$$

Circumferential Joint

$$t = \frac{PR_i}{2\sigma E + 0.4P} \qquad P = \frac{2\sigma Et}{R_i - 0.4t}$$

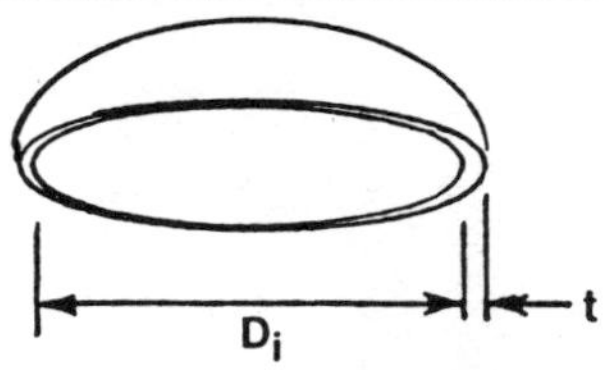

2:1 Ellipsoidal Head

$$t = \frac{PD_i}{2\sigma E - 0.2P} \qquad P = \frac{2\sigma Et}{D_i + 0.2t}$$

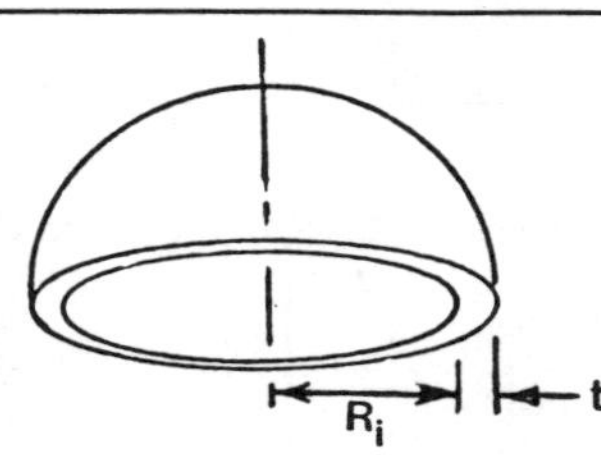

Sphere and Hemispherical Head

$$t = \frac{PR_i}{2SE - 0.2P} \qquad P = \frac{2\sigma Et}{R + 0.2t}$$

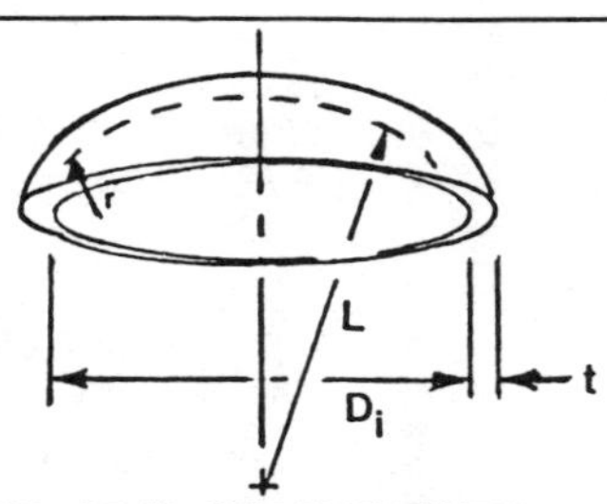

FOR VALUES OF M SEE SUPPLEMENT

ASME Flanged and Dished Head
when L/R = 16⅔

$$t = \frac{0.885PL}{\sigma E - 0.1P} \qquad P = \frac{\sigma Et}{0.885L + 0.1t}$$

When L/R < 16⅔

$$t = \frac{PLM}{2\sigma E - 0.2P} \qquad P = \frac{2\sigma Et}{LM + 0.2t}$$

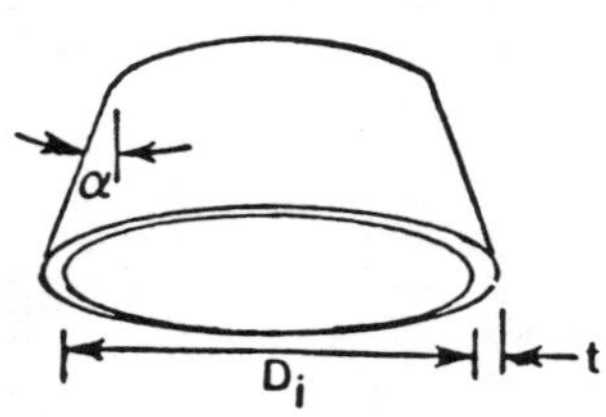

Conical Section

$$t = \frac{PD_i}{2\cos\alpha(\sigma E - 0.6P)} \qquad P = \frac{2\sigma Et\cos\alpha}{D_i + 1.2t\cos\alpha}$$

Supplement for ASME Formulations

1. For a cylindrical shell, when the wall thickness exceeds one half the inside radius or $P > 0.385\sigma E$, the formulas in ASME Code Appendix 1-2 shall be used.
2. For hemispherical heads without a straight flange, the efficiency of the head-to-shell joint is to be used if it is less than the efficiency of the seams in the head.
3. For ellipsoidal heads, where the ratio of the major axis is other than 2:1, refer to ASME Code Appendix 1-4(c).
4. To use the formulations for a conical section in the table, the half apex angle, α, shall not exceed 30°. If $\alpha > 30°$, then a special analysis is required per ASME Code Appendix 1-5(e).
5. For an ASME flanged and dished head (torispherical head) when $L/r < 16\frac{2}{3}$ the following values of M shall be used:

Values of Factor M

L/r	1.00	1.25	1.50	1.75	2.00	2.25	2.50	2.75	3.00	3.25	3.50	4.00	4.50	5.00	5.50	6.00	6.50
M	1.00	1.03	1.06	1.08	1.10	1.13	1.15	1.17	1.18	1.20	1.22	1.25	1.28	1.31	1.34	1.36	1.39
L/r	7.00	7.50	8.00	8.50	9.00	9.50	10.0	10.5	11.0	11.5	12.0	13.0	14.0	15.0	16.0	$16\frac{2}{3}$	*
M	1.41	1.44	1.46	1.48	1.50	1.52	1.54	1.56	1.58	1.60	1.62	1.65	1.69	1.72	1.75	1.77	

* The maximum allowed ratio: $L - t = D$. When $L/r > 16\frac{2}{3}$ (non-ASME Code construction), the values of M may be calculated by

$$M = \frac{1}{4\left(3 + \sqrt{\frac{L}{r}}\right)}$$

Appendix B

National Wind Design Standards

A standard is a collection of current practices, past experiences, and research knowledge. Standards that are developed by consensus groups (e.g., ASTM, ANSI), trade associations (e.g., AISC, ACI), or government groups (e.g., HUD, CPSC) carry more authority than other standards because they reflect wider ranges of materials.

The ANSI A58.1-1982 is a collection of information that is considered to be the state-of-the-art in the design of buildings and other structures. Local and regional building codes adopt portions of the ANSI standard for their own use. These local and regional codes are developed to represent the needs and interests of their respective areas and are written in legal language to be incorporated into state and local laws. Because these building codes are regional or local in scope, they often do not include everything in the ANSI standard, which is national in perspective. For this reason, one must be certain that a local code written for one area is applicable to the site being considered.

The ANSI standard does not have as much authority as the ASME vessel codes, and, unfortunately, does not have a referral committee or group to officially interpret the document. Therefore, one must make decisions based on past experience and accepted methods of design. The ANSI standard (Paragraph 6.6, p. 16) states that in determining the value for the gust response factor a rational analysis can be used. A note below the paragraph states that *one* such procedure for determining the gust response factor is in the standard's appendix. The note at the top of the appendix (p. 52) states clearly that it is not a part of the ANSI A58.1 miminum design standard. What all this implies is that one may follow the guide of the ANSI standard's appendix or use another rational analysis, which includes another wind standard. Thus, one *can* use another standard for design purposes. One of the most widely accepted international standards is the Australian Standard 1170, Part 2-1983, SAA Loading Code Part 2—Wind Forces.

The Australian Standard 1170 is more applicable to the process industries because in it are shape factors for geometries that are more common in that industry, e.g., circular shapes. However, before applying the shape factors of the Australian standard to the ANSI or any other national standard, one must be very careful to correctly convert the factors. This is because the codes have different basis upon which these factors are determined, and a direct application of other parameters is not possible. This is discussed later after we discuss the basis for the various standards.

CRITERIA FOR DETERMINING WIND SPEED

Wind is caused by differential heating of air masses by the sun. These masses of air at approximately one mile above the ground circulate air around their centers of pressure. At this altitude, the velocity and direction of the wind is almost entirely determined by macro-scale forces caused by large scale weather systems. Below this gradient height, the wind is modified by surface roughness, which reduces its velocity and changes its direction and turbulence. A secondary criterion, except for extreme wind conditions, is the temperature gradient, which affects the vertical mobility of turbulent eddies and therefore influences the surface velocity and the gradient height. Therefore, the exact nature of the surface wind at any point depends, first, on the general weather situation, which determines the gradient wind and the temperature gradient, and, second, on the surrounding topography and ground roughness which, together with the temperature gradient, modify the gradient wind to the surface wind.

Wind motion is further complicated by the rotation of the earth, which induces additional forces that cause the air moving across the earth's surface to be subjected to a force at right angles to the wind velocity vector. These additional forces are known as Coriolis forces.

Each country has adopted its own standard for measuring wind velocity. The U.S. National Weather Service and U.S. codes use the fastest-mile wind speed, which is defined as the average speed of one mile of air passing an anemometer. Thus, a fastest-mile wind speed of 120 mph means that a "mile" of wind passed the anemometer during a 30-second period. Other nations, namely Australia and Great Britain, use the two-second gust speed. This is based on the worst 2-second mean as measured by a cup anemometer. The mean gust speeds are recorded over a period of time such that a mean recurrence interval is determined. The mean recurrence interval is the reciprocal of the probability of exceeding a wind speed of a given magnitude at a particular location in one year. The risk, or probability, R, that the design wind speed will be equaled or surpassed at least once in the life of the tower is given by the expression

$$R = 1 - (1 - P_a)^n$$

where P_a = annual probability of exceedance (reciprocal of the mean recurrence interval)
n = life of the tower or stack

The risk that a given wind speed of specified magnitude will be equaled or exceeded increases with the period of time that the tower is exposed to the wind. Values of risk of exceeding design wind speed for a designated annual probability and a given design life of the structure are shown in Table B-1.

For example, if the design wind speed for a tower is based on an annual probability of 0.02 (mean recurrence interval of 50 years) and the projected tower life is 25 years, there is a 0.40 probability that the design wind

Table B-1
Probability of Exceeding Wind Design Speed
$P_A = 1 - (1 - P_A)^N$

Annual Probability	Design Life of Structure in N Years						
P_A	1	5	10	15	25	50	100
0.10	0.100	0.410	0.651	0.794	0.928	0.995	0.999
0.05	0.050	0.226	0.401	0.537	0.723	0.923	0.994
0.01	0.010	0.049	0.096	0.140	0.222	0.395	0.634
0.005	0.005	0.025	0.049	0.072	0.118	0.222	0.394

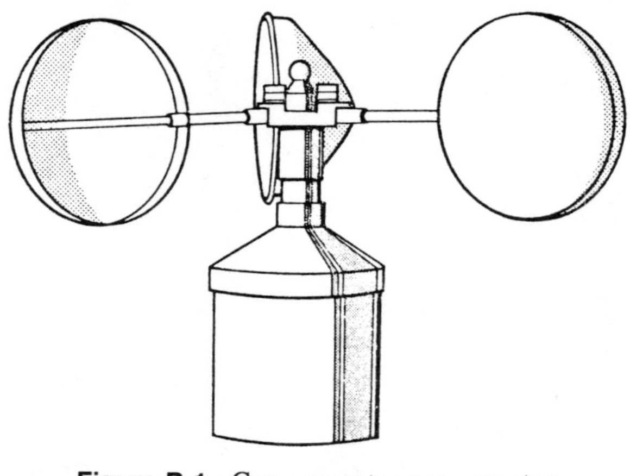

Figure B-1. Cup generator anemometer.

speed will be exceeded during the life of the structure. The United States and Australian wind codes use the 50-year recurrence interval.

The instrument for measuring the wind in the United States, Great Britain, and Australia is the cup-generator anemometer shown in Figure B-1. This device is operated by the wind striking the cups, which drive a small permanent alternator. The indicator, which incorporates a rectifier, is simply a voltmeter calibrated in miles per hour. In most recent cup-generator models the generator output is used to activate a pen-chart recorder which provides a record of continuous wind speed.

WIND SPEED RELATIONSHIPS

As stated previously, another method can be substituted for the appendix in ANSI A58.1. What this means is that another code could be used instead of the appendix. To do this one must be careful to utilize the correct conversion factors between standards. To accomplish this we refer to Figure B-2. For a 100-mph fastest mile wind speed in ANSI A58.1 we wish to determine the equivalent fastest mile wind speed for a 2-second gust using either the Australian or British code. From Figure B-2 we read from the ordinate 1.54 for 2 sec. Knowing that one mile of wind moving at 100 mph will pass the anemometer in 36 sec, we read 36 sec on the curve and arrive at $V_t/V_{3600} = 1.30$. Thus, the equivalent fastest mile wind speed is

$$V = \left(\frac{1.54}{1.30}\right)(100)\text{ mph} = 118.4\text{ mph}$$

for a 2-sec gust. For 110 mph, the values becomes

$$V = (1.18)(110)\text{ mph} = 129.8\text{ mph}$$

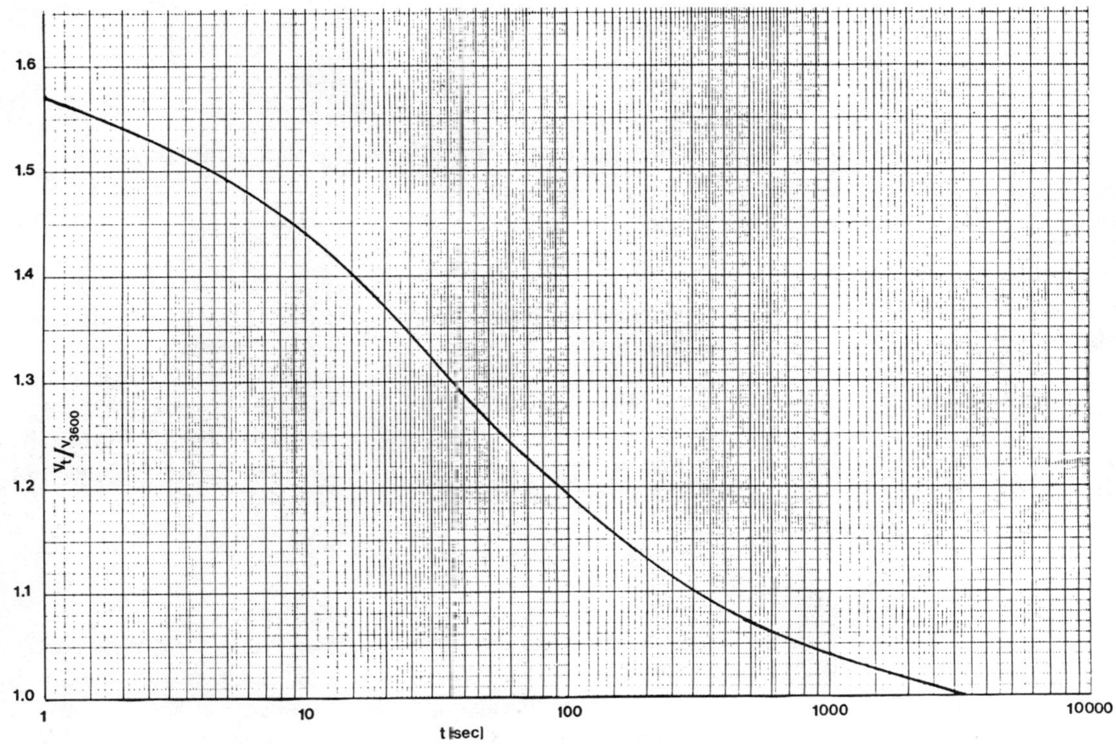

Figure B-2. Ratio of probable maximum wind speed averaged over t seconds to hourly mean speed.

Thus, the 1.18 factor would have to be used in the 2-sec gust code if that code were to be substituted for Appendix A of ANSI A58.1-1982.

Similarly, the Canadian code we must convert to obtain an equivalent fastest mile wind speed from the mean hourly. The mean hourly implies that the wind moves an average of 100 mph across the anemometer in a period of 3,600 sec. Reading Figure B-2 we have $V_t/V_{3600} = 1.0$. Thus

$$\frac{1.0}{1.3} = 0.769$$

which yields an equivalent velocity of 76.9 mph. With the Canadian code one must use 0.769 in use of shape constants and the various other parameters when using with ANSI A58.1. A comparison of the major wind codes is given in Tables B-2, B-3, B-4, and B-5.

ANSI A58.1-1982 WIND CATEGORIES

In the ANSI A58.1-1982 there are four wind categories—A, B, C, and D. The categories are described as follows:

Category A—A very restricted category in which the wind speed is drastically reduced. Most petrochemical and power facilities do not fall within this category. The wind force is reduced because the structure is considered to be among many tall structures. One example would be a ten-story building in downtown Manhattan, New York, where the taller buildings would block the stronger air currents.

Category B—A classification that encompasses some tall structures, but not enough to block the majority of wind gusts. An example of this category would be a tower in the midst of a large petrochemical facility where there were other towers that would block some of the wind force. A forest surrounding a tower is another example.

Category C—The most common classification for petrochemical applications. This category is open terrain where the tower would receive full impact from the wind with minimum ground resistance to the wind. An example of this category would be an open field or an airport.

Category D—A classification for wind moving over water. A beachhead, in which there is flat beach up to a row of buildings would be in Category D. Miami beach, from the ocean front up to the facade of hotels, is a good example. Behind the hotel fronts would be Category C. Another example of this classification would be a tall vertical vessel on an offshore structure.

Table B-2
Major U.S. and Foreign Building Codes and Standards Used in Wind Design

Code or Standard	Edition	Organization	Address
Australian Standard 1170, Part 2-Wind Forces	1983	Standards Association of Australia	Standards House 80 Arthur Street/North Sydney, N.S.W. Australia
British Code of Basic Data for Design of Buildings (CP3)	1972	British Standards Institution	British Standards Institution 2 Park Street London, W1A 285, England
Wind Loading Handbook (commentary on CP3)	1974	Building Research Establishment	Building Research Station Garston, Watford, WD2 7JR, England
National Building Code of Canada (NRCC No. 17303)	1980	National Research Council of Canada	National Research Council of Canada
The Supplement to the National Building Code of Canada (NRCC 17724)	1980	National Research Council of Canada	Ottawa, Ontario K1A OR6 Canada
ANSI A58.1-1982	1982	American National Standards Institute	1430 Broadway New York, New York 10018
Uniform Building Code	1982	International Conference of Building Officials	5360 South Workman Mill Road Whittier, California 90601
Standard Building Code	1982 with 1983 rev.	Southern Building Code Congress International	900 Montclair Road Birmingham, Alabama 35213
Basic Building Code	1984	Building Officials and Code Administrators International, Inc.	17926 South Halsted Street Homewood, Illinois 60430

Table B-3
Reference Wind Speed

Reference	Australian (SAA, 1983)	British (BSI, 1982)	Canadian (NRCC, 1980)	United States (ANSI, 1982)
Averaging time	2–3 second gust speed	2-second gust speed	Mean hourly	Fastest mile
Equivalent reference wind speed to fastest mile 100 mph	118.4	118.4	76.9	100

Table B-4
Parameters Used in the Major National Standards

Parameter	Australian (SAA, 1983)	British (BSI, 1972)	Canadian (NRCC, 1980)	United States (ANSI, 1982)
Wind Speed				
Terrain roughness	4	4	3	4
Local terrain	Yes	Yes	None	None
Height variation	Yes	Yes	Yes	Yes
Ref. speed	2-sec gusts	2-sec gusts	Mean hourly	Fastest mile
Wind Pressure				
Pressure coefficients	Tables in appendix includes figures	Tables, includes figures	Figures and tables in commentaries	Tables, figures and notes
Gusts				
Magnitude	Gust speed	Gust speed	Gust effect factor	Gust response factor
Spatial correlation	Reduction for large area	None	Gust effect factor	Area averaging
Gust frequency	Dynamic consideration for $h/b > 5$	Dynamic consideration not included	Dynamic consideration for $h/b > 4$ in. or for $h > 400$ ft	Dynamic consideration for $h/b > 5$
Analysis procedure	This standard is considered by many the best for use in the process industries. Figures and tables are easy to read. The standard actually provides the user with equations to curves. The analysis procedure is straight-forward.	Overall a very good code, its weakest part is the lack of dynamic consideration.	An excellent wind standard. The analysis procedure is straight-forward and the documents—code and supplement contain tables and figures easy to read.	Although the appendix is technically not considered a part of the standard, it contains figures difficult to read, namely Figure 6. For many structures the data extend beyond the limits of the curves in Figures 6 and 7. In the method in the appendix, one must assume an initial natural frequency, resulting in an iterative process. This method is extremely difficult in designing petrochemical towers without the use of a computer.

Table B-5
Limitations of Codes and Standards

Code or Standard	Statement of Limitation	Location
Australian Standard 1170, Part 2—1983	"Minimum Design Loads on Structures"	Title
National Building Code of Canada (NRCC, 1980)	". . . Essentially a Set of Minimum Regulations . . ."	Guide to the Use of the Code
British CP3	". . . Does Not Apply to Buildings . . . That Are of Unusual Shape or Location For Which Special Investigations May Be Necessary . . ."	Section 1 (Scope)
United States ANSI A58.1	"Minimum Design Loads . . ." "Specific Guidelines Are Given For . . . Wind Tunnel Investigations . . . For Buildings . . . Having Irregular Shapes . . ."	Title Paragraph 6.1
Uniform Building Code	"The purpose . . . is to provide minimum standards . . ."	Section 102
Basic Building Code (BOCA, 1984)	"The Basic Minimum Wind Speeds Are Shown in Figure 912.1 . . ."	Section 912.1
Standard Building Code, 1982 (SBCCI, 1982)	"The Purpose of This Code is to Provide Minimum Requirements . . ." "The Building Official May Require Evidence to Support the Design Pressures Used in the Design of Structures Not Included in This Section."	Preface Article 1205.2(a)

Appendix C

Properties of Pipe

PROPERTIES OF PIPE *

The following formulas are used in the computation of the values shown in the table:

† weight of pipe per foot (pounds)	=	$10.6802t(D-t)$
weight of water per foot (pounds)	=	$0.3405d^2$
square feet outside surface per foot	=	$0.2618D$
square feet inside surface per foot	=	$0.2618d$
inside area (square inches)	=	$0.785d^2$
area of metal (square inches)	=	$0.785(D^2-d^2)$
moment of inertia (inches⁴)	=	$0.0491(D^4-d^4)$
	=	$A_m R_g^{\,2}$
section modulus (inches³)	=	$\dfrac{0.0982(D^4-d^4)}{D}$
radius of gyration (inches)	=	$0.25\sqrt{D^2+d^2}$

A_m = area of metal (square inches)
d = inside diameter (inches)
D = outside diameter (inches)
R_g = radius of gyration (inches)
t = pipe wall thickness (inches)

† The ferritic steels may be about 5% less, and the austenitic stainless steels about 2% greater than the values shown in this table which are based on weights for carbon steel.

*** schedule numbers**

Standard weight pipe and schedule 40 are the same in all sizes through 10-inch; from 12-inch through 24-inch, standard weight pipe has a wall thickness of ⅜-inch.

Extra strong weight pipe and schedule 80 are the same in all sizes through 8-inch; from 8-inch through 24-inch, extra strong weight pipe has a wall thickness of ½-inch.

Double extra strong weight pipe has no corresponding schedule number.

a: ANSI B36.10 steel pipe schedule numbers

b: ANSI B36.10 steel pipe nominal wall thickness designation

c: ANSI B36.19 stainless steel pipe schedule numbers

nominal pipe size outside diameter, in.	schedule number* a	b	c	wall thickness, in.	inside diameter, in.	inside area, sq. in.	metal area, sq. in.	sq ft outside surface, per ft	sq ft inside surface, per ft	weight per ft, lb†	weight of water per ft, lb	moment of inertia, in.⁴	section modulus, in.³	radius gyration, in.
⅛	......	......	10S	0.049	0.307	0.0740	0.0548	0.106	0.0804	0.186	0.0321	0.00088	0.00437	0.1271
0.405	40	Std	40S	0.068	0.269	0.0568	0.0720	0.106	0.0705	0.245	0.0246	0.00106	0.00525	0.1215
	80	XS	80S	0.095	0.215	0.0364	0.0925	0.106	0.0563	0.315	0.0157	0.00122	0.00600	0.1146
¼	......	......	10S	0.065	0.410	0.1320	0.0970	0.141	0.1073	0.330	0.0572	0.00279	0.01032	0.1694
0.540	40	Std	40S	0.088	0.364	0.1041	0.1250	0.141	0.0955	0.425	0.0451	0.00331	0.01230	0.1628
	80	XS	80S	0.119	0.302	0.0716	0.1574	0.141	0.0794	0.535	0.0310	0.00378	0.01395	0.1547
⅜	......	......	SS	0.065	0.710	0.396	0.1582	0.220	0.1859	0.538	0.1716	0.01197	0.0285	0.2750
0.675	......	......	10S	0.065	0.545	0.2333	0.1246	0.177	0.1427	0.423	0.1011	0.00586	0.01737	0.2169
	40	Std	40S	0.091	0.493	0.1910	0.1670	0.177	0.1295	0.568	0.0827	0.00730	0.02160	0.2090
	80	XS	80S	0.126	0.423	0.1405	0.2173	0.177	0.1106	0.739	0.0609	0.00862	0.02554	0.1991
½	......	......	5S	0.065	0.710	0.3959	0.1583	0.220	0.1859	0.538	0.171	0.0120	0.0285	0.2750
0.840	......	......	10S	0.083	0.674	0.357	0.1974	0.220	0.1765	0.671	0.1547	0.01431	0.0341	0.2692
	40	Std	40S	0.109	0.622	0.304	0.2503	0.220	0.1628	0.851	0.1316	0.01710	0.0407	0.2613
	80	XS	80S	0.147	0.546	0.2340	0.320	0.220	0.1433	1.088	0.1013	0.02010	0.0478	0.2505
	160	......	......	0.187	0.466	0.1706	0.383	0.220	0.1220	1.304	0.0740	0.02213	0.0527	0.2402
	......	XXS	......	0.294	0.252	0.0499	0.504	0.220	0.0660	1.714	0.0216	0.02425	0.0577	0.2192
¾	......	......	5S	0.065	0.920	0.665	0.2011	0.275	0.2409	0.684	0.2882	0.02451	0.0467	0.349
1.050	......	......	10S	0.083	0.884	0.614	0.2521	0.275	0.2314	0.857	0.2661	0.02970	0.0566	0.343
	40	Std	40S	0.113	0.824	0.533	0.333	0.275	0.2157	1.131	0.2301	0.0370	0.0706	0.334
	80	XS	80S	0.154	0.742	0.432	0.435	0.275	0.1943	1.474	0.1875	0.0448	0.0853	0.321
	160	......	......	0.218	0.614	0.2961	0.570	0.275	0.1607	1.937	0.1284	0.0527	0.1004	0.304
	......	XXS	......	0.308	0.434	0.1479	0.718	0.275	0.1137	2.441	0.0641	0.0579	0.1104	0.2840
1	......	......	5S	0.065	1.185	1.103	0.2553	0.344	0.310	0.868	0.478	0.0500	0.0760	0.443
1.315	......	......	10S	0.109	1.097	0.945	0.413	0.344	0.2872	1.404	0.409	0.0757	0.1151	0.428
	40	Std	40S	0.133	1.049	0.864	0.494	0.344	0.2746	1.679	0.374	0.0874	0.1329	0.421
	80	XS	80S	0.179	0.957	0.719	0.639	0.344	0.2520	2.172	0.311	0.1056	0.1606	0.407
	160	......	......	0.250	0.815	0.522	0.836	0.344	0.2134	2.844	0.2261	0.1252	0.1903	0.387
	......	XXS	......	0.358	0.599	0.2818	1.076	0.344	0.1570	3.659	0.1221	0.1405	0.2137	0.361
1¼	......	......	5S	0.065	1.530	1.839	0.326	0.434	0.401	1.107	0.797	0.1038	0.1250	0.564
1.660	......	......	10S	0.109	1.442	1.633	0.531	0.434	0.378	1.805	0.707	0.1605	0.1934	0.550
	40	Std	40S	0.140	1.380	1.496	0.669	0.434	0.361	2.273	0.648	0.1948	0.2346	0.540
	80	XS	80S	0.191	1.278	1.283	0.881	0.434	0.335	2.997	0.555	0.2418	0.2913	0.524
	160	......	......	0.250	1.160	1.057	1.107	0.434	0.304	3.765	0.458	0.2839	0.342	0.506
	......	XXS	......	0.382	0.896	0.631	1.534	0.434	0.2346	5.214	0.2732	0.341	0.411	0.472
1½	......	......	5S	0.065	1.770	2.461	0.375	0.497	0.463	1.274	1.067	0.1580	0.1663	0.649
1.900	......	......	10S	0.109	1.682	2.222	0.613	0.497	0.440	2.085	0.962	0.2469	0.2599	0.634

Courtesy of ITT Grinnell.

PROPERTIES OF PIPE (Continued)

nominal pipe size *outside diameter,* in.	schedule number* a	schedule number* b	schedule number* c	wall thickness, in.	inside diameter, in.	inside area, sq. in.	metal area, sq. in.	sq ft outside surface, per ft	sq ft inside surface, per ft	weight per ft, lb†	weight of water per ft, lb	moment of inertia, in.⁴	section modulus, in.³	radius gyration, in.
	40	Std	40S	0.145	1.610	2.036	0.799	0.497	0.421	2.718	0.882	0.310	0.326	0.623
	80	XS	80S	0.200	1.500	1.767	1.068	0.497	0.393	3.631	0.765	0.391	0.412	0.605
1½	160	......	......	0.281	1.338	1.406	1.429	0.497	0.350	4.859	0.608	0.483	0.508	0.581
1.900	......	XXS	......	0.400	1.100	0.950	1.885	0.497	0.288	6.408	0.412	0.568	0.598	0.549
	......	......	......	0.525	0.850	0.567	2.267	0.497	0.223	7.710	0.246	0.6140	0.6470	0.5200
	......	......	......	0.650	0.600	0.283	2.551	0.497	0.157	8.678	0.123	0.6340	0.6670	0.4980
	......	......	5S	0.065	2.245	3.96	0.472	0.622	0.588	1.604	1.716	0.315	0.2652	0.817
	......	......	10S	0.109	2.157	3.65	0.776	0.622	0.565	2.638	1.582	0.499	0.420	0.802
2	40	Std	40S	0.154	2.067	3.36	1.075	0.622	0.541	3.653	1.455	0.666	0.561	0.787
2.375	80	XS	80S	0.218	1.939	2.953	1.477	0.622	0.508	5.022	1.280	0.868	0.731	0.766
	160	......	......	0.343	1.689	2.240	2.190	0.622	0.442	7.444	0.971	1.163	0.979	0.729
	......	XXS	......	0.436	1.503	1.774	2.656	0.622	0.393	9.029	0.769	1.312	1.104	0.703
	......	......	......	0.562	1.251	1.229	3.199	0.622	0.328	10.882	0.533	1.442	1.2140	0.6710
	......	......	......	0.687	1.001	0.787	3.641	0.622	0.262	12.385	0.341	1.5130	1.2740	0.6440
	......	......	5S	0.083	2.709	5.76	0.728	0.753	0.709	2.475	2.499	0.710	0.494	0.988
	......	......	10S	0.120	2.635	5.45	1.039	0.753	0.690	3.531	2.361	0.988	0.687	0.975
	40	Std	40S	0.203	2.469	4.79	1.704	0.753	0.646	5.793	2.076	1.530	1.064	0.947
2½	80	XS	80S	0.276	2.323	4.24	2.254	0.753	0.608	7.661	1.837	1.925	1.339	0.924
2.875	160	......	......	0.375	2.125	3.55	2.945	0.753	0.556	10.01	1.535	2.353	1.637	0.894
	......	XXS	......	0.552	1.771	2.464	4.03	0.753	0.464	13.70	1.067	2.872	1.998	0.844
	......	......	......	0.675	1.525	1.826	4.663	0.753	0.399	15.860	0.792	3.0890	2.1490	0.8140
	......	......	......	0.800	1.275	1.276	5.212	0.753	0.334	17.729	0.554	3.2250	2.2430	0.7860
	......	......	5S	0.083	3.334	8.73	0.891	0.916	0.873	3.03	3.78	1.301	0.744	1.208
	......	......	10S	0.120	3.260	8.35	1.274	0.916	0.853	4.33	3.61	1.822	1.041	1.196
	40	Std	40S	0.216	3.068	7.39	2.228	0.916	0.803	7.58	3.20	3.02	1.724	1.164
3	80	XS	80S	0.300	2.900	6.61	3.02	0.916	0.759	10.25	2.864	3.90	2.226	1.136
3.500	160	......	......	0.437	2.626	5.42	4.21	0.916	0.687	14.32	2.348	5.03	2.876	1.094
	......	XXS	......	0.600	2.300	4.15	5.47	0.916	0.602	18.58	1.801	5.99	3.43	1.047
	......	......	......	0.725	2.050	3.299	6.317	0.916	0.537	21.487	1.431	6.5010	3.7150	1.0140
	......	......	......	0.850	1.800	2.543	7.073	0.916	0.471	24.057	1.103	6.8530	3.9160	0.9840
	......	......	5S	0.083	3.834	11.55	1.021	1.047	1.004	3.47	5.01	1.960	0.980	1.385
3½	......	......	10S	0.120	3.760	11.10	1.463	1.047	0.984	4.97	4.81	2.756	1.378	1.372
4.000	40	Std	40S	0.226	3.548	9.89	2.680	1.047	0.929	9.11	4.28	4.79	2.394	1.337
	80	XS	80S	0.318	3.364	8.89	3.68	1.047	0.881	12.51	3.85	6.28	3.14	1.307
	......	XXS	......	0.636	2.728	5.845	6.721	1.047	0.716	22.850	2.530	9.8480	4.9240	1.2100
	......	......	5S	0.083	4.334	14.75	1.152	1.178	1.135	3.92	6.40	2.811	1.249	1.562
	......	......	10S	0.120	4.260	14.25	1.651	1.178	1.115	5.61	6.17	3.96	1.762	1.549
	......	......	......	0.188	4.124	13.357	2.547	1.178	1.082	8.560	5.800	5.8500	2.6000	1.5250
	40	Std	40S	0.237	4.026	12.73	3.17	1.178	1.054	10.79	5.51	7.23	3.21	1.510
4	80	XS	80S	0.337	3.826	11.50	4.41	1.178	1.002	14.98	4.98	9.61	4.27	1.477
4.500	120	......	......	0.437	3.626	10.33	5.58	1.178	0.949	18.96	4.48	11.65	5.18	1.445
	......	......	......	0.500	3.500	9.621	6.283	1.178	0.916	21.360	4.160	12.7710	5.6760	1.4250
	160	......	......	0.531	3.438	9.28	6.62	1.178	0.900	22.51	4.02	13.27	5.90	1.416
	......	XXS	......	0.674	3.152	7.80	8.10	1.178	0.825	27.54	3.38	15.29	6.79	1.374
	......	......	......	0.800	2.900	6.602	9.294	1.178	0.759	31.613	2.864	16.6610	7.4050	1.3380
	......	......	......	0.925	2.650	5.513	10.384	1.178	0.694	35.318	2.391	17.7130	7.8720	1.3060
	......	......	5S	0.109	5.345	22.44	1.868	1.456	1.399	6.35	9.73	6.95	2.498	1.929
	......	......	10S	0.134	5.295	22.02	2.285	1.456	1.386	7.77	9.53	8.43	3.03	1.920
	40	Std	40S	0.258	5.047	20.01	4.30	1.456	1.321	14.62	8.66	15.17	5.45	1.878
5	80	XS	80S	0.375	4.813	18.19	6.11	1.456	1.260	20.78	7.89	20.68	7.43	1.839
5.563	120	......	......	0.500	4.563	16.35	7.95	1.456	1.195	27.04	7.09	25.74	9.25	1.799
	160	......	......	0.625	4.313	14.61	9.70	1.456	1.129	32.96	6.33	30.0	10.80	1.760
	......	XXS	......	0.750	4.063	12.97	11.34	1.456	1.064	38.55	5.62	33.6	12.10	1.722
	......	......	......	0.875	3.813	11.413	12.880	1.456	0.998	43.810	4.951	36.6450	13.1750	1.6860
	......	......	......	1.000	3.563	9.966	14.328	1.456	0.933	47.734	4.232	39.1110	14.0610	1.6520

PROPERTIES OF PIPE (Continued)

nominal pipe size *outside diameter*, in.	schedule number* a	b	c	wall thick-ness, in.	inside diam-eter, in.	inside area, sq. in.	metal area, sq. in.	sq ft outside surface, per ft	sq ft inside surface, per ft	weight per ft, lb†	weight of water per ft, lb	moment of inertia, in.⁴	section modu-lus, in.³	radius gyra-tion, in.
	......	......	5S	0.109	6.407	32.2	2.231	1.734	1.677	5.37	13.98	11.85	3.58	2.304
	......	......	10S	0.134	6.357	31.7	2.733	1.734	1.664	9.29	13.74	14.40	4.35	2.295
	......	......	......	0.219	6.187	30.100	4.410	1.734	1.620	15.020	13.100	22.6600	6.8400	2.2700
	40	Std	40S	0.280	6.065	28.89	5.58	1.734	1.588	18.97	12.51	28.14	8.50	2.245
6	80	XS	80S	0.432	5.761	26.07	8.40	1.734	1.508	28.57	11.29	40.5	12.23	2.195
6.625	120	......	......	0.562	5.501	23.77	10.70	1.734	1.440	36.39	10.30	49.6	14.98	2.153
	160	......	......	0.718	5.189	21.15	13.33	1.734	1.358	45.30	9.16	59.0	17.81	2.104
	......	XXS	......	0.864	4.897	18.83	15.64	1.734	1.282	53.16	8.17	66.3	20.03	2.060
	......	......	......	1.000	4.625	16.792	17.662	1.734	1.211	60.076	7.284	72.1190	21.7720	2.0200
	......	......	......	1.125	4.375	15.025	19.429	1.734	1.145	66.084	6.517	76.5970	23.1240	1.9850
	......	......	5S	0.109	8.407	55.5	2.916	2.258	2.201	9.91	24.07	26.45	6.13	3.01
	......	......	10S	0.148	8.329	54.5	3.94	2.258	2.180	13.40	23.59	35.4	8.21	3.00
	......	......	......	0.219	8.187	52.630	5.800	2.258	2.150	19.640	22.900	51.3200	11.9000	2.9700
8	20	......	......	0.250	8.125	51.8	6.58	2.258	2.127	22.36	22.48	57.7	13.39	2.962
8.625	30	......	......	0.277	8.071	51.2	7.26	2.258	2.113	24.70	22.18	63.4	14.69	2.953
	40	Std	40S	0.322	7.981	50.0	8.40	2.258	2.089	28.55	21.69	72.5	16.81	2.938
	60	......	......	0.406	7.813	47.9	10.48	2.258	2.045	35.64	20.79	88.8	20.58	2.909
	80	XS	80S	0.500	7.625	45.7	12.76	2.258	1.996	43.39	19.80	105.7	24.52	2.878
	100	......	......	0.593	7.439	43.5	14.96	2.258	1.948	50.87	18.84	121.4	28.14	2.847
	120	......	......	0.718	7.189	40.6	17.84	2.258	1.882	60.63	17.60	140.6	32.6	2.807
8	140	......	......	0.812	7.001	38.5	19.93	2.258	1.833	67.76	16.69	153.8	35.7	2.777
8.625	160	......	......	0.906	6.813	36.5	21.97	2.258	1.784	74.69	15.80	165.9	38.5	2.748
	......	......	......	1.000	6.625	34.454	23.942	2.258	1.734	81.437	14.945	177.1320	41.0740	2.7190
	......	......	......	1.125	6.375	31.903	26.494	2.258	1.669	90.114	13.838	190.6210	44.2020	2.6810
	......	......	5S	0.134	10.482	86.3	4.52	2.815	2.744	15.15	37.4	63.7	11.85	3.75
	......	......	10S	0.165	10.420	85.3	5.49	2.815	2.728	18.70	36.9	76.9	14.30	3.74
	......	......	......	0.219	10.312	83.52	7.24	2.815	2.70	24.63	36.2	100.46	18.69	3.72
	20	......	......	0.250	10.250	82.5	8.26	2.815	2.683	28.04	35.8	113.7	21.16	3.71
	30	......	......	0.307	10.136	80.7	10.07	2.815	2.654	34.24	35.0	137.5	25.57	3.69
	40	Std	40S	0.365	10.020	78.9	11.91	2.815	2.623	40.48	34.1	160.8	29.90	3.67
10	60	XS	80S	0.500	9.750	74.7	16.10	2.815	2.553	54.74	32.3	212.0	39.4	3.63
10.750	80	......	......	0.593	9.564	71.8	18.92	2.815	2.504	64.33	31.1	244.9	45.6	3.60
	100	......	......	0.718	9.314	68.1	22.63	2.815	2.438	76.93	29.5	286.2	53.2	3.56
	120	......	......	0.843	9.064	64.5	26.24	2.815	2.373	89.20	28.0	324	60.3	3.52
	......	......	......	0.875	9.000	63.62	27.14	2.815	2.36	92.28	27.6	333.46	62.04	3.50
	140	......	......	1.000	8.750	60.1	30.6	2.815	2.291	104.13	26.1	368	68.4	3.47
	160	......	......	1.125	8.500	56.7	34.0	2.815	2.225	115.65	24.6	399	74.3	3.43
	......	......	......	1.250	8.250	53.45	37.31	2.815	2.16	126.82	23.2	428.17	79.66	3.39
	......	......	......	1.500	7.750	47.15	43.57	2.815	2.03	148.19	20.5	478.59	89.04	3.31
	......	......	5S	0.156	12.438	121.4	6.17	3.34	3.26	20.99	52.7	122.2	19.20	4.45
	......	......	10S	0.180	12.390	120.6	7.11	3.34	3.24	24.20	52.2	140.5	22.03	4.44
	20	......	......	0.250	12.250	117.9	9.84	3.34	3.21	33.38	51.1	191.9	30.1	4.42
	30	......	......	0.330	12.090	114.8	12.88	3.34	3.17	43.77	49.7	248.5	39.0	4.39
	......	Std	40S	0.375	12.000	113.1	14.58	3.34	3.14	49.56	49.0	279.3	43.8	4.38
	40	......	......	0.406	11.938	111.9	15.74	3.34	3.13	53.53	48.5	300	47.1	4.37
	......	XS	80S	0.500	11.750	108.4	19.24	3.34	3.08	65.42	47.0	362	56.7	4.33
12	60	......	......	0.562	11.626	106.2	21.52	3.34	3.04	73.16	46.0	401	62.8	4.31
12.750	80	......	......	0.687	11.376	101.6	26.04	3.34	2.978	88.51	44.0	475	74.5	4.27
	......	......	......	0.750	11.250	99.40	28.27	3.34	2.94	96.2	43.1	510.7	80.1	4.25
	100	......	......	0.843	11.064	96.1	31.5	3.34	2.897	107.20	41.6	562	88.1	4.22
	......	......	......	0.875	11.000	95.00	32.64	3.34	2.88	110.9	41.1	578.5	90.7	4.21
	120	......	......	1.000	10.750	90.8	36.9	3.34	2.814	125.49	39.3	642	100.7	4.17
	140	......	......	1.125	10.500	86.6	41.1	3.34	2.749	139.68	37.5	701	109.9	4.13
	......	......	......	1.250	10.250	82.50	45.16	3.34	2.68	153.6	35.8	755.5	118.5	4.09
	160	......	......	1.312	10.126	80.5	47.1	3.34	2.651	160.27	34.9	781	122.6	4.07

PROPERTIES OF PIPE (Continued)

nominal pipe size *outside diameter,* in.	schedule number* a	b	c	wall thick-ness, in.	inside diam-eter, in.	inside area, sq. in.	metal area, sq. in.	sq ft outside surface, per ft	sq ft inside surface, per ft	weight per ft, lb†	weight of water per ft, lb	moment of inertia, in.⁴	section modu-lus, in.³	radius gyra-tion, in.
	......	......	5S	0.156	13.688	147.20	6.78	3.67	3.58	23.0	63.7	162.6	23.2	4.90
	......	......	10S	0.188	13.624	145.80	8.16	3.67	3.57	27.7	63.1	194.6	27.8	4.88
	......	......	......	0.210	13.580	144.80	9.10	3.67	3.55	30.9	62.8	216.2	30.9	4.87
	......	......	......	0.219	13.562	144.50	9.48	3.67	3.55	32.2	62.6	225.1	32.2	4.87
	10	......	......	0.250	13.500	143.1	10.80	3.67	3.53	36.71	62.1	255.4	36.5	4.86
	......	......	......	0.281	13.438	141.80	12.11	3.67	3.52	41.2	61.5	285.2	40.7	4.85
	20	......	......	0.312	13.376	140.5	13.42	3.67	3.50	45.68	60.9	314	44.9	4.84
	......	......	......	0.344	13.312	139.20	14.76	3.67	3.48	50.2	60.3	344.3	49.2	4.83
14	30	Std	......	0.375	13.250	137.9	16.05	3.67	3.47	54.57	59.7	373	53.3	4.82
14.000	40	......	......	0.437	13.126	135.3	18.62	3.67	3.44	63.37	58.7	429	61.2	4.80
	......	......	......	0.469	13.062	134.00	19.94	3.67	3.42	67.8	58.0	456.8	65.3	4.79
	......	XS	......	0.500	13.000	132.7	21.21	3.67	3.40	72.09	57.5	484	69.1	4.78
	60	......	......	0.593	12.814	129.0	24.98	3.67	3.35	84.91	55.9	562	80.3	4.74
	......	......	......	0.625	12.750	127.7	26.26	3.67	3.34	89.28	55.3	589	84.1	4.73
	80	......	......	0.750	12.500	122.7	31.2	3.67	3.27	106.13	53.2	687	98.2	4.69
	100	......	......	0.937	12.126	115.5	38.5	3.67	3.17	130.73	50.0	825	117.8	4.63
	120	......	......	1.093	11.814	109.6	44.3	3.67	3.09	150.67	47.5	930	132.8	4.58
	140	......	......	1.250	11.500	103.9	50.1	3.67	3.01	170.22	45.0	1127	146.8	4.53
	160	......	......	1.406	11.188	98.3	55.6	3.67	2.929	189.12	42.6	1017	159.6	4.48
	......	......	5S	0.165	15.670	192.90	8.21	4.19	4.10	28	83.5	257	32.2	5.60
	......	......	10S	0.188	15.624	191.70	9.34	4.19	4.09	32	83.0	292	36.5	5.59
	10	......	......	0.250	15.500	188.7	12.37	4.19	4.06	42.05	81.8	384	48.0	5.57
	20	......	......	0.312	15.376	185.7	15.38	4.19	4.03	52.36	80.5	473	59.2	5.55
	30	Std	......	0.375	15.250	182.6	18.41	4.19	3.99	62.58	79.1	562	70.3	5.53
16	40	XS	......	0.500	15.000	176.7	24.35	4.19	3.93	82.77	76.5	732	91.5	5.48
16.000	60	......	......	0.656	14.688	169.4	31.6	4.19	3.85	107.50	73.4	933	116.6	5.43
	80	......	......	0.843	14.314	160.9	40.1	4.19	3.75	136.46	69.7	1157	144.6	5.37
	100	......	......	1.031	13.938	152.6	48.5	4.19	3.65	164.83	66.1	1365	170.6	5.30
	120	......	......	1.218	13.564	144.5	56.6	4.19	3.55	192.29	62.6	1556	194.5	5.24
	140	......	......	1.437	13.126	135.3	65.7	4.19	3.44	223.64	58.6	1760	220.0	5.17
	160	......	......	1.593	12.814	129.0	72.1	4.19	3.35	245.11	55.9	1894	236.7	5.12
	......	......	5S	0.165	17.670	245.20	9.24	4.71	4.63	31	106.2	368	40.8	6.31
	......	......	10S	0.188	17.624	243.90	10.52	4.71	4.61	36	105.7	417	46.4	6.30
	10	......	......	0.250	17.500	240.5	13.94	4.71	4.58	47.39	104.3	549	61.0	6.28
	20	......	......	0.312	17.376	237.1	17.34	4.71	4.55	59.03	102.8	678	75.5	6.25
	......	Std	......	0.375	17.250	233.7	20.76	4.71	4.52	70.59	101.2	807	89.6	6.23
18	30	......	......	0.437	17.126	230.4	24.11	4.71	4.48	82.06	99.9	931	103.4	6.21
	......	XS	......	0.500	17.00	227.0	27.49	4.71	4.45	93.45	98.4	1053	117.0	6.19
18.000	40	......	......	0.562	16.876	223.7	30.8	4.71	4.42	104.75	97.0	1172	130.2	6.17
	60	......	......	0.750	16.500	213.8	40.6	4.71	4.32	138.17	92.7	1515	168.3	6.10
	80	......	......	0.937	16.126	204.2	50.2	4.71	4.22	170.75	88.5	1834	203.8	6.04
	100	......	......	1.156	15.688	193.3	61.2	4.71	4.11	207.96	83.7	2180	242.2	5.97
	120	......	......	1.375	15.250	182.6	71.8	4.71	3.99	244.14	79.2	2499	277.6	5.90
	140	......	......	1.562	14.876	173.8	80.7	4.71	3.89	274.23	75.3	2750	306	5.84
	160	......	......	1.781	14.438	163.7	90.7	4.71	3.78	308.51	71.0	3020	336	5.77
	......	......	5S	0.188	19.634	302.40	11.70	5.24	5.14	40	131.0	574	57.4	7.00
	......	......	10S	0.218	19.564	300.60	13.55	5.24	5.12	46	130.2	663	66.3	6.99
	10	......	......	0.250	19.500	298.6	15.51	5.24	5.11	52.73	129.5	757	75.7	6.98
	20	Std	......	0.375	19.250	291.0	23.12	5.24	5.04	78.60	126.0	1114	111.4	6.94
20	30	XS	......	0.500	19.000	283.5	30.6	5.24	4.97	104.13	122.8	1457	145.7	6.90
20.000	40	......	......	0.593	18.814	278.0	36.2	5.24	4.93	122.91	120.4	1704	170.4	6.86
	60	......	......	0.812	18.376	265.2	48.9	5.24	4.81	166.40	115.0	2257	225.7	6.79
	......	......	......	0.875	18.250	261.6	52.6	5.24	4.78	178.73	113.4	2409	240.9	6.77
	80	......	......	1.031	17.938	252.7	61.4	5.24	4.70	208.87	109.4	2772	277.2	6.72
	100	......	......	1.281	17.438	238.8	75.3	5.24	4.57	256.10	103.4	3320	332	6.63

PROPERTIES OF PIPE (Continued)

nominal pipe size *outside diameter,* in.	schedule number* a	schedule number* b	schedule number* c	wall thick-ness, in.	inside diam-eter, in.	inside area, sq in.	metal area, sq in.	sq ft outside surface, per ft	sq ft inside surface, per ft	weight per ft, lb†	weight of water per ft, lb	moment of inertia, in.[4]	section modu-lus, in.[3]	radius gyra-tion, in.
20 *20.000*	120	......	......	1.500	17.000	227.0	87.2	5.24	4.45	296.37	98.3	3760	376	6.56
	140	......	......	1.750	16.500	213.8	100.3	5.24	4.32	341.10	92.6	4220	422	6.48
	160	......	......	1.968	16.064	202.7	111.5	5.24	4.21	379.01	87.9	4590	459	6.41
22 *22.000*	......	......	5S	0.188	21.624	367.3	12.88	5.76	5.66	44	159.1	766	69.7	7.71
	......	......	10S	0.218	21.564	365.2	14.92	5.76	5.65	51	158.2	885	80.4	7.70
	10	......	......	0.250	21.500	363.1	17.18	5.76	5.63	58	157.4	1010	91.8	7.69
	20	Std	......	0.375	21.250	354.7	25.48	5.76	5.56	87	153.7	1490	135.4	7.65
	30	XS	......	0.500	21.000	346.4	33.77	5.76	5.50	115	150.2	1953	177.5	7.61
	......	......	......	0.625	20.750	338.2	41.97	5.76	5.43	143	146.6	2400	218.2	7.56
	......	......	......	0.750	20.500	330.1	50.07	5.76	5.37	170	143.1	2829	257.2	7.52
	60	......	......	0.875	20.250	322.1	58.07	5.76	5.30	197	139.6	3245	295.0	7.47
	80	......	......	1.125	19.750	306.4	73.78	5.76	5.17	251	132.8	4029	366.3	7.39
	100	......	......	1.375	19.250	291.0	89.09	5.76	5.04	303	126.2	4758	432.6	7.31
	120	......	......	1.625	18.750	276.1	104.02	5.76	4.91	354	119.6	5432	493.8	7.23
	140	......	......	1.875	18.250	261.6	118.55	5.76	4.78	403	113.3	6054	550.3	7.15
	160	......	......	2.125	17.750	247.4	132.68	5.76	4.65	451	107.2	6626	602.4	7.07
24 *24.000*	10	......	......	0.250	23.500	434	18.65	6.28	6.15	63.41	188.0	1316	109.6	8.40
	20	Std	......	0.375	23.250	425	27.83	6.28	6.09	94.62	183.8	1943	161.9	8.35
	......	XS	......	0.500	23.000	415	36.9	6.28	6.02	125.49	180.1	2550	212.5	8.31
	30	......	......	0.562	22.876	411	41.4	6.28	5.99	140.80	178.1	2840	237.0	8.29
	......	......	......	0.625	22.750	406	45.9	6.28	5.96	156.03	176.2	3140	261.4	8.27
	40	......	......	0.687	22.626	402	50.3	6.28	5.92	171.17	174.3	3420	285.2	8.25
	......	......	......	0.750	22.500	398	54.8	6.28	5.89	186.24	172.4	3710	309	8.22
	......	......	5S	0.218	23.564	436.1	16.29	6.28	6.17	55	188.9	1152	96.0	8.41
	......	......	......	0.875	22.250	388.6	63.54	6.28	5.83	216	168.6	4256	354.7	8.18
	60	......	......	0.968	22.064	382	70.0	6.28	5.78	238.11	165.8	4650	388	8.15
	80	......	......	1.218	21.564	365	87.2	6.28	5.65	296.36	158.3	5670	473	8.07
	100	......	......	1.531	20.938	344	108.1	6.28	5.48	367.40	149.3	6850	571	7.96
	120	......	......	1.812	20.376	326	126.3	6.28	5.33	429.39	141.4	7830	652	7.87
	140	......	......	2.062	19.876	310	142.1	6.28	5.20	483.13	134.5	8630	719	7.79
	160	......	......	2.343	19.314	293	159.4	6.28	5.06	541.94	127.0	9460	788	7.70
26 *26.000*	......	......	......	0.250	25.500	510.7	19.85	6.81	6.68	67	221.4	1646	126.6	9.10
	10	......	......	0.312	25.376	505.8	25.18	6.81	6.64	86	219.2	2076	159.7	9.08
	......	Std	......	0.375	25.250	500.7	30.19	6.81	6.61	103	217.1	2478	190.6	9.06
	20	XS	......	0.500	25.000	490.9	40.06	6.81	6.54	136	212.8	3259	250.7	9.02
	......	......	......	0.625	24.750	481.1	49.82	6.81	6.48	169	208.6	4013	308.7	8.98
	......	......	......	0.750	24.500	471.4	59.49	6.81	6.41	202	204.4	4744	364.9	8.93
	......	......	......	0.875	24.250	461.9	69.07	6.81	6.35	235	200.2	5458	419.9	8.89
	......	......	......	1.000	24.000	452.4	78.54	6.81	6.28	267	196.1	6149	473.0	8.85
	......	......	......	1.125	23.750	443.0	87.91	6.81	6.22	299	192.1	6813	524.1	8.80
28 *28.000*	......	......	......	0.250	27.500	594.0	21.80	7.33	7.20	74	257.3	2098	149.8	9.81
	10	......	......	0.312	27.376	588.6	27.14	7.33	7.17	92	255.0	2601	185.8	9.79
	......	Std	......	0.375	27.250	583.2	32.54	7.33	7.13	111	252.6	3105	221.8	9.77
	20	XS	......	0.500	27.000	572.6	43.20	7.33	7.07	147	248.0	4085	291.8	9.72
	30	......	......	0.625	26.750	562.0	53.75	7.33	7.00	183	243.4	5038	359.8	9.68
	......	......	......	0.750	26.500	551.6	64.21	7.33	6.94	218	238.9	5964	426.0	9.64
	......	......	......	0.875	26.250	541.2	74.56	7.33	6.87	253	234.4	6865	490.3	9.60
	......	......	......	1.000	26.000	530.9	84.82	7.33	6.81	288	230.0	7740	552.8	9.55
	......	......	......	1.125	25.750	520.8	94.98	7.33	6.74	323	225.6	8590	613.6	9.51
30 *30.000*	......	......	5S	0.250	29.500	683.4	23.37	7.85	7.72	79	296.3	2585	172.3	10.52
	10	......	10S	0.312	29.376	677.8	29.19	7.85	7.69	99	293.7	3201	213.4	10.50
	......	Std	......	0.375	29.250	672.0	34.90	7.85	7.66	119	291.2	3823	254.8	10.48
	20	XS	......	0.500	29.000	660.5	46.34	7.85	7.59	158	286.2	5033	335.5	10.43
	30	......	......	0.625	28.750	649.2	57.68	7.85	7.53	196	281.3	6213	414.2	10.39

PROPERTIES OF PIPE (Continued)

nominal pipe size *outside diameter*, in.	schedule number* a	schedule number* b	schedule number* c	wall thick-ness, in.	inside diam-eter, in.	inside area, sq. in.	metal area, sq. in.	sq ft outside surface, per ft	sq ft inside surface, per ft	weight per ft, lb†	weight of water per ft lb	moment of inertia, in.[4]	section modu-lus, in.[3]	radius gyra-tion, in.
	40	......	......	0.750	28.500	637.9	68.92	7.85	7.46	234	276.6	7371	491.4	10.34
30	......	......	......	0.875	28.250	620.7	80.06	7.85	7.39	272	271.8	8494	566.2	10.30
30.000	......	......	......	1.000	28.000	615.7	91.11	7.85	7.33	310	267.0	9591	639.4	10.26
	......	......	......	1.125	27.750	604.7	102.05	7.85	7.26	347	262.2	10653	710.2	10.22
	......	......	......	0.250	31.500	779.2	24.93	8.38	8.25	85	337.8	3141	196.3	11.22
	10	......	......	0.312	31.376	773.2	31.02	8.38	8.21	106	335.2	3891	243.2	11.20
	......	Std	......	0.375	31.250	766.9	37.25	8.38	8.18	127	332.5	4656	291.0	11.18
	20	XS	......	0.500	31.000	754.7	49.48	8.38	8.11	168	327.2	6140	383.8	11.14
32	30	......	......	0.625	30.750	742.5	61.59	8.38	8.05	209	321.9	7578	473.6	11.09
32.000	40	......	......	0.688	30.624	736.6	67.68	8.38	8.02	230	319.0	8298	518.6	11.07
	......	......	......	0.750	30.500	730.5	73.63	8.38	7.98	250	316.7	8990	561.9	11.05
	......	......	......	0.875	30.250	718.3	85.52	8.38	7.92	291	311.6	10372	648.2	11.01
	......	......	......	1.000	30.000	706.8	97.38	8.38	7.85	331	306.4	11680	730.0	10.95
	......	......	......	1.125	29.750	694.7	109.0	8.38	7.79	371	301.3	13023	814.0	10.92
	......	......	......	0.250	33.500	881.2	26.50	8.90	8.77	90	382.0	3773	221.9	11.93
	10	......	......	0.312	33.376	874.9	32.99	8.90	8.74	112	379.3	4680	275.3	11.91
	......	Std	......	0.375	33.250	867.8	39.61	8.90	8.70	135	376.2	5597	329.2	11.89
	20	XS	......	0.500	33.000	855.3	52.62	8.90	8.64	179	370.8	7385	434.4	11.85
34	30	......	......	0.625	32.750	841.9	65.53	8.90	8.57	223	365.0	9124	536.7	11.80
34.000	40	......	......	0.688	32.624	835.9	72.00	8.90	8.54	245	362.1	9992	587.8	11.78
	......	......	......	0.750	32.500	829.3	78.34	8.90	8.51	266	359.5	10829	637.0	11.76
	......	......	......	0.875	32.250	816.4	91.01	8.90	8.44	310	354.1	12501	735.4	11.72
	......	......	......	1.000	32.000	804.2	103.67	8.90	8.38	353	348.6	14114	830.2	11.67
	......	......	......	1.125	31.750	791.3	116.13	8.90	8.31	395	343.2	15719	924.7	11.63
	......	......	......	0.250	35.500	989.7	28.11	9.42	9.29	96	429.1	4491	249.5	12.64
	10	......	......	0.312	35.376	982.9	34.95	9.42	9.26	119	426.1	5565	309.1	12.62
	......	Std	......	0.375	35.250	975.8	42.01	9.42	9.23	143	423.1	6664	370.2	12.59
	20	XS	......	0.500	35.000	962.1	55.76	9.42	9.16	190	417.1	8785	488.1	12.55
36	30	......	......	0.625	34.750	948.3	69.50	9.42	9.10	236	411.1	10872	604.0	12.51
36.000	40	......	......	0.750	34.500	934.7	83.01	9.42	9.03	282	405.3	12898	716.5	12.46
	......	......	......	0.875	34.250	920.6	96.50	9.42	8.97	328	399.4	14903	827.9	12.42
	......	......	......	1.000	34.000	907.9	109.96	9.42	8.90	374	393.6	16851	936.2	12.38
	......	......	......	1.125	33.750	894.2	123.19	9.42	8.89	419	387.9	18763	1042.4	12.34
	......	......	......	0.250	41.500	1352.6	32.82	10.99	10.86	112	586.4	7126	339.3	14.73
	......	Std	......	0.375	41.250	1336.3	49.08	10.99	10.80	167	579.3	10627	506.1	14.71
	20	XS	......	0.500	41.000	1320.2	65.18	10.99	10.73	222	572.3	14037	668.4	14.67
42	30	......	......	0.625	40.750	1304.1	81.28	10.99	10.67	276	565.4	17373	827.3	14.62
42.000	40	......	......	0.750	40.500	1288.2	97.23	10.99	10.60	330	558.4	20689	985.2	14.59
	......	......	......	1.000	40.000	1256.6	128.81	10.99	10.47	438	544.8	27080	1289.5	14.50
	......	......	......	1.250	39.500	1225.3	160.03	10.99	10.34	544	531.2	33233	1582.5	14.41
	......	......	......	1.500	39.000	1194.5	190.85	10.99	10.21	649	517.9	39181	1865.7	14.33

INSULATION WEIGHT FACTORS

To determine the weight per foot of any piping insulation, use the pipe size and nominal insulation thickness to find the insulation weight factor F in the chart shown below. Then multiply F by the density of the insulation in pounds per cubic foot.

Example. For 4″ pipe with 4″ nominal thickness insulation, $F = .77$. If the insulation density is 12 pounds per cubic foot, then the insulation weight is $.77 \times 12 = 9.24$ lb/ft.

Nominal Pipe Size	Nominal Insulation Thickness										
	1″	1½″	2″	2½″	3″	3½″	4″	4½″	5″	5½″	6″
1	.057	.10	.16	.23	.31	.40					
1¼	.051	.12	.15	.22	.30	.39					
1½	.066	.11	.21	.29	.38	.48					
2	.080	.14	.21	.29	.37	.47	.59				
2½	.091	.19	.27	.36	.46	.58	.70	.83			
3	.10	.17	.25	.34	.44	.56	.68	.81			
3½	.15	.23	.31	.41	.54	.66	.78	..	.97		
4	.13	.21	.30	.39	.51	.63	.77	.96	1.10		
5	.15	.24	.34	.45	.58	.71	.88	1.04	1.20		
6	.17	.27	.38	.51	.64	.83	.97	1.13	1.34		
8	..	.34	.47	.66	.80	.97	1.17	1.36	1.56	1.75	
10	..	.43	.59	.75	.93	1.12	1.32	1.54	1.76	1.99	
12	..	.50	.68	.88	1.07	1.28	1.52	1.74	1.99	2.24	2.50
14	..	.51	.70	.90	1.11	1.34	1.57	1.81	2.07	2.34	2.62
16	..	.57	.78	1.01	1.24	1.49	1.74	2.01	2.29	2.58	2.88
18	..	.64	.87	1.12	1.37	1.64	1.92	2.21	2.51	2.82	3.14
20	..	.70	.96	1.23	1.50	1.79	2.09	2.40	2.73	3.06	3.40
24	..	.83	1.13	1.44	1.77	2.10	2.44	2.80	3.16	3.54	3.92

LOAD CARRYING CAPACITIES OF THREADED HOT ROLLED STEEL ROD CONFORMING TO ASTM A-36

Nominal Rod Diameter, in.	⅜	½	⅝	¾	⅞	1	1⅛	1¼	1½	1¾	2	2¼	2½	2¾	3	3¼	3½
Root Area of Thread, sq. in.	.068	.126	.202	.302	.419	.552	.693	.889	1.293	1.744	2.300	3.023	3.719	4.619	5.621	6.720	7.918
Max. Safe Load, lbs. at Rod Temp. of 650°F	610	1130	1810	2710	3770	4960	6230	8000	11630	15700	20700	27200	33500	41580	50580	60480	71280

WEIGHTS OF PIPING MATERIALS

1″ PIPE 1.313″ O.D.

PIPE	Schedule No.	40	80	160	
	Wall Designation	Std.	XS		XXS
	Thickness—In.	.133	.179	.250	.358
	Pipe—Lbs/Ft	**1.68**	**2.17**	**2.84**	**3.66**
	Water—Lbs/Ft	**.37**	**.31**	**.23**	**.12**
WELDING FITTINGS	L.R. 90° Elbow	**.3** .3	**.4** .3	**.6** .3	**.7** .3
	S.R. 90° Elbow	**.2** .2			
	L.R. 45° Elbow	**.2** .2	**.3** .2	**.4** .2	**.4** .2
	Tee	**.8** .4	**.9** .4	**1.1** .4	**1.3** .4
	Lateral	**1.7** 1.1	**2.5** 1.1		
	Reducer	**.3** .2	**.4** .2	**.4** .2	**.5** .2
	Cap	**.2** .3	**.3** .3	**.4** .3	**.4** .3

INSULATION	Temperature Range °F		100-199	200-299	300-399	400-499	500-599	600-699	700-799	800-899	900-999	1000-1099	1100-1200
	85% Magnesia Calcium Silicate	Nom. Thick., In.	1	1	1½	2	2	2½	2½	2½	3	3	3
		Lbs/Ft.	**.72**	**.72**	**1.23**	**1.94**	**1.94**	**2.76**	**2.76**	**2.76**	**3.70**	**3.70**	**3.70**
	Combination	Nom. Thick., In.						2½	2½	2½	3	3	3
		Lbs/Ft						**3.30**	**3.30**	**3.30**	**4.70**	**4.70**	**4.70**
	*Asbestos Fiber—Sodium Silicate	Nom. Thick., In.	1	1	1	1	1	1½	1½	2	2	3	3
		Lbs/Ft	**.91**	**.91**	**.91**	**.91**	**.91**	**1.61**	**1.61**	**2.54**	**2.54**	**4.91**	**4.91**

	Pressure Rating psi	Cast Iron		Steel						
		125	250	150	300	400	600	900	1500	2500
FLANGES	Screwed or Slip-On	**2.5** 1.5	**4** 1.5	**2.5** 1.5	**4** 1.5	**5** 1.5	**5** 1.5	**12** 1.5	**12** 1.5	**15** 1.5
	Welding Neck			**2.3** 1.5	**5** 1.5	**7** 1.5	**7** 1.5	**12** 1.5	**12** 1.5	**16** 1.5
	Lap Joint			**2.5** 1.5	**4** 1.5	**5** 1.5	**5** 1.5	**12** 1.5	**12** 1.5	**15** 1.5
	Blind	**2.5** 1.5	**4** 1.5	**2.5** 1.5	**5** 1.5	**5** 1.5	**5** 1.5	**12** 1.5	**12** 1.5	**15** 1.5
FLANGED FITTINGS	S.R. 90° Elbow	**6** 3.6					**15** 3.7		**28** 3.8	
	L.R. 90° Elbow	**8** 3.8								
	45° Elbow	**5** 3.2					**14** 3.4		**26** 3.6	
	Tee	**11** 5.4					**20** 5.6		**39** 5.7	
VALVES	Flanged Bonnet Gate				**20** 1.2		**25** 1.5		**80** 4.3	
	Flanged Bonnet Globe or Angle								**84** 3.5	
	Flanged Bonnet Check									
	Pressure Seal Bonnet—Gate						**31** 1.7	**31** 1.7		
	Pressure Seal Bonnet—Globe									

Boldface type is weight in pounds. Lightface type beneath weight is weight factor for insulation.

Insulation thicknesses and weights are based on average conditions and do not constitute a recommendation for specific thicknesses of materials. Insulation weights are based on 85% magnesia and hydrous calcium silicate at 11 lbs/cubic foot. The listed thicknesses and weights of combination covering are the sums of the inner layer of diatomaceous earth at 21 lbs/cubic foot and the outer layer at 11 lbs/cubic foot.

Insulation weights include allowances for wire, cement, canvas, bands and paint, but not special surface finishes.

To find the weight of covering on flanges, valves or fittings, multiply the weight factor by the weight per foot of covering used on straight pipe.

Valve weights are approximate. When possible, obtain weights from the manufacturer.

Cast iron valve weights are for flanged end valves; steel weights for welding end valves.

All flanged fitting, flanged valve and flange weights include the proportional weight of bolts or studs to make up all joints.

* 16 lb cu. ft. density.

$1\frac{1}{4}''$ PIPE 1.660″ O.D. WEIGHTS OF PIPING MATERIALS

PIPE

Schedule No.	40	80	160	
Wall Designation	Std.	XS		XXS
Thickness—In.	.140	.191	.250	.382
Pipe—Lbs/Ft	**2.27**	**3.00**	**3.77**	**5.22**
Water—Lbs/Ft	**.65**	**.56**	**.46**	**.27**

WELDING FITTINGS

Fitting	40	80	160	XXS
L.R. 90° Elbow	**.6** .3	**.8** .3	**.9** .3	**1.2** .3
S.R. 90° Elbow	**.4** .2			
L.R. 45° Elbow	**.4** .2	**.5** .2	**.6** .2	**.8** .2
Tee	**1.3** .5	**1.6** .5	**1.9** .5	**2.4** .5
Lateral	**2.4** 1.2	**3.9** 1.2		
Reducer	**.5** .2	**.5** .2	**.6** .2	**.8** .2
Cap	**.3** .3	**.4** .3	**.5** .3	**.6** .3

INSULATION

Temperature Range °F		100-199	200-299	300-399	400-499	500-599	600-699	700-799	800-899	900-999	1000-1099	1100-1200
85% Magnesia Calcium Silicate	Nom. Thick., In.	1	1	1½	2	2	2½	2½	2½	3	3	3
	Lbs/Ft	**.65**	**.65**	**1.47**	**1.83**	**1.83**	**2.65**	**2.65**	**2.65**	**3.58**	**3.58**	**3.58**
Combination	Nom. Thick., In.						2½	2½	2½	3	3	3
	Lbs/Ft						**3.17**	**3.17**	**3.17**	**5.76**	**5.76**	**5.76**
*Asbestos Fiber—Sodium Silicate	Nom. Thick., In.	1	1	1	1½	1½	1½	2	2½	2½	3	3
	Lbs/Ft	**.82**	**.82**	**.82**	**1.93**	**1.93**	**1.93**	**2.45**	**3.58**	**3.58**	**4.82**	**4.82**

FLANGES, FLANGED FITTINGS, VALVES

Pressure Rating psi	Cast Iron		Steel						
	125	250	150	300	400	600	900	1500	2500
FLANGES									
Screwed or Slip-On	**2.5** 1.5	**5** 1.5	**3.5** 1.5	**5** 1.5	**7** 1.5	**7** 1.5	**13** 1.5	**13** 1.5	**23** 1.5
Welding Neck			**3.5** 1.5	**7** 1.5	**8** 1.5	**8** 1.5	**13** 1.5	**13** 1.5	**25** 1.5
Lap Joint			**3.5** 1.5	**5** 1.5	**7** 1.5	**7** 1.5	**13** 1.5	**13** 1.5	**22** 1.5
Blind	**3.5** 1.5	**5** 1.5	**3.5** 1.5	**7** 1.5	**7** 1.5	**7** 1.5	**13** 1.5	**13** 1.5	**23** 1.5
FLANGED FITTINGS									
S.R. 90° Elbow	**8** 3.6			**17** 3.7		**18** 3.8		**33** 3.9	
L.R. 90° Elbow	**10** 3.9			**18** 3.9					
45° Elbow	**7** 3.3			**15** 3.4		**16** 3.5		**31** 3.7	
Tee	**13** 5.4			**23** 5.6		**28** 5.7		**49** 5.9	
VALVES									
Flanged Bonnet Gate				**40** 4		**60** 4.2		**97** 4.6	
Flanged Bonnet Globe or Angle									
Flanged Bonnet Check				**21** 4					
Pressure Seal Bonnet—Gate							**38** 1.8	**38** 1.1	
Pressure Seal Bonnet—Globe									

Boldface type is weight in pounds. Lightface type beneath weight is weight factor for insulation.

Insulation thicknesses and weights are based on average conditions and do not constitute a recommendation for specific thicknesses of materials. Insulation weights are based on 85% magnesia and hydrous calcium silicate at 11 lbs/cubic foot. The listed thicknesses and weights of combination covering are the sums of the inner layer of diatomaceous earth at 21 lbs/cubic foot and the outer layer at 11 lbs/cubic foot.

Insulation weights include allowances for wire, cement, canvas, bands and paint, but not special surface finishes.

To find the weight of covering on flanges, valves or fittings, multiply the weight factor by the weight per foot of covering used on straight pipe.

Valve weights are approximate. When possible, obtain weights from the manufacturer.

Cast iron valve weights are for flanged end valves; steel weights for welding end valves.

All flanged fitting, flanged valve and flange weights include the proportional weight of bolts or studs to make up all joints.

* 16 lb cu. ft. density.

WEIGHTS OF PIPING MATERIALS

1.900″ O.D. 1½″ PIPE

PIPE	Schedule No.	40	80	160									
	Wall Designation	Std.	XS		XXS								
	Thickness—In.	.145	.200	.281	.400								
	Pipe—Lbs/Ft	**2.72**	**3.63**	**4.86**	**6.41**								
	Water—Lbs/Ft	**.88**	**.77**	**.61**	**.41**								
WELDING FITTINGS	L.R. 90° Elbow	**.8** .4	**1.1** .4	**1.4** .4	**1.8** .4								
	S.R. 90° Elbow	**.6** .3	**.7** .3										
	L.R. 45° Elbow	**.5** .2	**.7** .2	**.8** .2	**1** .2								
	Tee	**2** .6	**2.5** .6	**3.1** .6	**3.7** .6								
	Lateral	**3.3** 1.3	**5.4** 1.3										
	Reducer	**.6** .2	**.7** .2	**.9** .2	**1.2** .2								
	Cap	**.4** .3	**.5** .3	**.7** .3	**.7** .3								

	Temperature Range °F	100-199	200-299	300-399	400-499	500-599	600-699	700-799	800-899	900-999	1000-1099	1100-1200
INSULATION 85% Magnesia Calcium Silicate	Nom. Thick., In.	1	1	1½	2	2	2½	2½	2½	3	3	3
	Lbs/Ft	**.84**	**.84**	**1.35**	**2.52**	**2.52**	**3.47**	**3.47**	**3.47**	**4.52**	**4.52**	**4.52**
Combination	Nom. Thick., In.						2½	2½	2½	3	3	3
	Lbs/Ft						**4.20**	**4.20**	**4.20**	**5.62**	**5.62**	**5.62**
*Asbestos Fiber—Sodium Silicate	Nom. Thick., In.	1	1	1	1½	1½	2	2	2½	2½	3	3
	Lbs/Ft	**1.07**	**1.07**	**1.07**	**1.85**	**1.85**	**3.50**	**3.50**	**4.76**	**4.76**	**6.16**	**6.16**

	Pressure Rating psi	Cast Iron		Steel						
		125	250	150	300	400	600	900	1500	2500
FLANGES	Screwed or Slip-On	**3.5** 1.5	**7** 1.5	**3.5** 1.5	**8** 1.5	**9** 1.5	**9** 1.5	**19** 1.5	**19** 1.5	**31** 1.5
	Welding Neck			**4** 1.5	**9** 1.5	**12** 1.5	**12** 1.5	**19** 1.5	**19** 1.5	**34** 1.5
	Lap Joint			**3.5** 1.5	**8** 1.5	**9** 1.5	**9** 1.5	**19** 1.5	**19** 1.5	**31** 1.5
	Blind	**3.5** 1.5	**7** 1.5	**3.5** 1.5	**9** 1.5	**10** 1.5	**10** 1.5	**19** 1.5	**19** 1.5	**31** 1.5
FLANGED FITTINGS	S.R. 90° Elbow	**10** 3.7		**12** 3.7	**23** 3.8		**26** 3.9		**46** 4	
	L.R. 90° Elbow	**12** 4		**13** 4	**24** 4					
	45° Elbow	**9** 3.4		**11** 3.4	**21** 3.5		**23** 3.5		**39** 3.7	
	Tee	**17** 5.6		**20** 5.6	**30** 5.7		**37** 5.8		**70** 6	
VALVES	Flanged Bonnet Gate	**27** 6.8			**55** 4.2		**70** 4.5		**125** 5	
	Flanged Bonnet Globe or Angle				**40** 4.2		**45** 4.2		**170** 5	
	Flanged Bonnet Check			**30** 4.1	**35** 4.1		**40** 4.2		**110** 4.5	
	Pressure Seal Bonnet—Gate							**42** 1.9	**42** 1.2	
	Pressure Seal Bonnet—Globe									

* 16 lb cu. ft. density.

Boldface type is weight in pounds. Lightface type beneath weight is weight factor for insulation.

Insulation thicknesses and weights are based on average conditions and do not constitute a recommendation for specific thicknesses of materials. Insulation weights are based on 85% magnesia and hydrous calcium silicate at 11 lbs/cubic foot. The listed thicknesses and weights of combination covering are the sums of the inner layer of diatomaceous earth at 21 lbs/cubic foot and the outer layer at 11 lbs/cubic foot.

Insulation weights include allowances for wire, cement, canvas, bands and paint, but not special surface finishes.

To find the weight of covering on flanges, valves or fittings, multiply the weight factor by the weight per foot of covering used on straight pipe.

Valve weights are approximate. When possible, obtain weights from the manufacturer.

Cast iron valve weights are for flanged end valves; steel weights for welding end valves.

All flanged fitting, flanged valve and flange weights include the proportional weight of bolts or studs to make up all joints.

2″ PIPE 2.375″ O.D. WEIGHTS OF PIPING MATERIALS

PIPE	Schedule No.	40	80	160	
	Wall Designation	Std.	XS		XXS
	Thickness—In.	.154	.218	.343	.436
	Pipe—Lbs/Ft	**3.65**	**5.02**	**7.44**	**9.03**
	Water—Lbs/Ft	**1.46**	**1.28**	**.97**	**.77**
WELDING FITTINGS	L.R. 90° Elbow	**1.5** .5	**2** .5	**2.9** .5	**3.5** .5
	S.R. 90° Elbow	**1** .3	**1.3** .3		
	L.R. 45° Elbow	**.8** .2	**1.1** .2	**1.6** .2	**1.8** .2
	Tee	**3** .6	**3.7** .6	**5** .6	**5.7** .6
	Lateral	**5** 1.4	**7.8** 1.4		
	Reducer	**.9** .3	**1.2** .3	**1.6** .3	**1.9** .3
	Cap	**.5** .4	**.7** .4	**1.2** .4	**1.2** .4

INSULATION	Temperature Range °F	100-199	200-299	300-399	400-499	500-599	600-699	700-799	800-899	900-999	1000-1099	1100-1200
85% Magnesia Calcium Silicate	Nom. Thick., In.	1	1	1½	2	2	2½	2½	3	3	3	3½
	Lbs/Ft	**1.01**	**1.01**	**1.71**	**2.53**	**2.53**	**3.48**	**3.48**	**4.42**	**4.42**	**4.42**	**5.59**
Combination	Nom. Thick., In.						2½	2½	3	3	3	3½
	Lbs/Ft						**4.28**	**4.28**	**5.93**	**5.93**	**5.93**	**7.80**
*Asbestos Fiber—Sodium Silicate	Nom. Thick., In.	1	1	1	1½	1½	2	2	2½	2½	3	3
	Lbs/Ft	**1.26**	**1.26**	**1.26**	**2.20**	**2.20**	**3.32**	**3.32**	**4.57**	**4.57**	**5.99**	**5.99**

	Pressure Rating psi	Cast Iron		Steel						
		125	250	150	300	400	600	900	1500	2500
FLANGES	Screwed or Slip-On	**6** 1.5	**9** 1.5	**6** 1.5	**9** 1.5	**11** 1.5	**11** 1.5	**32** 1.5	**32** 1.5	**48** 1.5
	Welding Neck			**6** 1.5	**10** 1.5	**13** 1.5	**13** 1.5	**31** 1.5	**31** 1.5	**48** 1.5
	Lap Joint			**6** 1.5	**9** 1.5	**12** 1.5	**12** 1.5	**32** 1.5	**32** 1.5	**48** 1.5
	Blind	**6** 1.5	**10** 1.5	**4.8** 1.5	**10** 1.5	**12** 1.5	**12** 1.5	**31** 1.5	**31** 1.5	**49** 1.5
FLANGED FITTINGS	S.R. 90° Elbow	**16** 3.8	**24** 3.8	**19** 3.8	**29** 3.8		**35** 4		**83** 4.2	
	L.R. 90° Elbow	**18** 4.1	**27** 4.1	**22** 4.1	**31** 4.1					
	45° Elbow	**14** 3.4	**22** 3.5	**16** 3.4	**24** 3.5		**33** 3.7		**73** 3.9	
	Tee	**23** 5.7	**37** 5.7	**27** 5.7	**41** 5.7		**52** 6		**129** 6.3	
VALVES	Flanged Bonnet Gate	**37** 6.9	**52** 7.1	**40** 4	**65** 4.2		**80** 4.5		**190** 5	
	Flanged Bonnet Globe or Angle	**30** 7	**64** 7.3	**30** 3.8	**45** 4		**85** 4.5		**235** 5.5	
	Flanged Bonnet Check	**26** 7	**51** 7.3	**35** 3.8	**40** 4		**60** 4.2		**300** 5.8	
	Pressure Seal Bonnet—Gate								**150** 2.5	
	Pressure Seal Bonnet—Globe								**165** 3	

Boldface type is weight in pounds. Lightface type beneath weight is weight factor for insulation.

Insulation thicknesses and weights are based on average conditions and do not constitute a recommendation for specific thicknesses of materials. Insulation weights are based on 85% magnesia and hydrous calcium silicate at 11 lbs/cubic foot. The listed thicknesses and weights of combination covering are the sums of the inner layer of diatomaceous earth at 21 lbs/cubic foot and the outer layer at 11 lbs/cubic foot.

Insulation weights include allowances for wire, cement, canvas, bands and paint, but not special surface finishes.

To find the weight of covering on flanges, valves or fittings, multiply the weight factor by the weight per foot of covering used on straight pipe.

Valve weights are approximate. When possible, obtain weights from the manufacturer.

Cast iron valve weights are for flanged end valves; steel weights for welding end valves.

All flanged fitting, flanged valve and flange weights include the proportional weight of bolts or studs to make up all joints.

* 16 lb cu. ft. density.

WEIGHTS OF PIPING MATERIALS — 2.875″ O.D. 2½″ PIPE

PIPE	Schedule No.	40	80	160	
	Wall Designation	Std.	XS		XXS
	Thickness—In.	.203	.276	.375	.552
	Pipe—Lbs/Ft	**5.79**	**7.66**	**10.01**	**13.70**
	Water—Lbs/Ft	**2.08**	**1.84**	**1.54**	**1.07**
WELDING FITTINGS	L.R. 90° Elbow	**2.9** .6	**3.8** .6	**4.9** .6	**6.5** .6
	S.R. 90° Elbow	**1.9** .4	**2.5** .4		
	L.R. 45° Elbow	**1.6** .3	**2.1** .3	**2.7** .3	**3.5** .3
	Tee	**5.2** .8	**6.4** .8	**7.9** .8	**9.9** .8
	Lateral	9.2 1.5	14 1.5		
	Reducer	**1.6** .3	**2.1** .3	**2.7** .3	**3.4** .3
	Cap	**.8** .4	**1** .4	**2** .4	**2.1** .4

INSULATION	Temperature Range °F	100-199	200-299	300-399	400-499	500-599	600-699	700-799	800-899	900-999	1000-1099	1100-1200
85% Magnesia Calcium Silicate	Nom. Thick., In.	1	1	1½	2	2	2½	2½	3	3	3½	3½
	Lbs/Ft	**1.14**	**1.14**	**2.29**	**3.23**	**3.23**	**4.28**	**4.28**	**5.46**	**5.46**	**6.86**	**6.86**
Combination	Nom. Thick., In.						2½	2½	3	3	3½	3½
	Lbs/Ft						**5.20**	**5.20**	**7.36**	**7.36**	**9.58**	**9.58**
*Asbestos Fiber—Sodium Silicate	Nom. Thick., In.	1	1	1	1½	1½	2	2	2½	2½	3	3
	Lbs/Ft	**1.44**	**1.44**	**1.44**	**3.09**	**3.09**	**4.34**	**4.34**	**5.75**	**5.75**	**7.34**	**7.34**

	Pressure Rating psi	Cast Iron		Steel						
		125	250	150	300	400	600	900	1500	2500
FLANGES	Screwed or Slip-On	**8** 1.5	**14** 1.5	**9** 1.5	**14** 1.5	**17** 1.5	**17** 1.5	**46** 1.5	**46** 1.5	**69** 1.5
	Welding Neck			**9** 1.5	**14** 1.5	**20** 1.5	**20** 1.5	**46** 1.5	**46** 1.5	**66** 1.5
	Lap Joint			**9** 1.5	**14** 1.5	**18** 1.5	**18** 1.5	**45** 1.5	**45** 1.5	**67** 1.5
	Blind	**8** 1.5	**15** 1.5	**9** 1.5	**16** 1.5	**19** 1.5	**19** 1.5	**45** 1.5	**45** 1.5	**70** 1.5
FLANGED FITTINGS	S.R. 90° Elbow	**21** 3.8	**36** 3.9	**27** 3.8	**42** 3.9		**50** 4.1		**114** 4.4	
	L.R. 90° Elbow	**25** 4.2	**40** 4.2	**30** 4.2	**47** 4.2					
	45° Elbow	**19** 3.5	**34** 3.6	**22** 3.5	**35** 3.6		**46** 3.8		**99** 3.9	
	Tee	**32** 5.7	**55** 5.8	**42** 5.7	**61** 5.9		**77** 6.2		**169** 6.6	

VALVES	Flanged Bonnet Gate	**50** 7	**82** 7.1	**60** 4	**100** 4.2		**105** 4.6		**275** 5.2	
	Flanged Bonnet Globe or Angle	**43** 7.1	**87** 7.4	**50** 4	**70** 4.1		**120** 4.6		**325** 5.5	
	Flanged Bonnet Check	**36** 7.1	**71** 7.4	**40** 4	**50** 4		**105** 4.6		**320** 5.5	
	Pressure Seal Bonnet—Gate								**215** 2.5	
	Pressure Seal Bonnet—Globe								**230** 2.8	

* 16 lb cu. ft. density.

Boldface type is weight in pounds. Lightface type beneath weight is weight factor for insulation.

Insulation thicknesses and weights are based on average conditions and do not constitute a recommendation for specific thicknesses of materials. Insulation weights are based on 85% magnesia and hydrous calcium silicate at 11 lbs/cubic foot. The listed thicknesses and weights of combination covering are the sums of the inner layer of diatomaceous earth at 21 lbs/cubic foot and the outer layer at 11 lbs/cubic foot.

Insulation weights include allowances for wire, cement, canvas, bands and paint, but not special surface finishes.

To find the weight of covering on flanges, valves or fittings, multiply the weight factor by the weight per foot of covering used on straight pipe.

Valve weights are approximate. When possible, obtain weights from the manufacturer.

Cast iron valve weights are for flanged end valves; steel weights for welding end valves.

All flanged fitting, flanged valve and flange weights include the proportional weight of bolts or studs to make up all joints.

3″ PIPE 3.500″ O.D. WEIGHTS OF PIPING MATERIALS

PIPE	Schedule No.	40	80	160	
	Wall Designation	Std.	XS		XXS
	Thickness—In.	.216	.300	.438	.600
	Pipe—Lbs/Ft	**7.58**	**10.25**	**14.32**	**18.58**
	Water—Lbs/Ft	**3.20**	**2.86**	**2.35**	**1.80**
WELDING FITTINGS	L.R. 90° Elbow	**4.6** .8	**6.1** .8	**8.4** .8	**10.7** .8
	S.R. 90° Elbow	**3** .5	**4** .5		
	L.R. 45° Elbow	**2.4** .3	**3.2** .3	**4.4** .3	**5.4** .3
	Tee	**7.4** .8	**9.5** .8	**12.2** .8	**14.8** .8
	Lateral	**13** 1.8	**19** 1.8		
	Reducer	**2.2** .3	**2.9** .3	**3.7** .3	**4.7** .3
	Cap	**1.4** .5	**1.8** .5	**3.5** .5	**3.7** .5

INSULATION	Temperature Range °F	100-199	200-299	300-399	400-499	500-599	600-699	700-799	800-899	900-999	1000-1099	1100-1200
85% Magnesia Calcium Silicate	Nom. Thick., In.	1	1	1½	2	2	2½	3	3	3	3½	3½
	Lbs/Ft	**1.25**	**1.25**	**2.08**	**3.01**	**3.01**	**4.07**	**5.24**	**5.24**	**5.24**	**6.65**	**6.65**
Combination	Nom. Thick., In.						2½	3	3	3	3½	3½
	Lbs/Ft						**5.07**	**6.94**	**6.94**	**6.94**	**9.17**	**9.17**
*Asbestos Fiber—Sodium Silicate	Nom. Thick., In.	1	1	1	1½	1½	2	2	3	3	3½	3½
	Lbs/Ft	**1.61**	**1.61**	**1.61**	**2.74**	**2.74**	**3.98**	**3.98**	**6.99**	**6.99**	**8.99**	**8.99**

	Pressure Rating psi	Cast Iron		Steel						
		125	250	150	300	400	600	900	1500	2500
FLANGES	Screwed or Slip-On	**9** 1.5	**17** 1.5	**9** 1.5	**17** 1.5	**20** 1.5	**20** 1.5	**37** 1.5	**61** 1.5	**102** 1.5
	Welding Neck			**11** 1.5	**19** 1.5	**27** 1.5	**27** 1.5	**38** 1.5	**61** 1.5	**113** 1.5
	Lap Joint			**9** 1.5	**17** 1.5	**19** 1.5	**19** 1.5	**36** 1.5	**60** 1.5	**99** 1.5
	Blind	**10** 1.5	**19** 1.5	**10** 1.5	**20** 1.5	**24** 1.5	**24** 1.5	**38** 1.5	**61** 1.5	**105** 1.5
FLANGED FITTINGS	S.R. 90° Elbow	**26** 3.9	**46** 4	**32** 3.9	**53** 4		**67** 4.1	**98** 4.3	**150** 4.6	
	L.R. 90° Elbow	**30** 4.3	**50** 4.3	**40** 4.3	**63** 4.3					
	45° Elbow	**22** 3.5	**41** 3.6	**28** 3.5	**46** 3.6		**60** 3.8	**93** 3.9	**135** 4	
	Tee	**39** 5.9	**67** 6	**52** 5.9	**81** 6		**102** 6.2	**151** 6.5	**238** 6.9	
VALVES	Flanged Bonnet Gate	**66** 7	**112** 7.4	**70** 4	**125** 4.4		**155** 4.8	**260** 5	**410** 5.5	
	Flanged Bonnet Globe or Angle	**56** 7.2	**121** 7.6	**60** 4.3	**95** 4.5		**155** 4.8	**225** 5	**495** 5.5	
	Flanged Bonnet Check	**46** 7.2	**100** 7.6	**60** 4.3	**70** 4.4		**120** 4.8	**150** 4.9	**440** 5.8	
	Pressure Seal Bonnet—Gate							**208** 3	**235** 3.2	
	Pressure Seal Bonnet—Globe							**135** 2.5	**180** 3	

Boldface type is weight in pounds. Lightface type beneath weight is weight factor for insulation.

Insulation thicknesses and weights are based on average conditions and do not constitute a recommendation for specific thicknesses of materials. Insulation weights are based on 85% magnesia and hydrous calcium silicate at 11 lbs/cubic foot. The listed thicknesses and weights of combination covering are the sums of the inner layer of diatomaceous earth at 21 lbs/cubic foot and the outer layer at 11 lbs/cubic foot.

Insulation weights include allowances for wire, cement, canvas, bands and paint, but not special surface finishes.

To find the weight of covering on flanges, valves or fittings, multiply the weight factor by the weight per foot of covering used on straight pipe.

Valve weights are approximate. When possible, obtain weights from the manufacturer.

Cast iron valve weights are for flanged end valves; steel weights for welding end valves.

All flanged fitting, flanged valve and flange weights include the proportional weight of bolts or studs to make up all joints.

*** 16 lb cu. ft. density.**

WEIGHTS OF PIPING MATERIALS — 4.000″ O.D. 3½″ PIPE

PIPE	Schedule No.	40	80	
	Wall Designation	Std.	XS	XXS
	Thickness—In.	.226	.318	.636
	Pipe—Lbs/Ft	**9.11**	**12.51**	**22.85**
	Water—Lbs/Ft	**4.28**	**3.85**	**2.53**
WELDING FITTINGS	L.R. 90° Elbow	**6.4** .9	**8.7** .9	**15.4** .9
	S.R. 90° Elbow	**4.3** .6	**5.8** .6	
	L.R. 45° Elbow	**3.3** .4	**4.4** .4	**7.5** .4
	Tee	**9.9** .9	**12.6** .9	**20** .9
	Lateral	**17** 1.8	**26** 1.8	
	Reducer	**3.1** .3	**4.1** .3	**6.9** .3
	Cap	**2.1** .6	**2.8** .6	**5.5** .6

INSULATION	Temperature Range °F	100-199	200-299	300-399	400-499	500-599	600-699	700-799	800-899	900-999	1000-1099	1100-1200
85% Magnesia Calcium Silicate	Nom. Thick., In.	1	1	1½	2	2½	2½	3	3	3½	3½	3½
	Lbs/Ft	**1.83**	**1.83**	**2.77**	**3.71**	**4.88**	**4.88**	**6.39**	**6.39**	**7.80**	**7.80**	**7.80**
Combination	Nom. Thick., In.						2½	3	3	3½	3½	3½
	Lbs/Ft						**6.49**	**8.71**	**8.71**	**10.8**	**10.8**	**10.8**
*Asbestos Fiber—Sodium Silicate	Nom. Thick., In.	1	1	1	1½	1½	2	2	3	3	3½	3½
	Lbs/Ft	**2.41**	**2.41**	**2.41**	**3.65**	**3.65**	**5.07**	**5.07**	**8.66**	**8.66**	**10.62**	**10.62**

	Pressure Rating psi	Cast Iron		Steel						
		125	250	150	300	400	600	900	1500	2500
FLANGES	Screwed or Slip-On	**13** 1.5	**21** 1.5	**13** 1.5	**21** 1.5	**27** 1.5	**27** 1.5			
	Welding Neck			**14** 1.5	**22** 1.5	**32** 1.5	**32** 1.5			
	Lap Joint			**13** 1.5	**21** 1.5	**26** 1.5	**26** 1.5			
	Blind	**14** 1.5	**23** 1.5	**15** 1.5	**25** 1.5	**35** 1.5	**35** 1.5			
FLANGED FITTINGS	S.R. 90° Elbow	**35** 4	**56** 4.1	**49** 4			**82** 4.3			
	L.R. 90° Elbow	**40** 4.4	**62** 4.4	**54** 4.4						
	45° Elbow	**31** 3.6	**51** 3.7	**39** 3.6			**75** 3.9			
	Tee	**54** 6	**86** 6.2	**70** 6			**133** 6.4			
VALVES	Flanged Bonnet Gate	**82** 7.1	**143** 7.5	**90** 4.1	**155** 4.5		**180** 4.8	**360** 5	**510** 5.5	
	Flanged Bonnet Globe or Angle	**74** 7.3	**137** 7.7				**160** 4.7			
	Flanged Bonnet Check	**71** 7.3	**125** 7.7				**125** 4.7			
	Pressure Seal Bonnet—Gate						**140** 2.5	**295** 2.8	**380** 3	
	Pressure Seal Bonnet—Globe									

* 16 lb cu. ft. density.

Boldface type is weight in pounds. Lightface type beneath weight is weight factor for insulation.

Insulation thicknesses and weights are based on average conditions and do not constitute a recommendation for specific thicknesses of materials. Insulation weights are based on 85% magnesia and hydrous calcium silicate at 11 lbs/cubic foot. The listed thicknesses and weights of combination covering are the sums of the inner layer of diatomaceous earth at 21 lbs/cubic foot and the outer layer at 11 lbs/cubic foot.

Insulation weights include allowances for wire, cement, canvas, bands and paint, but not special surface finishes.

To find the weight of covering on flanges, valves or fittings, multiply the weight factor by the weight per foot of covering used on straight pipe.

Valve weights are approximate. When possible, obtain weights from the manufacturer.

Cast iron valve weights are for flanged end valves; steel weights for welding end valves.

All flanged fitting, flanged valve and flange weights include the proportional weight of bolts or studs to make up all joints.

4″ PIPE 4.500″ O.D. WEIGHTS OF PIPING MATERIALS

PIPE	Schedule No.	40	80	120	160								
	Wall Designation	Std.	XS			XXS							
	Thickness—In.	.237	.337	.438	.531	.674							
	Pipe—Lbs/Ft	**10.79**	**14.98**	**18.96**	**22.51**	**27.54**							
	Water—Lbs/Ft	**5.51**	**4.98**	**4.48**	**4.02**	**3.38**							
WELDING FITTINGS	L.R. 90° Elbow	**8.7** 1	**11.9** 1		**17.6** 1	**21** 1							
	S.R. 90° Elbow	**5.8** .7	**7.9** .7										
	L.R. 45° Elbow	**4.3** .4	**5.9** .4		**8.5** .4	**10.1** .4							
	Tee	**12.6** 1	**16.4** 1		**23** 1	**27** 1							
	Lateral	**21** 2.1	**33** 2.1										
	Reducer	**3.6** .3	**4.9** .3		**6.6** .3	**8.2** .3							
	Cap	**2.6** .6	**3.4** .6		**6.5** .6	**6.7** .6							

INSULATION	Temperature Range °F	100-199	200-299	300-399	400-499	500-599	600-699	700-799	800-899	900-999	1000-1099	1100-1200
85% Magnesia Calcium Silicate	Nom. Thick., In.	1	1	1½	2	2½	2½	3	3	3½	3½	4
	Lbs/Ft	**1.62**	**1.62**	**2.55**	**3.61**	**4.66**	**4.66**	**6.07**	**6.07**	**7.48**	**7.48**	**9.10**
Combination	Nom. Thick., In.						2½	3	3	3½	3½	3½
	Lbs/Ft						**6.07**	**8.30**	**8.30**	**10.6**	**10.6**	**10.6**
*Asbestos Fiber—Sodium Silicate	Nom. Thick., In.	1	1	1	1½	1½	2	2	3	3	3½	3½
	Lbs/Ft	**2.04**	**2.04**	**2.04**	**3.28**	**3.28**	**4.70**	**4.70**	**8.29**	**8.29**	**10.25**	**10.25**

	Pressure Rating psi	Cast Iron		Steel						
		125	250	150	300	400	600	900	1500	2500
FLANGES	Screwed or Slip-On	**16** 1.5	**26** 1.5	**15** 1.5	**26** 1.5	**32** 1.5	**43** 1.5	**66** 1.5	**90** 1.5	**158** 1.5
	Welding Neck			**17** 1.5	**29** 1.5	**41** 1.5	**48** 1.5	**64** 1.5	**90** 1.5	**177** 1.5
	Lap Joint			**15** 1.5	**26** 1.5	**31** 1.5	**42** 1.5	**64** 1.5	**92** 1.5	**153** 1.5
	Blind	**18** 1.5	**29** 1.5	**19** 1.5	**31** 1.5	**39** 1.5	**47** 1.5	**67** 1.5	**90** 1.5	**164** 1.5
FLANGED FITTINGS	S.R. 90° Elbow	**45** 4.1	**72** 4.2	**59** 4.1	**85** 4.2	**99** 4.3	**128** 4.4	**185** 4.5	**254** 4.8	
	L.R. 90° Elbow	**52** 4.5	**79** 4.5	**72** 4.5	**98** 4.5					
	45° Elbow	**40** 3.7	**65** 3.8	**51** 3.7	**78** 3.8	**82** 3.9	**119** 4	**170** 4.1	**214** 4.2	
	Tee	**70** 6.1	**109** 6.3	**86** 6.1	**121** 6.3	**153** 6.4	**187** 6.6	**262** 6.8	**386** 7.2	
VALVES	Flanged Bonnet Gate	**109** 7.2	**188** 7.5	**100** 4.2	**175** 4.5	**195** 5	**255** 5.1	**455** 5.4	**735** 6	
	Flanged Bonnet Globe or Angle	**97** 7.4	**177** 7.8	**95** 4.3	**145** 4.8	**215** 5	**230** 5.1	**415** 5.5	**800** 6	
	Flanged Bonnet Check	**80** 7.4	**146** 7.8	**80** 4.3	**105** 4.5	**160** 4.8	**195** 5	**320** 5.6	**780** 6	
	Pressure Seal Bonnet—Gate						**215** 2.8	**380** 3	**520** 4	
	Pressure Seal Bonnet—Globe							**240** 2.7	**290** 3	

Boldface type is weight in pounds. Lightface type beneath weight is weight factor for insulation.

Insulation thicknesses and weights are based on average conditions and do not constitute a recommendation for specific thicknesses of materials. Insulation weights are based on 85% magnesia and hydrous calcium silicate at 11 lbs/cubic foot. The listed thicknesses and weights of combination covering are the sums of the inner layer of diatomaceous earth at 21 lbs/cubic foot and the outer layer at 11 lbs/cubic foot.

Insulation weights include allowances for wire, cement, canvas, bands and paint, but not special surface finishes.

To find the weight of covering on flanges, valves or fittings, multiply the weight factor by the weight per foot of covering used on straight pipe.

Valve weights are approximate. When possible, obtain weights from the manufacturer.

Cast iron valve weights are for flanged end valves; steel weights for welding end valves.

All flanged fitting, flanged valve and flange weights include the proportional weight of bolts or studs to make up all joints.

* 16 lb cu. ft. density.

WEIGHTS OF PIPING MATERIALS

5.563″ O.D. **5″** PIPE

PIPE	Schedule No.	40	80	120	160	
	Wall Designation	Std.	XS			XXS
	Thickness—In.	.258	.375	.500	.625	.750
	Pipe—Lbs/Ft	**14.62**	**20.78**	**27.04**	**32.96**	**38.55**
	Water—Lbs/Ft	**8.66**	**7.89**	**7.09**	**6.33**	**5.62**
WELDING FITTINGS	L.R., 90° Elbow	**14.7** 1.3	**21** 1.3		**32** 1.3	**37** 1.3
	S.R. 90° Elbow	**9.8** .8	**13.7** .8			
	L.R. 45° Elbow	**7.3** .5	**10.2** .5		**15.6** .5	**17.7** .5
	Tee	**19.8** 1.2	**26** 1.2		**39** 1.2	**43** 1.2
	Lateral	**31** 2.5	**50** 2.5			
	Reducer	**6** .4	**8.3** .4		**12.4** .4	**14.2** .4
	Cap	**4.2** .7	**5.7** .7		**11** .7	**11** .7

INSULATION	Temperature Range °F	100-199	200-299	300-399	400-499	500-599	600-699	700-799	800-899	900-999	1000-1099	1100-1200
*Asbestos Fiber—Sodium Silicate	Nom. Thick., In.	1	1½	1½	2	2½	2½	3	3½	3½	4	4
	Lbs/Ft	**1.86**	**2.92**	**2.92**	**4.08**	**5.38**	**5.38**	**6.90**	**8.41**	**8.41**	**10.4**	**10.4**
Combination	Nom. Thick., In.						2½	3	3½	3½	4	4
	Lbs/Ft						**7.01**	**9.30**	**11.8**	**11.8**	**14.9**	**14.9**
85% Magnesia Calcium Silicate	Nom. Thick., In.	1	1	1	1½	1½	2½	2½	3	3	4	4
	Lbs/Ft	**2.34**	**2.34**	**2.34**	**3.76**	**3.76**	**7.35**	**7.35**	**9.31**	**9.31**	**14.37**	**14.37**

	Pressure Rating psi	Cast Iron		Steel						
		125	250	150	300	400	600	900	1500	2500
FLANGES	Screwed or Slip-On	**20** 1.5	**32** 1.5	**18** 1.5	**32** 1.5	**37** 1.5	**73** 1.5	**100** 1.5	**162** 1.5	**259** 1.5
	Welding Neck			**22** 1.5	**36** 1.5	**49** 1.5	**78** 1.5	**103** 1.5	**162** 1.5	**293** 1.5
	Lap Joint			**18** 1.5	**32** 1.5	**35** 1.5	**71** 1.5	**98** 1.5	**168** 1.5	**253** 1.5
	Blind	**23** 1.5	**37** 1.5	**23** 1.5	**39** 1.5	**50** 1.5	**78** 1.5	**104** 1.5	**172** 1.5	**272** 1.5
FLANGED FITTINGS	S.R. 90° Elbow	**58** 4.3	**94** 4.3	**80** 4.3	**113** 4.3	**123** 4.5	**205** 4.7	**268** 4.8	**435** 5.2	
	L.R. 90° Elbow	**68** 4.7	**105** 4.7	**91** 4.7	**128** 4.7					
	45° Elbow	**51** 3.8	**83** 3.8	**66** 3.8	**98** 3.8	**123** 4	**180** 4.2	**239** 4.3	**350** 4.5	
	Tee	**90** 6.4	**145** 6.5	**119** 6.4	**172** 6.4	**179** 6.8	**304** 7	**415** 7.2	**665** 7.8	
VALVES	Flanged Bonnet Gate	**138** 7.3	**264** 7.9	**150** 4.3	**265** 4.9	**310** 5.3	**455** 5.5	**615** 6	**1340** 7	
	Flanged Bonnet Globe or Angle	**138** 7.6	**247** 8	**155** 4.3	**215** 5	**355** 5.2	**515** 5.8	**555** 5.8	**950** 6	
	Flanged Bonnet Check	**118** 7.6	**210** 8	**110** 4.3	**165** 5	**185** 5	**350** 5.8	**560** 6	**1150** 7	
	Pressure Seal Bonnet—Gate						**350** 3.1	**520** 3.8	**865** 4.5	
	Pressure Seal Bonnet—Globe							**280** 4	**450** 4.5	

Boldface type is weight in pounds. Lightface type beneath weight is weight factor for insulation.

Insulation thicknesses and weights are based on average conditions and do not constitute a recommendation for specific thicknesses of materials. Insulation weights are based on 85% magnesia and hydrous calcium silicate at 11 lbs/cubic foot. The listed thicknesses and weights of combination covering are the sums of the inner layer of diatomaceous earth at 21 lbs/cubic foot and the outer layer at 11 lbs/cubic foot.

Insulation weights include allowances for wire, cement, canvas, bands and paint, but not special surface finishes.

To find the weight of covering on flanges, valves or fittings, multiply the weight factor by the weight per foot of covering used on straight pipe.

Valve weights are approximate. When possible, obtain weights from the manufacturer.

Cast iron valve weights are for flanged end valves; steel weights for welding end valves.

All flanged fitting, flanged valve and flange weights include the proportional weight of bolts or studs to make up all joints.

* 16 lb cu. ft. density.

6″ PIPE 6.625″ O.D. WEIGHTS OF PIPING MATERIALS

PIPE	Schedule No.	40	80	120	160	
	Wall Designation	Std.	XS			XXS
	Thickness—In.	.280	.432	.562	.718	.864
	Pipe—Lbs/Ft	**18.97**	**28.57**	**36.39**	**45.3**	**53.2**
	Water—Lbs/Ft	**12.51**	**11.29**	**10.30**	**9.2**	**8.2**
WELDING FITTINGS	L.R. 90° Elbow	**23** 1.5	**34** 1.5		**53** 1.5	**62** 1.5
	S.R. 90° Elbow	**15.2** 1	**23** 1			
	L.R. 45° Elbow	**11.3** .6	**16.7** .6		**26** .6	**30** .6
	Tee	**29.3** 1.4	**42** 1.4		**60** 1.4	**68** 1.4
	Lateral	**42** 2.9	**79** 2.9			
	Reducer	**8.7** .5	**12.6** .5		**18.8** .5	**21** .5
	Cap	**6.4** .9	**9.2** .9		**17.5** .9	**17.5** .9

INSULATION	Temperature Range °F	100-199	200-299	300-399	400-499	500-599	600-699	700-799	800-899	900-999	1000-1099	1100-1200
85% Magnesia Calcium Silicate	Nom. Thick., In.	1	1½	2	2	2½	3	3	3½	3½	4	4
	Lbs/Ft	**2.11**	**3.28**	**4.57**	**4.57**	**6.09**	**7.60**	**7.60**	**9.82**	**98.2**	**11.5**	**11.4**
Combination	Nom. Thick., In.						3	3	3½	3½	4	4
	Lbs/Ft						**10.3**	**10.3**	**13.4**	**13.4**	**16.6**	**16.6**
*Asbestos Fiber—Sodium Silicate	Nom. Thick., In.	1	1	1	1½	1½	2½	2½	3½	3½	4	4
	Lbs/Ft	**2.57**	**2.57**	**2.57**	**4.18**	**4.18**	**8.10**	**8.10**	**13.31**	**13.31**	**15.85**	**15.85**

	Pressure Rating psi	Cast Iron		Steel						
		125	250	150	300	400	600	900	1500	2500
FLANGES	Screwed or Slip-On	**25** 1.5	**42** 1.5	**22** 1.5	**45** 1.5	**54** 1.5	**95** 1.5	**128** 1.5	**202** 1.5	**396** 1.5
	Welding Neck			**27** 1.5	**48** 1.5	**67** 1.5	**96** 1.5	**130** 1.5	**202** 1.5	**451** 1.5
	Lap Joint			**22** 1.5	**45** 1.5	**52** 1.5	**93** 1.5	**125** 1.5	**208** 1.5	**387** 1.5
	Blind	**28** 1.5	**51** 1.5	**29** 1.5	**56** 1.5	**71** 1.5	**101** 1.5	**133** 1.5	**197** 1.5	**418** 1.5
FLANGED FITTINGS	S.R. 90° Elbow	**74** 4.3	**125** 4.4	**90** 4.3	**147** 4.4	**184** 4.6	**275** 4.8	**375** 5	**566** 5.3	
	L.R. 90° Elbow	**91** 4.9	**145** 4.9	**126** 4.9	**182** 4.9					
	45° Elbow	**66** 3.8	**115** 3.9	**82** 3.8	**132** 3.9	**149** 4.1	**240** 4.3	**320** 4.3	**476** 4.6	
	Tee	**114** 6.5	**195** 6.6	**149** 6.5	**217** 6.6	**279** 6.9	**400** 7.2	**565** 7.5	**839** 8	
VALVES	Flanged Bonnet Gate	**172** 7.3	**359** 8	**190** 4.3	**360** 5	**435** 5.5	**620** 5.8	**835** 6	**1595** 7	
	Flanged Bonnet Globe or Angle	**184** 7.8	**345** 8.2	**185** 4.4	**275** 5	**415** 5.3	**645** 5.8	**765** 6	**1800** 7	
	Flanged Bonnet Check	**154** 7.8	**286** 8.2	**150** 4.8	**200** 5	**360** 5.4	**445** 6	**800** 6.4	**1630** 7	
	Pressure Seal Bonnet—Gate						**580** 3.5	**750** 4	**1215** 5	
	Pressure Seal Bonnet—Globe							**730** 4	**780** 5	

Boldface type is weight in pounds. Lightface type beneath weight is weight factor for insulation.

Insulation thicknesses and weights are based on average conditions and do not constitute a recommendation for specific thicknesses of materials. Insulation weights are based on 85% magnesia and hydrous calcium silicate at 11 lbs/cubic foot. The listed thicknesses and weights of combination covering are the sums of the inner layer of diatomaceous earth at 21 lbs/cubic foot and the outer layer at 11 lbs/cubic foot.

Insulation weights include allowances for wire, cement, canvas, bands and paint, but not special surface finishes.

To find the weight of covering on flanges, valves or fittings, multiply the weight factor by the weight per foot of covering used on straight pipe.

Valve weights are approximate. When possible, obtain weights from the manufacturer.

Cast iron valve weights are for flanged end valves; steel weights for welding end valves.

All flanged fitting, flanged valve and flange weights include the proportional weight of bolts or studs to make up all joints.

* 16 lb cu. ft. density.

WEIGHTS OF PIPING MATERIALS

8.625″ O.D. 8″ PIPE

PIPE	Schedule No.	20	30	40	60	80	100	120	140		160		
	Wall Designation			Std.		XS				XXS			
	Thickness—In.	.250	.277	.322	.406	.500	.593	.718	.812	.875	.906		
	Pipe—Lbs/Ft	**22.36**	**24.70**	**28.55**	**35.64**	**43.4**	**50.9**	**60.6**	**67.8**	**72.4**	**74.7**		
	Water—Lbs/Ft	22.48	22.18	21.69	20.79	19.8	18.8	17.6	16.7	16.1	15.8		
WELDING FITTINGS	L.R. 90° Elbow			**46** 2		**69** 2				**114** 2	**117** 2		
	S.R. 90° Elbow			**31** 1.3		**46** 1.3							
	L.R. 45° Elbow			**23** .8		**34** .8				**55** .8	**56** .8		
	Tee			**54** 1.8		**76** 1.8				**118** 1.8	**120** 1.8		
	Lateral			**76** 3.8		**140** 3.8							
	Reducer			**13.9** .5		**20** .5				**32** .5	**33** .5		
	Cap			**11.3** 1		**16.3** 1				**31** 1	**32** 1		

	Temperature Range °F	100-199	200-299	300-399	400-499	500-599	600-699	700-799	800-899	900-999	1000-1099	1100-1200
INSULATION: 85% Magnesia Calcium Silicate	Nom. Thick., In.	1½	1½	2	2	2½	3	3½	3½	4	4	4½
	Lbs/Ft	**4.13**	**4.13**	**5.64**	**5.64**	**7.85**	**9.48**	**11.5**	**11.5**	**13.8**	**13.8**	**16.0**
Combination	Nom. Thick., In.						3	3½	3½	4	4	4½
	Lbs/Ft						**12.9**	**16.2**	**16.2**	**20.4**	**20.4**	**23.8**
*Asbestos Fiber—Sodium Silicate	Nom. Thick., In.	1½	1½	1½	1½	1½	2½	2½	3½	3½	4½	4½
	Lbs/Ft	**5.38**	**5.38**	**5.38**	**5.38**	**5.38**	**10.60**	**10.60**	**15.85**	**15.85**	**20.85**	**20.85**

	Pressure Rating psi	Cast Iron		Steel						
		125	250	150	300	400	600	900	1500	2500
FLANGES	Screwed or Slip-On	**34** 1.5	**64** 1.5	**33** 1.5	**67** 1.5	**82** 1.5	**135** 1.5	**207** 1.5	**319** 1.5	**601** 1.5
	Welding Neck			**42** 1.5	**76** 1.5	**104** 1.5	**137** 1.5	**222** 1.5	**334** 1.5	**692** 1.5
	Lap Joint			**33** 1.5	**67** 1.5	**79** 1.5	**132** 1.5	**223** 1.5	**347** 1.5	**587** 1.5
	Blind	**45** 1.5	**83** 1.5	**48** 1.5	**90** 1.5	**115** 1.5	**159** 1.5	**232** 1.5	**363** 1.5	**649** 1.5
FLANGED FITTINGS	S.R. 90° Elbow	**117** 4.5	**201** 4.7	**157** 4.5	**238** 4.7	**310** 5	**435** 5.2	**639** 5.4	**995** 5.7	
	L.R. 90° Elbow	**152** 5.3	**236** 5.3	**202** 5.3	**283** 5.3					
	45° Elbow	**101** 3.9	**171** 4	**127** 3.9	**203** 4	**215** 4.1	**360** 4.4	**507** 4.5	**870** 4.8	
	Tee	**175** 6.8	**304** 7.1	**230** 6.8	**337** 7.1	**445** 7.5	**610** 7.8	**978** 8.1	**1465** 8.6	
VALVES	Flanged Bonnet Gate	**251** 7.5	**583** 8.1	**305** 4.5	**505** 5.1	**730** 6	**960** 6.3	**1180** 6.6	**2740** 7	
	Flanged Bonnet Globe or Angle	**317** 8.4	**554** 8.6	**475** 5.4	**505** 5.5	**610** 5.9	**1130** 6.3	**1160** 6.3	**2865** 7	
	Flanged Bonnet Check	**302** 8.4	**454** 8.6	**235** 5.2	**310** 5.3	**475** 5.6	**725** 6	**1140** 6.4	**2075** 7	
	Pressure Seal Bonnet—Gate						**925** 4.5	**1185** 4.7	**2345** 5.5	
	Pressure Seal Bonnet—Globe							**1550** 4	**1680** 5	

Boldface type is weight in pounds. Lightface type beneath weight is weight factor for insulation.

Insulation thicknesses and weights are based on average conditions and do not constitute a recommendation for specific thicknesses of materials. Insulation weights are based on 85% magnesia and hydrous calcium silicate at 11 lbs/cubic foot. The listed thicknesses and weights of combination covering are the sums of the inner layer of diatomaceous earth at 21 lbs/cubic foot and the outer layer at 11 lbs/cubic foot.

Insulation weights include allowances for wire, cement, canvas, bands and paint, but not special surface finishes.

To find the weight of covering on flanges, valves or fittings, multiply the weight factor by the weight per foot of covering used on straight pipe.

Valve weights are approximate. When possible, obtain weights from the manufacturer.

Cast iron valve weights are for flanged end valves; steel weights for welding end valves.

All flanged fitting, flanged valve and flange weights include the proportional weight of bolts or studs to make up all joints.

* 16 lb cu. ft. density.

10″ PIPE 10.750″ O.D. WEIGHTS OF PIPING MATERIALS

PIPE											
Schedule No.	20	30	40	60	80	100	120	140	160		
Wall Designation			Std.	XS							
Thickness—In	.250	.307	.365	.500	.593	.718	.843	1.000	1.125		
Pipe—Lbs/Ft	**28.04**	**34.24**	**40.5**	**54.7**	**64.3**	**76.9**	**89.2**	**104.1**	**115.7**		
Water—Lbs/Ft	**35.77**	**34.98**	**34.1**	**32.3**	**31.1**	**29.5**	**28.0**	**26.1**	**24.6**		

WELDING FITTINGS	20	30	40	60	80	100	120	140	160		
L.R. 90° Elbow			**82** 2.5	**109** 2.5					**226** 2.5		
S.R. 90° Elbow			**54** 1.7	**73** 1.7							
L.R. 45° Elbow			**40** 1	**54** 1					**109** 1		
Tee			**91** 2.1	**118** 2.1					**222** 2.1		
Lateral			**124** 4.4	**202** 4.4							
Reducer			**23** .6	**31** .6					**58** .6		
Cap			**20** 1.3	**26** 1.3					**54** 1.3		

INSULATION		100-199	200-299	300-399	400-499	500-599	600-699	700-799	800-899	900-999	1000-1099	1100-1200
	Temperature Range °F											
85% Magnesia Calcium Silicate	Nom. Thick., In.	1½	1½	2	2½	2½	3	3½	3½	4	4	4½
	Lbs/Ft	**5.20**	**5.20**	**7.07**	**8.93**	**8.93**	**11.0**	**13.2**	**13.2**	**15.5**	**15.5**	**18.1**
Combination	Nom. Thick., In.						3	3½	3½	4	4	4½
	Lbs/Ft						**15.4**	**19.3**	**19.3**	**23**	**23**	**27.2**
*Asbestos Fiber—Sodium Silicate	Nom. Thick., In.	1½	1½	1½	1½	1½	2½	2½	3½	3½	4½	4½
	Lbs/Ft	**6.77**	**6.77**	**6.77**	**6.77**	**6.77**	**12.03**	**12.03**	**18.39**	**18.39**	**25.21**	**25.21**

	Pressure Rating psi	Cast Iron		Steel						
		125	250	150	300	400	600	900	1500	2500
FLANGES	Screwed or Slip-On	**53** 1.5	**97** 1.5	**51** 1.5	**100** 1.5	**117** 1.5	**213** 1.5	**293** 1.5	**528** 1.5	**1148** 1.5
	Welding Neck			**60** 1.5	**110** 1.5	**152** 1.5	**225** 1.5	**316** 1.5	**546** 1.5	**1291** 1.5
	Lap Joint			**51** 1.5	**110** 1.5	**138** 1.5	**231** 1.5	**325** 1.5	**577** 1.5	**1120** 1.5
	Blind	**71** 1.5	**136** 1.5	**78** 1.5	**146** 1.5	**181** 1.5	**267** 1.5	**338** 1.5	**599** 1.5	**1248** 1.5
FLANGED FITTINGS	S.R. 90° Elbow	**190** 4.8	**323** 4.9	**240** 4.8	**343** 4.9	**462** 5.2	**747** 5.6	**995** 5.8		
	L.R. 90° Elbow	**245** 5.8	**383** 5.8	**290** 5.8	**438** 5.8					
	45° Elbow	**160** 4.1	**273** 4.2	**185** 4.1	**288** 4.2	**332** 4.3	**572** 4.6	**732** 4.7		
	Tee	**293** 7.2	**479** 7.4	**353** 7.2	**527** 7.4	**578** 7.8	**1007** 8.4	**1417** 8.7		
VALVES	Flanged Bonnet Gate	**471** 7.7	**899** 8.3	**455** 4.5	**750** 5	**1035** 6	**1575** 6.9	**2140** 7.1	**3690** 8	
	Flanged Bonnet Globe or Angle	**541** 9.1	**943** 9.1	**485** 4.5	**855** 5.5	**1070** 6	**1500** 6.3	**2500** 6.8	**4160** 8	
	Flanged Bonnet Check	**453** 9.1	**751** 9.1	**370** 6	**485** 6.1	**605** 6.3	**1030** 6.8	**1350** 7	**2280** 7.5	
	Pressure Seal Bonnet—Gate						**1450** 4.9	**1860** 5.5	**3150** 6	
	Pressure Seal Bonnet—Globe							**1800** 5	**1910** 6	

Boldface type is weight in pounds. Lightface type beneath weight is weight factor for insulation.

Insulation thicknesses and weights are based on average conditions and do not constitute a recommendation for specific thicknesses of materials. Insulation weights are based on 85% magnesia and hydrous calcium silicate at 11 lbs/cubic foot. The listed thicknesses and weights of combination covering are the sums of the inner layer of diatomaceous earth at 21 lbs/cubic foot and the outer layer at 11 lbs/cubic foot.

Insulation weights include allowances for wire, cement, canvas, bands and paint, but not special surface finishes.

To find the weight of covering on flanges, valves or fittings, multiply the weight factor by the weight per foot of covering used on straight pipe.

Valve weights are approximate. When possible, obtain weights from the manufacturer.

Cast iron valve weights are for flanged end valves; steel weights for welding end valves.

All flanged fitting, flanged valve and flange weights include the proportional weight of bolts or studs to make up all joints.

* 16 lb cu. ft. density.

WEIGHTS OF PIPING MATERIALS 12.750″ O.D. 12″ PIPE

PIPE	Schedule No.	20	30		40		60	80	100	120	140	160	
	Wall Designation			Std.		XS							
	Thickness—In.	.250	.330	.375	.406	.500	.562	.687	.843	1.000	1.125	1.312	
	Pipe—Lbs/Ft	**33.38**	**43.8**	**49.6**	**53.5**	**65.4**	**73.2**	**88.5**	**107.2**	**125.5**	**139.7**	**160.3**	
	Water—Lbs/Ft	**51.10**	**49.7**	**49.0**	**48.5**	**47.0**	**46.0**	**44.0**	**41.6**	**39.3**	**37.5**	**34.9**	
WELDING FITTINGS	L.R. 90° Elbow			**119** 3		**157** 3						**375** 3	
	S.R. 90° Elbow			**80** 2		**104** 2							
	L.R. 45° Elbow			**60** 1.3		**78** 1.3						**181** 1.3	
	Tee			**132** 2.5		**167** 2.5						**360** 2.5	
	Lateral			**180** 5.4		**273** 5.4							
	Reducer			**33** .7		**44** .7						**94** .7	
	Cap			**30** 1.5		**38** 1.5						**89** 1.5	

INSULATION	Temperature Range °F		100-199	200-299	300-399	400-499	500-599	600-699	700-799	800-899	900-999	1000-1099	1100-1200
	85% Magnesia Calcium Silicate	Nom. Thick., In.	1½	1½	2	2½	3	3	3½	4	4	4½	4½
		Lbs/Ft	**6.04**	**6.04**	**8.13**	**10.5**	**12.7**	**12.7**	**15.1**	**17.9**	**17.9**	**20.4**	**20.4**
	Combination	Nom. Thick., In.						3	3½	4	4	4½	4½
		Lbs/Ft						**17.7**	**21.9**	**26.7**	**26.7**	**31.1**	**31.1**
	*Asbestos Fiber—Sodium Silicate	Nom. Thick., In.	1½	1½	1½	1½	1½	2½	2½	4	4	5	5
		Lbs/Ft	**5.22**	**5.22**	**5.22**	**5.22**	**5.22**	**14.20**	**14.20**	**24.64**	**24.64**	**32.40**	**32.40**

	Pressure Rating psi	Cast Iron		Steel						
		125	250	150	300	400	600	900	1500	2500
FLANGES	Screwed or Slip-On	**71** 1.5	**137** 1.5	**72** 1.5	**140** 1.5	**164** 1.5	**261** 1.5	**388** 1.5	**820** 1.5	**1611** 1.5
	Welding Neck			**88** 1.5	**163** 1.5	**212** 1.5	**272** 1.5	**434** 1.5	**843** 1.5	**1919** 1.5
	Lap Joint			**72** 1.5	**164** 1.5	**187** 1.5	**286** 1.5	**433** 1.5	**902** 1.5	**1573** 1.5
	Blind	**96** 1.5	**177** 1.5	**118** 1.5	**209** 1.5	**261** 1.5	**341** 1.5	**475** 1.5	**928** 1.5	**1775** 1.5
FLANGED FITTINGS	S.R. 90° Elbow	**265** 5	**453** 5.2	**345** 5	**509** 5.2	**669** 5.5	**815** 5.8	**1474** 6.2		
	L.R. 90° Elbow	**375** 6.2	**553** 6.2	**485** 6.2	**624** 6.2			**1598** 6.2		
	45° Elbow	**235** 4.3	**383** 4.3	**282** 4.3	**414** 4.3	**469** 4.5	**705** 4.7	**1124** 4.8		
	Tee	**403** 7.5	**684** 7.8	**513** 7.5	**754** 7.8	**943** 8.3	**1361** 8.7	**1928** 9.3		
VALVES	Flanged Bonnet Gate	**687** 7.8	**1298** 8.5	**635** 4	**1015** 5	**1420** 5.5	**2155** 7	**2770** 7.2	**4650** 8	
	Flanged Bonnet Globe or Angle	**808** 9.4	**1200** 9.5	**710** 5	**1410** 5.5					
	Flanged Bonnet Check	**674** 9.4	**1160** 9.5	**560** 6	**720** 6.5		**1410** 7.2	**2600** 8	**3370** 8	
	Pressure Seal Bonnet—Gate						**1975** 5.5	**2560** 6	**4515** 7	
	Pressure Seal Bonnet—Globe									

Boldface type is weight in pounds. Lightface type beneath weight is weight factor for insulation.

Insulation thicknesses and weights are based on average conditions and do not constitute a recommendation for specific thicknesses of materials. Insulation weights are based on 85% magnesia and hydrous calcium silicate at 11 lbs/cubic foot. The listed thicknesses and weights of combination covering are the sums of the inner layer of diatomaceous earth at 21 lbs/cubic foot and the outer layer at 11 lbs/cubic foot.

Insulation weights include allowances for wire, cement, canvas, bands and paint, but not special surface finishes.

To find the weight of covering on flanges, valves or fittings, multiply the weight factor by the weight per foot of covering used on straight pipe.

Valve weights are approximate. When possible, obtain weights from the manufacturer.

Cast iron valve weights are for flanged end valves; steel weights for welding end valves.

All flanged fitting, flanged valve and flange weights include the proportional weight of bolts or studs to make up all joints.

* 16 lb cu. ft. density.

14″ PIPE 14″ O.D. WEIGHTS OF PIPING MATERIALS

PIPE	Schedule No.	10	20	30	40		60	80	100	120	140	160
	Wall Designation			Std.		XS						
	Thickness—In.	.250	.312	.375	.438	.500	.593	.750	.937	1.093	1.250	1.406
	Pipe—Lbs/Ft	**36.71**	**45.7**	**54.6**	**63.4**	**72.1**	**84.9**	**106.1**	**130.7**	**150.7**	**170.2**	**189.1**
	Water—Lbs/Ft	**62.06**	**60.92**	**59.7**	**58.7**	**57.5**	**55.9**	**53.2**	**50.0**	**47.5**	**45.0**	**42.6**
WELDING FITTINGS	L.R. 90° Elbow			**154** 3.5		**202** 3.5						
	S.R. 90° Elbow			**102** 2.3		**135** 2.3						
	L.R. 45° Elbow			**77** 1.5		**100** 1.5						
	Tee			**159** 2.8		**203** 2.8						
	Lateral			**218** 5.8		**340** 5.8						
	Reducer			**63** 1.1		**83** 1.1						
	Cap			**35** 1.7		**46** 1.7						

INSULATION	Temperature Range °F		100-199	200-299	300-399	400-499	500-599	600-699	700-799	800-899	900-999	1000-1099	1100-1200
	85% Magnesia Calcium Silicate	Nom. Thick., In.	1½	1½	2	2½	3	3	3½	4	4	4½	4½
		Lbs/Ft	**6.16**	**6.16**	**8.38**	**10.7**	**13.1**	**13.1**	**15.8**	**18.5**	**18.5**	**21.3**	**21.3**
	Combination	Nom. Thick., In.						3	3½	4	4	4½	4½
		Lbs/Ft						**18.2**	**22.8**	**27.5**	**27.5**	**32.4**	**32.4**
	*Asbestos Fiber—Sodium Silicate	Nom. Thick., In.	1½	1½	1½	2	2	3	3	4	4	5	5
		Lbs/Ft	**7.90**	**7.90**	**7.90**	**11.18**	**11.18**	**18.00**	**18.00**	**25.42**	**25.42**	**33.53**	**33.53**

	Pressure Rating psi	Cast Iron		Steel						
		125	250	150	300	400	600	900	1500	2500
FLANGES	Screwed or Slip-On	**93** 1.5	**184** 1.5	**96** 1.5	**195** 1.5	**235** 1.5	**318** 1.5	**460** 1.5	**1016** 1.5	
	Welding Neck			**113** 1.5	**217** 1.5	**277** 1.5	**406** 1.5	**642** 1.5	**1241** 1.5	
	Lap Joint			**110** 1.5	**220** 1.5	**254** 1.5	**349** 1.5	**477** 1.5	**1076** 1.5	
	Blind	**126** 1.5	**239** 1.5	**142** 1.5	**267** 1.5	**354** 1.5	**437** 1.5	**574** 1.5		
FLANGED FITTINGS	S.R. 90° Elbow	**372** 5.3	**617** 5.5	**497** 5.3	**632** 5.5	**664** 5.7	**918** 5.9	**1549** 6.4		
	L.R. 90° Elbow	**492** 6.6	**767** 6.6	**622** 6.6	**772** 6.6					
	45° Elbow	**292** 4.3	**497** 4.4	**377** 4.3	**587** 4.4	**638** 4.6	**883** 4.8	**1246** 4.9		
	Tee	**563** 8	**956** 8.4	**683** 8	**968** 8.3	**1131** 8.6	**1652** 8.9	**2318** 9.6		
VALVES	Flanged Bonnet Gate	**921** 7.9	**1762** 8.8	**905** 4.9	**1525** 6	**1920** 6.3	**2960** 7	**4170** 8	**6425** 8.8	
	Flanged Bonnet Globe or Angle	**1171** 9.9								
	Flanged Bonnet Check	**885** 9.9		**1010** 5	**1155** 5.2					
	Pressure Seal Bonnet—Gate						**2620** 6	**3475** 6.5	**6380** 7.5	
	Pressure Seal Bonnet—Globe									

Boldface type is weight in pounds. Lightface type beneath weight is weight factor for insulation.

Insulation thicknesses and weights are based on average conditions and do not constitute a recommendation for specific thicknesses of materials. Insulation weights are based on 85% magnesia and hydrous calcium silicate at 11 lbs/cubic foot. The listed thicknesses and weights of combination covering are the sums of the inner layer of diatomaceous earth at 21 lbs/cubic foot and the outer layer at 11 lbs/cubic foot.

Insulation weights include allowances for wire, cement, canvas, bands and paint, but not special surface finishes.

To find the weight of covering on flanges, valves or fittings, multiply the weight factor by the weight per foot of covering used on straight pipe.

Valve weights are approximate. When possible, obtain weights from the manufacturer.

Cast iron valve weights are for flanged end valves; steel weights for welding end valves.

All flanged fitting, flanged valve and flange weights include the proportional weight of bolts or studs to make up all joints.

* 16 lb cu. ft. density

WEIGHTS OF PIPING MATERIALS

16″ O.D. 16″ PIPE

	Schedule No.	10	20	30	40	60	80	100	120	140	160		
PIPE	Wall Designation			Std.	XS								
	Thickness—In.	.250	.312	.375	.500	.656	.843	1.031	1.218	1.438	1.593		
	Pipe—Lbs/Ft	**42.1**	**52.4**	**62.6**	**82.8**	**107.5**	**136.5**	**164.8**	**192.3**	**223.6**	**245.1**		
	Water—Lbs/Ft	**81.8**	**80.5**	**79.1**	**76.5**	**73.4**	**69.7**	**66.1**	**62.6**	**58.6**	**55.9**		
WELDING FITTINGS	L.R. 90° Elbow			**201** 4	**265** 4								
	S.R. 90° Elbow			**135** 2.5	**177** 2.5								
	L.R. 45° Elbow			**100** 1.7	**132** 1.7								
	Tee			**202** 3.2	**257** 3.2								
	Lateral			**275** 6.7	**433** 6.7								
	Reducer			**78** 1.2	**102** 1.2								
	Cap			**44** 1.8	**58** 1.8								

		Temperature Range °F	100-199	200-299	300-399	400-499	500-599	600-699	700-799	800-899	900-999	1000-1099	1100-1200
INSULATION	85% Magnesia Calcium Silicate	Nom. Thick., In.	1½	1½	2	2½	3	3	3½	4	4	4½	4½
		Lbs/Ft	**6.90**	**6.90**	**9.33**	**12.0**	**14.6**	**14.6**	**17.5**	**20.5**	**20.5**	**23.6**	**23.6**
	Combination	Nom. Thick., In.						3	3½	4	4	4½	4½
		Lbs/Ft						**20.3**	**25.2**	**30.7**	**30.7**	**36.0**	**36.0**
	*Asbestos Fiber—Sodium Silicate	Nom. Thick., In.	1½	1½	1½	1½	1½	2½	2½	3½	3½	4½	4½
		Lbs/Ft	**9.26**	**9.26**	**9.26**	**9.26**	**9.26**	**16.35**	**16.35**	**24.11**	**24.11**	**32.57**	**32.57**

	Pressure Rating psi	Cast Iron		Steel						
		125	250	150	300	400	600	900	1500	2500
FLANGES	Screwed or Slip-On	**120** 1.5	**233** 1.5	**108** 1.5	**262** 1.5	**310** 1.5	**442** 1.5	**559** 1.5	**1297** 1.5	
	Welding Neck			**142** 1.5	**288** 1.5	**351** 1.5	**577** 1.5	**785** 1.5	**1597** 1.5	
	Lap Joint			**143** 1.5	**282** 1.5	**337** 1.5	**476** 1.5	**588** 1.5	**1372** 1.5	
	Blind	**175** 1.5	**308** 1.5	**185** 1.5	**349** 1.5	**455** 1.5	**603** 1.5	**719** 1.5		
FLANGED FITTINGS	S.R. 90° Elbow	**501** 5.5	**826** 5.8	**656** 5.5	**958** 5.8	**1014** 6	**1402** 6.3	**1886** 6.7		
	L.R. 90° Elbow	**701** 7	**1036** 7	**781** 7	**1058** 7					
	45° Elbow	**391** 4.3	**696** 4.6	**481** 4.3	**708** 4.6	**839** 4.7	**1212** 5	**1586** 5		
	Tee	**746** 8.3	**1263** 8.7	**961** 8.3	**1404** 8.6	**1671** 9	**2128** 9.4	**3054** 10		
VALVES	Flanged Bonnet Gate	**1254** 8	**2321** 9	**1190** 5	**2015** 7	**2300** 7.2	**3675** 7.9	**4950** 8.2	**7875** 9	
	Flanged Bonnet Globe or Angle									
	Flanged Bonnet Check	**1166** 10.5			**1225** 6					
	Pressure Seal Bonnet—Gate						**3230** 7		**8130** 8	
	Pressure Seal Bonnet—Globe									

Boldface type is weight in pounds. Lightface type beneath weight is weight factor for insulation.

Insulation thicknesses and weights are based on average conditions and do not constitute a recommendation for specific thicknesses of materials. Insulation weights are based on 85% magnesia and hydrous calcium silicate at 11 lbs/cubic foot. The listed thicknesses and weights of combination covering are the sums of the inner layer of diatomaceous earth at 21 lbs/cubic foot and the outer layer at 11 lbs/cubic foot.

Insulation weights include allowances for wire, cement, canvas, bands and paint, but not special surface finishes.

To find the weight of covering on flanges, valves or fittings, multiply the weight factor by the weight per foot of covering used on straight pipe.

Valve weights are approximate. When possible, obtain weights from the manufacturer.

Cast iron valve weights are for flanged end valves; steel weights for welding end valves.

All flanged fitting, flanged valve and flange weights include the proportional weight of bolts or studs to make up all joints.

* 16 lb cu. ft. density.

18″ PIPE 18″ O.D. WEIGHTS OF PIPING MATERIALS

PIPE	Schedule No.	10	20		30		40	60	80	100	120	140	160
	Wall Designation			Std.		XS							
	Thickness—In.	.250	.312	.375	.438	.500	.562	.750	.937	1.156	1.375	1.562	1.781
	Pipe—Lbs/Ft	**47.4**	**59.0**	**70.6**	**82.1**	**93.5**	**104.8**	**138.2**	**170.8**	**208.0**	**244.1**	**274.2**	**308.5**
	Water—Lbs/Ft	**104.3**	**102.8**	**101.2**	**99.9**	**98.4**	**97**	**92.7**	**88.5**	**83.7**	**79.2**	**75.3**	**71.0**
WELDING FITTINGS	L.R. 90° Elbow			**256** 4.5		**338** 4.5							
	S.R. 90° Elbow			**171** 2.8		**225** 2.8							
	L.R. 45° Elbow			**128** 1.9		**168** 1.9							
	Tee			**258** 3.6		**328** 3.6							
	Lateral			**326** 7.5		**526** 7.5							
	Reducer			**94** 1.3		**123** 1.3							
	Cap			**57** 2.1		**75** 2.1							

INSULATION	Temperature Range °F		100-199	200-299	300-399	400-499	500-599	600-699	700-799	800-899	900-999	1000-1099	1100-1200
	85% Magnesia Calcium Silicate	Nom. Thick., In.	1½	1½	2	2½	3	3	3½	4	4	4½	4½
		Lbs/Ft	**7.73**	**7.73**	**10.4**	**13.3**	**16.3**	**16.3**	**19.3**	**22.6**	**22.6**	**25.9**	**25.9**
	Combination	Nom. Thick., In.						3	3½	4	4	4½	4½
		Lbs/Ft						**22.7**	**28.0**	**33.8**	**33.8**	**39.5**	**39.5**
	*Asbestos Fiber—Sodium Silicate	Nom. Thick., In.	1½	1½	1½	2	2	3	3	4	4	5	5
		Lbs/Ft	**9.93**	**9.93**	**9.93**	**13.72**	**13.72**	**21.84**	**21.84**	**31.22**	**31.22**	**40.77**	**40.77**

	Pressure Rating psi	Cast Iron		Steel						
		125	250	150	300	400	600	900	1500	2500
FLANGES	Screwed or Slip-On	**140** 1.5		**140** 1.5	**331** 1.5	**380** 1.5	**573** 1.5	**797** 1.5	**1694** 1.5	
	Welding Neck			**160** 1.5	**355** 1.5	**430** 1.5	**652** 1.5	**1074** 1.5	**2069** 1.5	
	Lap Joint			**166** 1.5	**355** 1.5	**415** 1.5	**566** 1.5	**820** 1.5	**1769** 1.5	
	Blind	**210** 1.5	**396** 1.5	**229** 1.5	**440** 1.5	**572** 1.5	**762** 1.5	**1030** 1.5		
FLANGED FITTINGS	S.R. 90° Elbow	**621** 5.8	**1060** 6	**711** 5.8	**1126** 6	**1340** 6.2	**1793** 6.6	**2817** 7		
	L.R. 90° Elbow	**881** 7.4	**1350** 7.4	**941** 7.4	**1426** 7.4					
	45° Elbow	**461** 4.4	**870** 4.7	**521** 4.4	**901** 4.7	**1040** 4.8	**1543** 5	**2252** 5.2		
	Tee	**921** 8.6	**1625** 9	**1010** 8.6	**1602** 9	**1909** 9.3	**2690** 9.9	**4327** 10.5		
VALVES	Flanged Bonnet Gate	**1629** 8.2	**2578** 9.3	**1510** 6	**2505** 6.5	**3765** 7	**4460** 7.8	**6675** 8.5		
	Flanged Bonnet Globe or Angle									
	Flanged Bonnet Check	**1371** 10.5								
	Pressure Seal Bonnet—Gate						**3100** 5.5	**3400** 5.6	**4200** 6	
	Pressure Seal Bonnet—Globe									

Boldface type is weight in pounds. Lightface type beneath weight is weight factor for insulation.

Insulation thicknesses and weights are based on average conditions and do not constitute a recommendation for specific thicknesses of materials. Insulation weights are based on 85% magnesia and hydrous calcium silicate at 11 lbs/cubic foot. The listed thicknesses and weights of combination covering are the sums of the inner layer of diatomaceous earth at 21 lbs/cubic foot and the outer layer at 11 lbs/cubic foot.

Insulation weights include allowances for wire, cement, canvas, bands and paint, but not special surface finishes.

To find the weight of covering on flanges, valves or fittings, multiply the weight factor by the weight per foot of covering used on straight pipe.

Valve weights are approximate. When possible, obtain weights from the manufacturer.

Cast iron valve weights are for flanged end valves; steel weights for welding end valves.

All flanged fitting, flanged valve and flange weights include the proportional weight of bolts or studs to make up all joints.

*** 16 lb cu. ft. density.**

WEIGHTS OF PIPING MATERIALS

20″ O.D. **20″** PIPE

PIPE	Schedule No.	10	20	30	40	60	80	100	120	140	160		
	Wall Designation		Std.	XS									
	Thickness—In.	.250	.375	.500	.593	.812	1.031	1.281	1.500	1.750	1.968		
	Pipe—Lbs/Ft	**52.7**	**78.6**	**104.1**	**122.9**	**166.4**	**208.9**	**256.1**	**296.4**	**341.1**	**379.0**		
	Water—Lbs/Ft	**129.5**	**126.0**	**122.8**	**120.4**	**115.0**	**109.4**	**103.4**	**98.3**	**92.6**	**87.9**		
WELDING FITTINGS	L.R. 90° Elbow		**317** 5	**419** 5									
	S.R. 90° Elbow		**212** 3.4	**278** 3.4									
	L.R. 45° Elbow		**158** 2.1	**208** 2.1									
	Tee		**321** 4	**407** 4									
	Lateral		**396** 8.3	**628** 8.3									
	Reducer		**142** 1.7	**186** 1.7									
	Cap		**72** 2.3	**94** 2.3									

INSULATION	Temperature Range °F	100-199	200-299	300-399	400-499	500-599	600-699	700-799	800-899	900-999	1000-1099	1100-1200
85% Magnesia Calcium Silicate	Nom. Thick., In.	1½	1½	2	2½	3	3	3½	4	4	4½	4½
	Lbs/Ft	**8.45**	**8.45**	**11.6**	**14.6**	**17.7**	**17.7**	**21.1**	**24.6**	**24.6**	**28.1**	**28.1**
Combination	Nom. Thick., In.						3	3½	4	4	4½	4½
	Lbs/Ft						**24.7**	**30.7**	**37.0**	**37.0**	**43.1**	**43.1**
*Asbestos Fiber—Sodium Silicate	Nom. Thick., In.	1½	1½	1½	2	2	3	3	4	4	5	5
	Lbs/Ft	**10.96**	**10.96**	**10.96**	**14.86**	**14.86**	**24.24**	**24.24**	**33.79**	**33.79**	**44.03**	**44.03**

	Pressure Rating psi	Cast Iron		Steel						
		125	250	150	300	400	600	900	1500	2500
FLANGES	Screwed or Slip-On	**176** 1.5		**181** 1.5	**378** 1.5	**468** 1.5	**733** 1.5	**972** 1.5	**2114** 1.5	
	Welding Neck			**196** 1.5	**431** 1.5	**535** 1.5	**811** 1.5	**1344** 1.5	**2614** 1.5	
	Lap Joint			**211** 1.5	**428** 1.5	**510** 1.5	**725** 1.5	**1048** 1.5	**2189** 1.5	
	Blind	**276** 1.5	**487** 1.5	**298** 1.5	**545** 1.5	**711** 1.5	**976** 1.5	**1287** 1.5		
FLANGED FITTINGS	S.R. 90° Elbow	**792** 6	**1315** 6.3	**922** 6	**1375** 6.3	**1680** 6.5	**2314** 6.9	**3610** 7.3		
	L.R. 90° Elbow	**1132** 7.8	**1725** 7.8	**1352** 7.8	**1705** 7.8					
	45° Elbow	**592** 4.6	**1055** 4.8	**652** 4.6	**1105** 4.8	**1330** 4.9	**1917** 5.2	**2848** 5.4		
	Tee	**1178** 9	**2022** 9.5	**1378** 9	**1908** 9.5	**2370** 9.7	**3463** 10.1	**5520** 11		
VALVES	Flanged Bonnet Gate	**1934** 8.3	**3823** 9.5	**1855** 6	**3370** 7	**5700** 8	**5755** 8			
	Flanged Bonnet Globe or Angle									
	Flanged Bonnet Check	**1772** 11								
	Pressure Seal Bonnet—Gate									
	Pressure Seal Bonnet—Globe									

Boldface type is weight in pounds. Lightface type beneath weight is weight factor for insulation.

Insulation thicknesses and weights are based on average conditions and do not constitute a recommendation for specific thicknesses of materials. Insulation weights are based on 85% magnesia and hydrous calcium silicate at 11 lbs/cubic foot. The listed thicknesses and weights of combination covering are the sums of the inner layer of diatomaceous earth at 21 lbs/cubic foot and the outer layer at 11 lbs/cubic foot.

Insulation weights include allowances for wire, cement, canvas, bands and paint, but not special surface finishes.

To find the weight of covering on flanges, valves or fittings, multiply the weight factor by the weight per foot of covering used on straight pipe.

Valve weights are approximate. When possible, obtain weights from the manufacturer.

Cast iron valve weights are for flanged end valves; steel weights for welding end valves.

All flanged fitting, flanged valve and flange weights include the proportional weight of bolts or studs to make up all joints.

* 16 lb cu. ft. density.

24″ PIPE 24″ O.D. WEIGHTS OF PIPING MATERIALS

PIPE												
	Schedule No.	10	20		30	40	60	80	100	120	140	160
	Wall Designation		Std.	XS								
	Thickness—In.	.250	.375	.500	.562	.687	.968	1.218	1.531	1.812	2.062	2.343
	Pipe—Lbs/Ft	**63.4**	**94.6**	**125.5**	**140.8**	**171.2**	**238.1**	**296.4**	**367.4**	**429.4**	**483.1**	**541.9**
	Water—Lbs/Ft	**188.0**	**183.8**	**180.1**	**178.1**	**174.3**	**165.8**	**158.3**	**149.3**	**141.4**	**134.5**	**127.0**
WELDING FITTINGS	L.R. 90° Elbow		**458** 6	**606** 6								
	S.R. 90° Elbow		**305** 3.7	**404** 3.7								
	L.R. 45° Elbow		**229** 2.5	**302** 2.5								
	Tee		**445** 4.9	**563** 4.9								
	Lateral		**544** 10	**882** 10								
	Reducer		**167** 1.7	**220** 1.7								
	Cap		**102** 2.8	**134** 2.8								

INSULATION		Temperature Range °F	100-199	200-299	300-399	400-499	500-599	600-699	700-799	800-899	900-999	1000-1099	1100-1200
	85% Magnesia Calcium Silicate	Nom. Thick., In.	1½	1½	2	2½	3	3	3½	4	4	4½	4½
		Lbs/Ft	**10.0**	**10.0**	**13.4**	**17.0**	**21.0**	**21.0**	**24.8**	**28.7**	**28.7**	**32.9**	**32.9**
	Combination	Nom. Thick., In.						3	3½	4	4	4½	4½
		Lbs/Ft						**29.2**	**36.0**	**43.1**	**43.1**	**50.6**	**50.6**
	*Asbestos Fiber—Sodium Silicate	Nom. Thick., In.	1½	1½	1½	2	2	3	3	4½	4½	5	5
		Lbs/Ft	**13.55**	**13.55**	**13.55**	**18.44**	**18.44**	**28.38**	**28.38**	**45.06**	**45.06**	**50.97**	**50.97**

	Pressure Rating psi	Cast Iron		Steel						
		125	250	150	300	400	600	900	1500	2500
FLANGES	Screwed or Slip-On	**255** 1.5		**245** 1.5	**577** 1.5	**676** 1.5	**1056** 1.5	**1823** 1.5	**3378** 1.5	
	Welding Neck			**295** 1.5	**632** 1.5	**777** 1.5	**1157** 1.5	**2450** 1.5	**4153** 1.5	
	Lap Joint			**295** 1.5	**617** 1.5	**752** 1.5	**1046** 1.5	**2002** 1.5	**3478** 1.5	
	Blind	**405** 1.5	**757** 1.5	**446** 1.5	**841** 1.5	**1073** 1.5	**1355** 1.5	**2442** 1.5		
FLANGED FITTINGS	S.R. 90° Elbow	**1231** 6.7	**2014** 6.8	**1671** 6.7	**2174** 6.8	**2474** 7.1	**3506** 7.6	**6155** 8.1		
	L.R. 90° Elbow	**1711** 8.7	**2644** 8.7	**1821** 8.7	**2874** 8.7					
	45° Elbow	**871** 4.8	**1604** 5	**1121** 4.8	**1634** 5	**1974** 5.1	**2831** 5.5	**5124** 6		
	Tee	**1836** 10	**3061** 10.2	**2276** 10	**3161** 10.2	**3811** 10.6	**5184** 11.4	**9387** 12.1		
VALVES	Flanged Bonnet Gate	**3062** 8.5	**6484** 9.8	**2500** 5	**4675** 7	**6995** 8.7	**8020** 9.5			
	Flanged Bonnet Globe or Angle									
	Flanged Bonnet Check	**2956** 12								
	Pressure Seal Bonnet—Gate									
	Pressure Seal Bonnet—Globe									

Boldface type is weight in pounds. Lightface type beneath weight is weight factor for insulation.

Insulation thicknesses and weights are based on average conditions and do not constitute a recommendation for specific thicknesses of materials. Insulation weights are based on 85% magnesia and hydrous calcium silicate at 11 lbs/cubic foot. The listed thicknesses and weights of combination covering are the sums of the inner layer of diatomaceous earth at 21 lbs/cubic foot and the outer layer at 11 lbs/cubic foot.

Insulation weights include allowances for wire, cement, canvas, bands and paint, but not special surface finishes.

To find the weight of covering on flanges, valves or fittings, multiply the weight factor by the weight per foot of covering used on straight pipe.

Valve weights are approximate. When possible, obtain weights from the manufacturer.

Cast iron valve weights are for flanged end valves; steel weights for welding end valves.

All flanged fitting, flanged valve and flange weights include the proportional weight of bolts or studs to make up all joints.

* 16 lb cu. ft. density.

WEIGHTS OF PIPING MATERIALS

26″ O.D **26″** PIPE

PIPE	Schedule No.	10		20
	Wall Designation		Std.	XS
	Thickness—In.	.312	.375	.500
	Pipe—Lbs/Ft	**85.7**	**102.6**	**136.2**
	Water—Lbs/Ft	**219.2**	**216.8**	**212.5**
WELDING FITTINGS	L.R. 90° Elbow		**602** 8.5	**713** 8.5
	S.R. 90° Elbow		**359** 5	**474** 5
	L.R. 45° Elbow		**269** 3.5	**355** 3.5
	Tee		**634** 6.8	**794** 6.8
	Lateral			
	Reducer		**200** 2.5	**272** 2.5
	Cap		**110** 4.3	**145** 4.3

	Temperature Range °F	100-199	200-299	300-399	400-499	500-599	600-699	700-799	800-899	900-999	1000-1099	1100-1200
INSULATION: 85% Magnesia Calcium Silicate	Nom. Thick., In.	1½	1½	2	2½	3	3½	4	4½	5	5	6
	Lbs/Ft	**10.4**	**10.4**	**14.1**	**18.0**	**21.9**	26.0	30.2	34.6	39.1	39.1	48.4
Combination	Nom. Thick., In.						3½	4½	5½	6	6½	7
	Lbs/Ft						**37.0**	**51.9**	**67.8**	**76.0**	**84.5**	**93.2**
*Asbestos Fiber—Sodium Silicate	Nom. Thick., In.	2	2	2	2	2	3	3	4½	4½	5	5
	Lbs/Ft	18.93	18.93	18.93	18.93	18.93	29.87	29.87	47.60	47.60	53.85	53.85

	Pressure Rating psi	Cast Iron		Steel						
		125	250	150	300	400	600	900	1500	2500
FLANGES	Screwed or Slip-On			**250** 1.5	**570** 1.5	**650** 1.5	**950** 1.5	**1525** 1.5		
	Welding Neck			**300** 1.5	**670** 1.5	**750** 1.5	**1025** 1.5	**1575** 1.5		
	Lap Joint									
	Blind			**525** 1.5	**1050** 1.5	**1125** 1.5	**1525** 1.5	**2200** 1.5		
FLANGED FITTINGS	S.R. 90° Elbow									
	L.R. 90° Elbow									
	45° Elbow									
	Tee									
VALVES	Flanged Bonnet Gate									
	Flanged Bonnet Globe or Angle									
	Flanged Bonnet Check									
	Pressure Seal Bonnet—Gate									
	Pressure Seal Bonnet—Globe									

Boldface type is weight in pounds. Lightface type beneath weight is weight factor for insulation.

Insulation thicknesses and weights are based on average conditions and do not constitute a recommendation for specific thicknesses of materials. Insulation weights are based on 85% magnesia and hydrous calcium silicate at 11 lbs/cubic foot. The listed thicknesses and weights of combination covering are the sums of the inner layer of diatomaceous earth at 21 lbs/cubic foot and the outer layer at 11 lbs/cubic foot.

Insulation weights include allowances for wire, cement, canvas, bands and paint, but not special surface finishes.

To find the weight of covering on flanges, valves or fittings, multiply the weight factor by the weight per foot of covering used on straight pipe.

Valve weights are approximate. When possible, obtain weights from manufacturer. Cast iron valve weights are for flanged end valves; steel weights for welding end valves.

All flanged fitting, flanged valve and flange weights include the proportional weight of bolts or studs to make up all joints.

* 16 lb cu. ft. density.

28″ PIPE 28″ O.D.

WEIGHTS OF PIPING MATERIALS

PIPE	Schedule No.	10		20	30								
	Wall Designation		Std.	XS									
	Thickness—In.	.312	.375	.500	.625								
	Pipe—Lbs/Ft	**92.4**	**110.6**	**146.9**	**182.7**								
	Water—Lbs/Ft	**255.0**	**252.7**	**248.1**	**243.6**								
WELDING FITTINGS	L.R. 90° Elbow		**626** 9	**829** 9									
	S.R. 90° Elbow		**415** 5.4	**551** 5.4									
	L.R. 45° Elbow		**312** 3.6	**413** 3.6									
	Tee		**729** 7	**910** 7									
	Lateral												
	Reducer		**210** 2.7	**290** 2.7									
	Cap		**120** 4.5	**160** 4.5									

INSULATION	Temperature Range °F	100-199	200-299	300-399	400-499	500-599	600-699	700-799	800-899	900-999	1000-1099	1100-1200
85% Magnesia Calcium Silicate	Nom. Thick., In.	1½	1½	2	2½	3	3½	4	4½	5	5	6
	Lbs/Ft	**11.2**	**11.2**	**15.1**	**19.2**	**23.4**	27.8	32.3	36.9	41.6	41.6	51.4
Combination	Nom. Thick., In.						3½	4½	5½	6	6½	7
	Lbs/Ft						**39.5**	**55.4**	**72.2**	**80.9**	**89.8**	**99.0**
*Asbestos Fiber—Sodium Silicate	Nom. Thick., In.	2	2	2	2	2	3	3	4½	4½	5	5
	Lbs/Ft	20.26	20.26	20.26	20.26	20.26	31.90	31.90	52.51	52.51	59.17	59.17

	Pressure Rating psi	Cast Iron		Steel						
		125	250	150	300	400	600	900	1500	2500
FLANGES	Screwed or Slip-On			**285** 1.5	**720** 1.5	**780** 1.5	**1075** 1.5	**1800** 1.5		
	Welding Neck			**315** 1.5	**810** 1.5	**880** 1.5	**1175** 1.5	**1850** 1.5		
	Lap Joint									
	Blind			**620** 1.5	**1275** 1.5	**1425** 1.5	**1750** 1.5	**2575** 1.5		
FLANGED FITTINGS	S.R. 90° Elbow									
	L.R. 90° Elbow									
	45° Elbow									
	Tee									
VALVES	Flanged Bonnet Gate									
	Flanged Bonnet Globe or Angle									
	Flanged Bonnet Check									
	Pressure Seal Bonnet—Gate									
	Pressure Seal Bonnet—Globe									

Boldface type is weight in pounds. Lightface type beneath weight is weight factor for insulation.

Insulation thicknesses and weights are based on average conditions and do not constitute a recommendation for specific thicknesses of materials. Insulation weights are based on 85% magnesia and hydrous calcium silicate at 11 lbs/cubic foot. The listed thicknesses and weights of combination covering are the sums of the inner layer of diatomaceous earth at 21 lbs/cubic foot and the outer layer at 11 lbs/cubic foot.

Insulation weights include allowances for wire, cement, canvas, bands and paint, but not special surface finishes.

To find the weight of covering on flanges, valves or fittings, multiply the weight factor by the weight per foot of covering used on straight pipe.

Valve weights are approximate. When possible, obtain weights from manufacturer.

Cast iron valve weights are for flanged end valves; steel weights for welding end valves.

All flanged fitting, flanged valve and flange weights include the proportional weight of bolts or studs to make up all joints.

* 16 lb cu. ft. density.

WEIGHTS OF PIPING MATERIALS

30″ O.D. 30″ PIPE

PIPE	Schedule No.	10			20		30					
	Wall Designation		Std.		XS							
	Thickness—In.	.312	.375	.438	.500	.562	.625					
	Pipe—Lbs/Ft	**98.9**	**118.7**	**138.0**	**157.6**	**176.8**	**196.1**					
	Water—Lbs/Ft	**293.5**	**291.0**	**288.4**	**286.0**	**283.6**	**281.1**					
WELDING FITTINGS	L.R. 90° Elbow		**775** 10		**953** 10							
	S.R. 90° Elbow		**470** 5.9		**644** 5.9							
	L.R. 45° Elbow		**358** 3.9		**475** 3.9							
	Tee		**855** 7.8		**1065** 7.8							
	Lateral											
	Reducer		**220** 3.9		**315** 3.9							
	Cap		**125** 4.8		**175** 4.8							
	Temperature Range °F	100-199	200-299	300-399	400-499	500-599	600-699	700-799	800-899	900-999	1000-1099	1100-1200
INSULATION: 85% Magnesia Calcium Silicate	Nom. Thick., In.	1½	1½	2	2½	3	3½	4	4½	5	5	6
	Lbs/Ft	**11.9**	**11.9**	**16.1**	**20.5**	**25.0**	**29.5**	**34.3**	**39.1**	**44.1**	**44.1**	**54.4**
INSULATION: Combination	Nom. Thick., In.						3½	4½	5½	6	6½	7
	Lbs/Ft						**42.1**	**58.9**	**76.5**	**85.7**	**95.1**	**104.7**
INSULATION: *Asbestos Fiber—Sodium Silicate	Nom. Thick., In.	2½	2½	2½	2½	2½	3	3	4½	4½	5	5
	Lbs/Ft	**27.17**	**27.17**	**27.17**	**27.17**	**27.17**	**33.43**	**33.43**	**55.18**	**55.18**	**62.18**	**62.18**

	Pressure Rating psi	Cast Iron 125	Cast Iron 250	Steel 150	Steel 300	Steel 400	Steel 600	Steel 900	Steel 1500	Steel 2500
FLANGES	Screwed or Slip-On			**315** 1.5	**810** 1.5	**900** 1.5	**1175** 1.5	**2075** 1.5		
	Welding Neck			**360** 1.5	**930** 1.5	**1000** 1.5	**1300** 1.5	**2150** 1.5		
	Lap Joint									
	Blind			**720** 1.5	**1500** 1.5	**1675** 1.5	**2000** 1.5	**3025** 1.5		
FLANGED FITTINGS	S.R. 90° Elbow									
	L.R. 90° Elbow									
	45° Elbow									
	Tee									
VALVES	Flanged Bonnet Gate									
	Flanged Bonnet Globe or Angle									
	Flanged Bonnet Check									
	Pressure Seal Bonnet—Gate									
	Pressure Seal Bonnet—Globe									

* 16 lb cu. ft. density.

Boldface type is weight in pounds. Lightface type beneath weight is weight factor for insulation.

Insulation thicknesses and weights are based on average conditions and do not constitute a recommendation for specific thicknesses of materials. Insulation weights are based on 85% magnesia and hydrous calcium silicate at 11 lbs/cubic foot. The listed thicknesses and weights of combination covering are the sums of the inner layer of diatomaceous earth at 21 lbs/cubic foot and the outer layer at 11 lbs/cubic foot.

Insulation weights include allowances for wire, cement, canvas, bands and paint, but not special surface finishes.

To find the weight of covering on flanges, valves or fittings, multiply the weight factor by the weight per foot of covering used on straight pipe.

Valve weights are approximate. When possible, obtain weights from manufacturer.

Cast iron valve weights are for flanged end valves; steel weights for welding end valves.

All flanged fitting, flanged valve and flange weights include the proportional weight of bolts or studs to make up all joints.

32″ PIPE 32″ O.D. WEIGHTS OF PIPING MATERIALS

PIPE	Schedule No.	10		20	30	40
	Wall Designation		Std.	XS		
	Thickness—In.	.312	.375	.500	.625	.688
	Pipe—Lbs/Ft	**105.8**	**126.7**	**168.2**	**209.4**	**229.9**
	Water—Lbs/Ft	**335.0**	**332.3**	**327.0**	**321.8**	**319.2**
WELDING FITTINGS	L.R. 90° Elbow		**818** 10.5	**1090** 10.5		
	S.R. 90° Elbow		**546** 6.3	**722** 6.3		
	L.R. 45° Elbow		**408** 4.2	**541** 4.2		
	Tee		**991** 8.4	**1230** 8.4		
	Lateral					
	Reducer		**255** 3.1	**335** 3.1		
	Cap		**145** 5.2	**190** 5.2		

INSULATION	Temperature Range °F	100-199	200-299	300-399	400-499	500-599	600-699	700-799	800-899	900-999	1000-1099	1100-1200
85% Magnesia Calcium Silicate	Nom. Thick., In.	1½	1½	2	2½	3	3½	4	4½	5	5	6
	Lbs/Ft	**12.7**	**12.7**	**17.1**	**21.7**	**26.5**	**31.3**	**36.3**	**41.4**	**46.6**	**46.6**	**57.5**
Combination	Nom. Thick., In.						3½	4½	5½	6	6½	7
	Lbs/Ft						**44.7**	**62.3**	**80.9**	**90.5**	**100.4**	**110.5**
*Asbestos Fiber—Sodium Silicate	Nom. Thick., In.	3	3	3	3	3	3	3	4½	4½	5	5
	Lbs/Ft	**41.50**	**41.50**	**41.50**	**41.50**	**41.50**	**41.50**	**41.50**	**57.27**	**57.27**	**64.49**	**64.49**

	Pressure Rating psi	Cast Iron		Steel						
		125	250	150	300	400	600	900	1500	2500
FLANGES	Screwed or Slip-On			**395** 1.5	**890** 1.5	**1025** 1.5	**1375** 1.5	**2500** 1.5		
	Welding Neck			**435** 1.5	**1025** 1.5	**1150** 1.5	**1500** 1.5	**2575** 1.5		
	Lap Joint									
	Blind			**870** 1.5	**1775** 1.5	**1975** 1.5	**2300** 1.5	**3650** 1.5		
FLANGED FITTINGS	S.R. 90° Elbow									
	L.R. 90° Elbow									
	45° Elbow									
	Tee									
VALVES	Flanged Bonnet Gate									
	Flanged Bonnet Globe or Angle									
	Flanged Bonnet Check									
	Pressure Seal Bonnet—Gate									
	Pressure Seal Bonnet—Globe									

* 16 lb cu. ft. density.

Boldface type is weight in pounds. Lightface type beneath weight is weight factor for insulation.

Insulation thicknesses and weights are based on average conditions and do not constitute a recommendation for specific thicknesses of materials. Insulation weights are based on 85% magnesia and hydrous calcium silicate at 11 lbs/cubic foot. The listed thicknesses and weights of combination covering are the sums of the inner layer of diatomaceous earth at 21 lbs/cubic foot and the outer layer at 11 lbs/cubic foot.

Insulation weights include allowances for wire, cement, canvas, bands and paint, but not special surface finishes.

To find the weight of covering on flanges, valves or fittings, multiply the weight factor by the weight per foot of covering used on straight pipe.

Valve weights are approximate. When possible, obtain weights from manufacturer.

Cast iron valve weights are for flanged end valves; steel weights for welding end valves.

All flanged fitting, flanged valve and flange weights include the proportional weight of bolts or studs to make up all joints.

WEIGHTS OF PIPING MATERIALS

34″ O.D. 34″ PIPE

PIPE	Schedule No.	10		20	30	40
	Wall Designation		Std.	XS		
	Thickness—In.	.312	.375	.500	.625	.688
	Pipe—Lbs/Ft	**112.4**	**134.7**	**178.9**	**222.8**	**244.6**
	Water—Lbs/Ft	**379.1**	**376.0**	**370.3**	**365.0**	**362.2**
WELDING FITTINGS	L.R. 90° Elbow		**926** 11	**1230** 11		
	S.R. 90° Elbow		**617** 5.5	**817** 5.5		
	L.R. 45° Elbow		**463** 4.4	**615** 4.4		
	Tee		**1136** 8.9	**1420** 8.9		
	Lateral					
	Reducer		**270** 3.3	**355** 3.3		
	Cap		**160** 5.6	**210** 5.6		

INSULATION	Temperature Range °F	100-199	200-299	300-399	400-499	500-599	600-699	700-799	800-899	900-999	1000-1099	1100-1200
85% Magnesia Calcium Silicate	Nom. Thick., In.	1½	1½	2	2½	3	3½	4	4½	5	5	6
	Lbs/Ft	**13.4**	**13.4**	**18.2**	**23.0**	**28.0**	33.1	38.3	43.7	49.1	49.1	60.5
Combination	Nom. Thick., In.						3½	4½	5½	6	6½	7
	Lbs/Ft						**47.2**	**65.8**	**85.3**	**95.4**	**105.7**	**116.3**
*Asbestos Fiber—Sodium Silicate	Nom. Thick., In.	3	3	3	3	3	3	3	4½	4½	5	5
	Lbs/Ft	38.74	38.74	38.74	38.74	38.74	38.74	38.74	60.50	60.50	68.04	68.04

	Pressure Rating psi	Cast Iron		Steel						
		125	250	150	300	400	600	900	1500	2500
FLANGES	Screwed or Slip-On			**420** 1.5	**1075** 1.5	**1150** 1.5	**1500** 1.5	**2950** 1.5		
	Welding Neck			**465** 1.5	**1200** 1.5	**1300** 1.5	**1650** 1.5	**3025** 1.5		
	Lap Joint									
	Blind			**990** 1.5	**2025** 1.5	**2250** 1.5	**2575** 1.5	**4275** 1.5		
FLANGED FITTINGS	S.R. 90° Elbow									
	L.R. 90° Elbow									
	45° Elbow									
	Tee									
VALVES	Flanged Bonnet Gate									
	Flanged Bonnet Globe or Angle									
	Flanged Bonnet Check									
	Presssure Seal Bonnet—Gate									
	Pressure Seal Bonnet—Globe									

Boldface type is weight in pounds. Lightface type beneath weight is weight factor for insulation.

Insulation thicknesses and weights are based on average conditions and do not constitute a recommendation for specific thicknesses of materials. Insulation weights are based on 85% magnesia and hydrous calcium silicate at 11 lbs/cubic foot. The listed thicknesses and weights of combination covering are the sums of the inner layer of diatomaceous earth at 21 lbs/cubic foot and the outer layer at 11 lbs/cubic foot.

Insulation weights include allowances for wire, cement, canvas, bands and paint, but not special surface finishes.

To find the weight of covering on flanges, valves or fittings, multiply the weight factor by the weight per foot of covering used on straight pipe.

Valve weights are approximate. When possible, obtain weights from manufacturer.

Cast iron valve weights are for flanged end valves; steel weights for welding end valves.

All flanged fitting, flanged valve and flange weights include the proportional weight of bolts or studs to make up all joints.

* 16 lb cu. ft. density.

36″ PIPE 36″ O.D. WEIGHTS OF PIPING MATERIALS

PIPE											
Schedule No.	10		20	30	40						
Wall Designation		Std.	XS								
Thickness—In.	.312	.375	.500	.625	.750	1.125					
Pipe—Lbs/Ft	**119.1**	**142.7**	**189.6**	**236.1**	**282.4**	**419.03**					
Water—Lbs/Ft	**425.9**	**422.6**	**416.6**	**411.0**	**405.1**	**387.67**					

WELDED FITTINGS											
L.R. 90° Elbow		**1040** 12	**1380** 12								
S.R. 90° Elbow		**692** 5	**913** 5								
L.R. 45° Elbow		**518** 4.8	**686** 4.8								
Tee		**1294** 9.5	**1610** 9.5								
Lateral											
Reducer		**340** 3.6	**360** 3.6								
Cap		**175** 6	**235** 6								

INSULATION		100-199	200-299	300-399	400-499	500-599	600-699	700-799	800-899	900-999	1000-1099	1100-1200
Temperature Range °F												
85% Magnesia Calcium Silicate	Nom. Thick., In.	1½	1½	2	2½	3	3½	4	4½	5	5	6
	Lbs/Ft	**14.2**	**14.2**	**19.2**	**24.2**	**29.5**	**34.8**	**40.3**	**45.9**	**51.7**	**51.7**	**63.5**
Combination	Nom. Thick., In.						3½	4½	5½	6	6½	7
	Lbs/Ft						**49.8**	**69.3**	**89.7**	**100.2**	**111.0**	**122.0**
*Asbestos Fiber—Sodium Silicate	Nom. Thick., In.	3	3	3	3	3	3	3	4½	4½	5	5
	Lbs/Ft	**40.84**	**40.84**	**40.84**	**40.84**	**40.84**	**40.84**	**40.84**	**55.83**	**55.83**	**71.48**	**71.48**

	Pressure Rating psi	Cast Iron		Steel						
		125	250	150	300	400	600	900	1500	2500
FLANGES	Screwed or Slip-On			**480** 1.5	**1200** 1.5	**1325** 1.5	**1600** 1.5	**3350** 1.5		
	Welding Neck			**520** 1.5	**1300** 1.5	**1475** 1.5	**1750** 1 5	**3450** 1.5		
	Lap Joint									
	Blind			**1125** 1.5	**2275** 1.5	**2525** 1.5	**2950** 1.5	**4900** 1.5		
FLANGED FITTINGS	S.R. 90° Elbow									
	L.R. 90° Elbow									
	45° Elbow									
	Tee									
VALVES	Flanged Bonnet Gate									
	Flanged Bonnet Globe or Angle									
	Flanged Bonnet Check									
	Pressure Seal Bonnet—Gate									
	Pressure Seal Bonnet—Globe									

Boldface type is weight in pounds. Lightface type beneath weight is weight factor for insulation.

Insulation thicknesses and weights are based on average conditions and do not constitute a recommendation for specific thicknesses of materials. Insulation weights are based on 85% magnesia and hydrous calcium silicate at 11 lbs/cubic foot. The listed thicknesses and weights of combination covering are the sums of the inner layer of diatomaceous earth at 21 lbs/cubic foot and the outer layer at 11 lbs/cubic foot.

Insulation weights include allowances for wire, cement, canvas, bands and paint, but not special surface finishes.

To find the weight of covering on flanges, valves or fittings, multiply the weight factor by the weight per foot of covering used on straight pipe.

Valve weights are approximate. When possible, obtain weights from manufacturer.

Cast iron valve weights are for flanged end valves; steel weights for welding end valves.

All flanged fitting, flanged valve and flange weights include the proportional weight of bolts or studs to make up all joints.

* 16 lb cu. ft. density.

Appendix D

Conversion Factors

Alphabetical Conversion Factors

A

TO CONVERT	INTO	MULTIPLY BY
Abcoulomb	Statcoulombs	2.998×10^{10}
Acre	Sq. chain (Gunters)	10
Acre	Rods	160
Acre	Square links (Gunters)	1×10^{5}
Acre	Hectare or sq. hectometer	.4047
acres	sq feet	43,560.0
acres	sq meters	4,047.
acres	sq miles	1.562×10^{-3}
acres	sq yards	4,840.
acre-feet	cu feet	43,560.0
acre-feet	gallons	3.259×10^{5}
amperes/sq cm	amps/sq in.	6.452
amperes/sq cm	amps/sq meter	10^{4}
amperes/sq in.	amps/sq cm	0.1550
amperes/sq in.	amps/sq meter	1,550.0
amperes/sq meter	amps/sq cm	10^{-4}
amperes/sq meter	amps/sq in.	6.452×10^{-4}
ampere-hours	coulombs	3,600.0
ampere-hours	faradays	0.03731
ampere-turns	gilberts	1.257
ampere-turns/cm	amp-turns/in.	2.540
ampere-turns/cm	amp-turns/meter	100.0
ampere-turns/cm	gilberts/cm	1.257
ampere-turns/in.	amp-turns/cm	0.3937
ampere-turns/in.	amp-turns/meter	39.37
ampere-turns/in.	gilberts/cm	0.4950
ampere-turns/meter	amp/turns/cm	0.01
ampere-turns/meter	amp-turns/in.	0.0254
ampere-turns/meter	gilberts/cm	0.01257
Angstrom unit	Inch	3937×10^{-9}
Angstrom unit	Meter	1×10^{-10}
Angstrom unit	Micron or (Mu)	1×10^{-4}
Are	Acre (US)	.02471
Ares	sq. yards	119.60
ares	acres	0.02471
ares	sq meters	100.0
Astronomical Unit	Kilometers	1.495×10^{8}
Atmospheres	Ton/sq. inch	.007348
atmospheres	cms of mercury	76.0
atmospheres	ft of water (at 4°C)	33.90
atmospheres	in. of mercury (at 0°C)	29.92
atmospheres	kgs/sq cm	1.0333
atmospheres	kgs/sq meter	10,332.
atmospheres	pounds/sq in.	14.70
atmospheres	tons/sq ft	1.058

B

TO CONVERT	INTO	MULTIPLY BY
Barrels (U.S., dry)	cu. inches	7056.
Barrels (U.S., dry)	quarts (dry)	105.0
Barrels (U.S., liquid)	gallons	31.5
barrels (oil)	gallons (oil)	42.0
bars	atmospheres	0.9869
bars	dynes/sq cm	10^{6}
bars	kgs/sq meter	1.020×10^{4}
bars	pounds/sq ft	2,089.
bars	pounds/sq in.	14.50
Baryl	Dyne/sq. cm.	1.000
Bolt (US Cloth)	Meters	36.576
BTU	Liter–Atmosphere	10.409
Btu	ergs	1.0550×10^{10}
Btu	foot-lbs	778.3
Btu	gram-calories	252.0
Btu	horsepower-hrs	3.931×10^{-4}
Btu	joules	1,054.8
Btu	kilogram-calories	0.2520
Btu	kilogram-meters	107.5
Btu	kilowatt-hrs	2.928×10^{-4}
Btu/hr	foot-pounds/sec	0.2162
Btu/hr	gram-cal/sec	0.0700
Btu/hr	horsepower-hrs	3.929×10^{-4}
Btu/hr	watts	0.2931
Btu/min	foot-lbs/sec	12.96
Btu/min	horsepower	0.02356
Btu/min	kilowatts	0.01757
Btu/min	watts	17.57
Btu/sq ft/min	watts/sq in.	0.1221
Bucket (Br. dry)	Cubic Cm.	1.818×10^{4}
bushels	cu ft	1.2445
bushels	cu in.	2,150.4
bushels	cu meters	0.03524
bushels	liters	35.24
bushels	pecks	4.0
bushels	pints (dry)	64.0
bushels	quarts (dry)	32.0

C

TO CONVERT	INTO	MULTIPLY BY
Calories, gram (mean)	B.T.U. (mean)	3.9685×10^{-3}
Candle/sq. cm	Lamberts	3.142
Candle/sq. inch	Lamberts	.4870
centares (centiares)	sq meters	1.0
Centigrade	Fahrenheit	$(C° \times 9/5)+32$
centigrams	grams	0.01
Centiliter	Ounce fluid (US)	.3382
Centiliter	Cubic inch	.6103
Centiliter	drams	2.705
centiliters	liters	0.01
centimeters	feet	3.281×10^{-2}
centimeters	inches	0.3937
centimeters	kilometers	10^{-5}
centimeters	meters	0.01
centimeters	miles	6.214×10^{-6}
centimeters	millimeters	10.0
centimeters	mils	393.7
centimeters	yards	1.094×10^{-2}
centimeter-dynes	cm-grams	1.020×10^{-3}
centimeter-dynes	meter-kgs	1.020×10^{-8}
centimeter-dynes	pound-feet	7.376×10^{-8}
centimeter-grams	cm-dynes	980.7
centimeter-grams	meter-kgs	10^{-5}
centimeter-grams	pound-feet	7.233×10^{-5}
centimeters of mercury	atmospheres	0.01316
centimeters of mercury	feet of water	0.4461
centimeters of mercury	kgs/sq meter	136.0
centimeters of mercury	pounds/sq ft	27.85
centimeters of mercury	pounds/sq in.	0.1934
centimeters/sec	feet/min	1.1969
centimeters/sec	feet/sec	0.03281
centimeters/sec	kilometers/hr	0.036
centimeters/sec	knots	0.1943
centimeters/sec	meters/min	0.6
centimeters/sec	miles/hr	0.02237
centimeters/sec	miles/min	3.728×10^{-4}
centimeters/sec/sec	feet/sec/sec	0.03281
centimeters/sec/sec	kms/hr/sec	0.036
centimeters/sec/sec	meters/sec/sec	0.01
centimeters/sec/sec	miles/hr/sec	0.02237
Chain	Inches	792.00
Chain	meters	20.12
Chains (surveyors' or Gunter's)	yards	22.00
circular mils	sq cms	5.067×10^{-6}
circular mils	sq mils	0.7854
Circumference	Radians	6.283
circular mils	sq inches	7.854×10^{-7}
Cords	cord feet	8
Cord feet	cu. feet	16
Coulomb	Statcoulombs	2.998×10^{9}
coulombs	faradays	1.036×10^{-5}

(Continued). Alphabetical Conversion Factors

TO CONVERT	INTO	MULTIPLY BY
coulombs/sq cm	coulombs/sq in.	64.52
coulombs/sq cm	coulombs/sq meter	10^{4}
coulombs/sq in.	coulombs/sq cm	0.1550
coulombs/sq in.	coulombs/sq meter	1,550.
coulombs/sq meter	coulombs/sq cm	10^{-4}
coulombs/sq meter	coulombs/sq in.	6.452×10^{-4}
cubic centimeters	cu feet	3.531×10^{-5}
cubic centimeters	cu inches	0.06102
cubic centimeters	cu meters	10^{-6}
cubic centimeters	cu yards	1.308×10^{-6}
cubic centimeters	gallons (U. S. liq.)	2.642×10^{-4}
cubic centimeters	liters	0.001
cubic centimeters	pints (U.S. liq.)	2.113×10^{-3}
cubic centimeters	quarts (U.S. liq.)	1.057×10^{-3}
cubic feet	bushels (dry)	0.8036
cubic feet	cu cms	28,320.0
cubic feet	cu inches	1,728.0
cubic feet	cu meters	0.02832
cubic feet	cu yards	0.03704
cubic feet	gallons (U.S. liq.)	7.48052
cubic feet	liters	28.32
cubic feet	pints (U.S. liq.)	59.84
cubic feet	quarts (U.S. liq.)	29.92
cubic feet/min	cu cms/sec	472.0
cubic feet/min	gallons/sec	0.1247
cubic feet/min	liters/sec	0.4720
cubic feet/min	pounds of water/min	62.43
cubic feet/sec	million gals/day	0.646317
cubic feet/sec	gallons/min	448.831
cubic inches	cu cms	16.39
cubic inches	cu feet	5.787×10^{-4}
cubic inches	cu meters	1.639×10^{-5}
cubic inches	cu yards	2.143×10^{-5}
cubic inches	gallons	4.329×10^{-3}
cubic inches	liters	0.01639
cubic inches	mil-feet	1.061×10^{5}
cubic inches	pints (U.S. liq.)	0.03463
cubic inches	quarts (U.S. liq.)	0.01732
cubic meters	bushels (dry)	28.38
cubic meters	cu cms	10^{6}
cubic meters	cu feet	35.31
cubic meters	cu inches	61,023.0
cubic meters	cu yards	1.308
cubic meters	gallons (U.S. liq.)	264.2
cubic meters	liters	1,000.0
cubic meters	pints (U.S. liq.)	2,113.0
cubic meters	quarts (U.S. liq.)	1,057.
cubic yards	cu cms	7.646×10^{5}
cubic yards	cu feet	27.0
cubic yards	cu inches	46,656.0
cubic yards	cu meters	0.7646
cubic yards	gallons (U.S. liq.)	202.0
cubic yards	liters	764.6
cubic yards	pints (U.S. liq.)	1,615.9
cubic yards	quarts (U.S. liq.)	807.9
cubic yards/min	cubic ft/sec	0.45
cubic yards/min	gallons/sec	3.367
cubic yards/min	liters/sec	12.74
D		
Dalton	Gram	1.650×10^{-24}
days	seconds	86,400.0
decigrams	grams	0.1
deciliters	liters	0.1
decimeters	meters	0.1
degrees (angle)	quadrants	0.01111
degrees (angle)	radians	0.01745
degrees (angle)	seconds	3,600.0
degrees/sec	radians/sec	0.01745
degrees/sec	revolutions/min	0.1667
degrees/sec	revolutions/sec	2.778×10^{-3}
dekagrams	grams	10.0
dekaliters	liters	10.0
dekameters	meters	10.0
Drams (apothecaries' or troy)	ounces (avoidupois)	0.1371429
Drams (apothecaries' or troy)	ounces (troy)	0.125
Drams (U.S., fluid or apoth.)	cubic cm.	3.6967
drams	grams	1.7718
drams	grains	27.3437
drams	ounces	0.0625
Dyne/cm	Erg/sq. millimeter	.01
Dyne/sq. cm.	Atmospheres	9.869×10^{-7}
Dyne/sq. cm.	Inch of Mercury at 0°C	2.953×10^{-5}
Dyne/sq. cm.	Inch of Water at 4°C	4.015×10^{-4}
dynes	grams	1.020×10^{-3}
dynes	joules/cm	10^{-7}
dynes	joules/meter (newtons)	10^{-5}
dynes	kilograms	1.020×10^{-6}
dynes	poundals	7.233×10^{-5}
dynes	pounds	2.248×10^{-6}
dynes/sq cm	bars	10^{-6}
E		
Ell	Cm.	114.30
Ell	Inches	45
Em, Pica	Inch	.167
Em, Pica	Cm.	.4233
Erg/sec	Dyne – cm/sec	1.000
ergs	Btu	9.480×10^{-11}
ergs	dyne-centimeters	1.0
ergs	foot-pounds	7.367×10^{-8}
ergs	gram-calories	0.2389×10^{-7}
ergs	gram-cms	1.020×10^{-3}
ergs	horsepower-hrs	3.7250×10^{-14}
ergs	joules	10^{-7}
ergs	kg-calories	2.389×10^{-11}
ergs	kg-meters	1.020×10^{-8}
ergs	kilowatt-hrs	0.2778×10^{-13}
ergs	watt-hours	0.2778×10^{-10}
ergs/sec	Btu/min	$5{,}688 \times 10^{-9}$
ergs/sec	ft-lbs/min	4.427×10^{-6}
ergs/sec	ft-lbs/sec	7.3756×10^{-8}
ergs/sec	horsepower	1.341×10^{-10}
ergs/sec	kg-calories/min	1.433×10^{-9}
ergs/sec	kilowatts	10^{-10}
F		
farads	microfarads	10^{6}
Faraday/sec	Ampere (absolute)	9.6500×10^{4}
faradays	ampere-hours	26.80
faradays	coulombs	9.649×10^{4}
Fathom	Meter	1.828804
fathoms	feet	6.0
feet	centimeters	30.48
feet	kilometers	3.048×10^{-4}
feet	meters	0.3048
feet	miles (naut.)	1.645×10^{-4}
feet	miles (stat.)	1.894×10^{-4}
feet	millimeters	304.8
feet	mils	1.2×10^{4}
feet of water	atmospheres	0.02950
feet of water	in. of mercury	0.8826
feet of water	kgs/sq cm	0.03048

(Continued). Alphabetical Conversion Factors

TO CONVERT	INTO	MULTIPLY BY
feet of water	kgs/sq meter	304.8
feet of water	pounds/sq ft	62.43
feet of water	pounds/sq in.	0.4335
feet/min	cms/sec	0.5080
feet/min	feet/sec	0.01667
feet/min	kms/hr	0.01829
feet/min	meters/min	0.3048
feet/min	miles/hr	0.01136
feet/sec	cms/sec	30.48
feet/sec	kms/hr	1.097
feet/sec	knots	0.5921
feet/sec	meters/min	18.29
feet/sec	miles/hr	0.6818
feet/sec	miles/min	0.01136
feet/sec/sec	cms/sec/sec	30.48
feet/sec/sec	kms/hr/sec	1.097
feet/sec/sec	meters/sec/sec	0.3048
feet/sec/sec	miles/hr/sec	0.6818
feet/100 feet	per cent grade	1.0
Foot – candle	Lumen/sq. meter	10.764
foot-pounds	Btu	1.286×10^{-3}
foot-pounds	ergs	1.356×10^{7}
foot-pounds	gram-calories	0.3238
foot-pounds	hp-hrs	5.050×10^{-7}
foot-pounds	joules	1.356
foot-pounds	kg-calories	3.24×10^{-4}
foot-pounds	kg-meters	0.1383
foot-pounds	kilowatt-hrs	3.766×10^{-7}
foot-pounds/min	Btu/min	1.286×10^{-3}
foot-pounds/min	foot-pounds/sec	0.01667
foot-pounds/min	horsepower	3.030×10^{-5}
foot-pounds/min	kg-calories/min	3.24×10^{-4}
foot-pounds/min	kilowatts	2.260×10^{-5}
foot-pounds/sec	Btu/hr	4.6263
foot-pounds/sec	Btu/min	0.07717
foot-pounds/sec	horsepower	1.818×10^{-3}
foot-pounds/sec	kg-calories/min	0.01945
foot-pounds/sec	kilowatts	1.356×10^{-3}
Furlongs	miles (U.S.)	0.125
furlongs	rods	40.0
furlongs	feet	660.0

G

TO CONVERT	INTO	MULTIPLY BY
gallons	cu cms	3,785.0
gallons	cu feet	0.1337
gallons	cu inches	231.0
gallons	cu meters	3.785×10^{-3}
gallons	cu yards	4.951×10^{-3}
gallons	liters	3.785
gallons (liq. Br. Imp.)	gallons (U.S. liq.)	1.20095
gallons (U.S.)	gallons (Imp.)	0.83267
gallons of water	pounds of water	8.3453
gallons/min	cu ft/sec	2.228×10^{-3}
gallons/min	liters/sec	0.06308
gallons/min	cu ft/hr	8.0208
gausses	lines/sq in.	6.452
gausses	webers/sq cm	10^{-8}
gausses	webers/sq in.	6.452×10^{-8}
gausses	webers/sq meter	10^{-4}
gilberts	ampere-turns	0.7958
gilberts/cm	amp-turns/cm	0.7958
gilberts/cm	amp-turns/in	2.021
gilberts/cm	amp-turns/meter	79.58
Gills (British)	cubic cm.	142.07
gills	liters	0.1183
gills	pints (liq.)	0.25
Grade	Radian	.01571
Grains	drams (avoirdupois)	0.03657143
grains (troy)	grains (avdp)	1.0
grains (troy)	grams	0.06480
grains (troy)	ounces (avdp)	2.0833×10^{-3}
grains (troy)	pennyweight (troy)	0.04167
grains/U.S. gal	parts/million	17.118
grains/U.S. gal	pounds/million gal	142.86
grains/Imp. gal	parts/million	14.286
grams	dynes	980.7
grams	grains	15.43
grams	joules/cm	9.807×10^{-5}
grams	joules/meter (newtons)	9.807×10^{-3}
grams	kilograms	0.001
grams	milligrams	1,000.
grams	ounces (avdp)	0.03527
grams	ounces (troy)	0.03215
grams	poundals	0.07093
grams	pounds	2.205×10^{-3}
grams/cm	pounds/inch	5.600×10^{-3}
grams/cu cm	pounds/cu ft	62.43
grams/cu cm	pounds/cu in	0.03613
grams/cu cm	pounds/mil-foot	3.405×10^{-7}
grams/liter	grains/gal	58.417
grams/liter	pounds/ gal	8.345
grams/liter	pounds/cu ft	0.062427
grams/liter	parts/million	1,000.0
grams/sq cm	pounds/sq ft	2.0481
gram-calories	Btu	3.9683×10^{-3}
gram-calories	ergs	4.1868×10^{7}
gram-calories	foot-pounds	3.0880
gram-calories	horsepower-hrs	1.5596×10^{-6}
gram-calories	kilowatt-hrs	1.1630×10^{-6}
gram-calories	watt-hrs	1.1630×10^{-3}
gram-calories/sec	Btu/hr	14.286
gram-centimeters	Btu	9.297×10^{-8}
gram-centimeters	ergs	980.7
gram-centimeters	joules	9.807×10^{-5}
gram-centimeters	kg-cal	2.343×10^{-8}
gram-centimeters	kg-meters	10^{-5}

H

TO CONVERT	INTO	MULTIPLY BY
Hand	Cm.	10.16
hectares	acres	2.471
hectares	sq feet	1.076×10^{5}
hectograms	grams	100.0
hectoliters	liters	100.0
hectometers	meters	100.0
hectowatts	watts	100.0
henries	millihenries	1,000.0
Hogsheads (British)	cubic ft.	10.114
Hogsheads (U.S.)	cubic ft.	8.42184
Hogsheads (U.S.)	gallons (U.S.)	63
horsepower	Btu/min	42.44
horsepower	foot-lbs/min	33,000.
horsepower	foot-lbs/sec	550.0
horsepower (metric) (542.5 ft lb/sec)	horsepower (550 ft lb/sec)	0.9863
horsepower (550 ft lb/sec)	horsepower (metric) (542.5 ft lb/sec)	1.014
horsepower	kg-calories/min	10.68
horsepower	kilowatts	0.7457
horsepower	watts	745.7
horsepower (boiler)	Btu/hr	33.479
horsepower (boiler)	kilowatts	9.803
horsepower-hrs	Btu	2,547.
horsepower-hrs	ergs	2.6845×10^{13}
horsepower-hrs	foot-lbs	1.98×10^{6}
horsepower-hrs	gram-calories	641,190.
horsepower-hrs	joules	2.684×10^{6}

(Continued). Alphabetical Conversion Factors

TO CONVERT	INTO	MULTIPLY BY
horsepower-hrs	kg-calories	641.1
horsepower-hrs	kg-meters	2.737×10^{5}
horsepower-hrs	kilowatt-hrs	0.7457
hours	days	4.167×10^{-2}
hours	weeks	5.952×10^{-3}
Hundredweights (long)	pounds	112
Hundredweights (long)	tons (long)	0.05
Hundredweights (short)	ounces (avoirdupois)	1600
Hundredweights (short)	pounds	100
Hundredweights (short)	tons (metric)	0.0453592
Hundredweights (short)	tons (long)	0.0446429
	I	
inches	centimeters	2.540
inches	meters	2.540×10^{-2}
inches	miles	1.578×10^{-5}
inches	millimeters	25.40
inches	mils	1,000.0
inches	yards	2.778×10^{-2}
inches of mercury	atmospheres	0.03342
inches of mercury	feet of water	1.133
inches of mercury	kgs/sq cm	0.03453
inches of mercury	kgs/sq meter	345.3
inches of mercury	pounds/sq ft	70.73
inches of mercury	pounds/sq in.	0.4912
inches of water (at 4°C)	atmospheres	2.458×10^{-3}
inches of water (at 4°C)	inches of mercury	0.07355
inches of water (at 4°C)	kgs/sq cm	2.540×10^{-3}
inches of water (at 4°C)	ounces/sq in.	0.5781
inches of water (at 4°C)	pounds/sq ft	5.204
inches of water (at 4°C)	pounds/sq in.	0.03613
International Ampere	Ampere (absolute)	.9998
International Volt	Volts (absolute)	1.0003
International volt	Joules (absolute)	$1\text{-}593 \times 10^{-19}$
International volt	Joules	9.654×10^{4}
	J	
joules	Btu	9.480×10^{-4}
joules	ergs	10^{7}
joules	foot-pounds	0.7376
joules	kg-calories	2.389×10^{-4}
joules	kg-meters	0.1020
joules	watt-hrs	2.778×10^{-4}
joules/cm	grams	1.020×10^{4}
joules/cm	dynes	10^{7}
joules/cm	joules/meter (newtons)	100.0
joules/cm	poundals	723.3
joules/cm	pounds	22.48
	K	
kilograms	dynes	980,665.
kilograms	grams	1,000.0
kilograms	joules/cm	0.09807
kilograms	joules/meter (newtons)	9.807
kilograms	poundals	70.93
kilograms	pounds	2.205
kilograms	tons (long)	9.842×10^{-4}
kilograms	tons (short)	1.102×10^{-3}
kilograms/cu meter	grams/cu cm	0.001
kilograms/cu meter	pounds/cu ft	0.06243
kilograms/cu meter	pounds/cu in.	3.613×10^{-5}
kilograms/cu meter	pounds/mil-foot	3.405×10^{-10}
kilograms/meter	pounds/ft	0.6720
Kilogram/sq. cm.	Dynes	980,665
kilograms/sq cm	atmospheres	0.9678
kilograms/sq cm	feet of water	32.81
kilograms/sq cm	inches of mercury	28.96
kilograms/sq cm	pounds/sq ft	2,048.
kilograms/sq cm	pounds/sq in.	14.22
kilograms/sq meter	atmospheres	9.678×10^{-5}
kilograms/sq meter	bars	98.07×10^{-6}
kilograms/sq meter	feet of water	3.281×10^{-3}
kilograms/sq meter	inches of mercury	2.896×10^{-3}
kilograms/sq meter	pounds/sq ft	0.2048
kilograms/sq meter	pounds/sq in.	1.422×10^{-3}
kilograms/sq mm	kgs/sq meter	10^{6}
kilogram-calories	Btu	3.968
kilogram-calories	foot-pounds	3,088.
kilogram-calories	hp-hrs	1.560×10^{-3}
kilogram-calories	joules	4,186.
kilogram-calories	kg-meters	426.9
kilogram-calories	kilojoules	4.186
kilogram-calories	kilowatt-hrs	1.163×10^{-3}
kilogram meters	Btu	9.294×10^{-3}
kilogram meters	ergs	9.804×10^{7}
kilogram meters	foot-pounds	7.233
kilogram meters	joules	9.804
kilogram meters	kg-calories	2.342×10^{-3}
kilogram meters	kilowatt-hrs	2.723×10^{-6}
kilolines	maxwells	1,000.0
kiloliters	liters	1,000.0
kilometers	centimeters	10^{5}
kilometers	feet	3,281.
kilometers	inches	3.937×10^{4}
kilometers	meters	1,000.0
kilometers	miles	0.6214
kilometers	millimeters	10^{6}
kilometers	yards	1,094.
kilometers/hr	cms/sec	27.78
kilometers/hr	feet/min	54.68
kilometers/hr	feet/sec	0.9113
kilometers/hr	knots	0.5396
kilometers/hr	meters/min	16.67
kilometers/hr	miles/hr	0.6214
kilometers/hr/sec	cms/sec/sec	27.78
kilometers/hr/sec	ft/sec/sec	0.9113
kilometers/hr/sec	meters/sec/sec	0.2778
kilometers/hr/sec	miles/hr/sec	0.6214
kilowatts	Btu/min	56.92
kilowatts	foot-lbs/min	4.426×10^{4}
kilowatts	foot-lbs/sec	737.6
kilowatts	horsepower	1.341
kilowatts	kg-calories/min	14.34
kilowatts	watts	1,000.0
kilowatt-hrs	Btu	3,413.
kilowatt-hrs	ergs	3.600×10^{13}
kilowatt-hrs	foot-lbs	2.655×10^{6}
kilowatt-hrs	gram-calories	859,850.
kilowatt-hrs	horsepower-hrs	1.341
kilowatt-hrs	joules	3.6×10^{6}
kilowatt-hrs	kg-calories	860.5
kilowatt-hrs	kg-meters	3.671×10^{5}
kilowatt-hrs	pounds of water evaporated from and at 212° F.	3.53
kilowatt-hrs	pounds of water raised from 62° to 212° F.	22.75
knots	feet/hr	6,080.
knots	kilometers/hr	1.8532
knots	nautical miles/hr	1.0
knots	statute miles/hr	1.151

(Continued). Alphabetical Conversion Factors

TO CONVERT	INTO	MULTIPLY BY
knots	yards/hr	2,027.
knots	feet/sec	1.689
	L	
league	miles (approx.)	3.0
Light year	Miles	5.9 x 10^{12}
Light year	Kilometers	9.46091 x 10^{12}
lines/sq cm	gausses	1.0
lines/sq in.	gausses	0.1550
lines/sq in.	webers/sq cm	1.550 x 10^{-9}
lines/sq in.	webers/sq in.	10^{-8}
lines/sq in.	webers/sq meter	1.550 x 10^{-5}
links (engineer's)	inches	12.0
links (surveyor's)	inches	7.92
liters	bushels (U.S. dry)	0.02838
liters	cu cm	1,000.0
liters	cu feet	0.03531
liters	cu inches	61.02
liters	cu meters	0.001
liters	cu yards	1.308 x 10^{-3}
liters	gallons (U.S. liq.)	0.2642
liters	pints (U.S. liq.)	2.113
liters	quarts (U.S. liq.)	1.057
liters/min	cu ft/sec	5.886 x 10^{-4}
liters/min	gals/sec	4.403 x 10^{-3}
lumens/sq ft	foot-candles	1.0
Lumen	Spherical candle power	.07958
Lumen	Watt	.001496
Lumen/sq. ft.	Lumen/sq. meter	10.76
lux	foot-candles	0.0929
	M	
maxwells	kilolines	0.001
maxwells	webers	10^{-8}
megalines	maxwells	10^{6}
megohms	microhms	10^{12}
megohms	ohms	10^{6}
meters	centimeters	100.0
meters	feet	3.281
meters	inches	39.37
meters	kilometers	0.001
meters	miles (naut.)	5.396 x 10^{-4}
meters	miles (stat.)	6.214 x 10^{-4}
meters	millimeters	1,000.0
meters	yards	1.094
meters	varas	1.179
meters/min	cms/sec	1.667
meters/min	feet/min	3.281
meters/min	feet/sec	0.05468
meters/min	kms/hr	0.06
meters/min	knots	0.03238
meters/min	miles/hr	0.03728
meters/sec	feet/min	196.8
meters/sec	feet/sec	3.281
meters/sec	kilometers/hr	3.6
meters/sec	kilometers/min	0.06
meters/sec	miles/hr	2.237
meters/sec	miles/min	0.03728
meters/sec/sec	cms/sec/sec	100.0
meters/sec/sec	ft/sec/sec	3.281
meters/sec/sec	kms/hr/sec	3.6
meters/sec/sec	miles/hr/sec	2.237
meter-kilograms	cm-dynes	9.807 x 10^{7}
meter-kilograms	cm-grams	10^{5}
meter-kilograms	pound-feet	7.233
microfarad	farads	10^{-6}
micrograms	grams	10^{-6}
microhms	megohms	10^{-12}
microhms	ohms	10^{-6}
microliters	liters	10^{-6}
Microns	meters	1 x 10^{-6}
miles (naut.)	feet	6,080.27
miles (naut.)	kilometers	1.853
miles (naut.)	meters	1,853.
miles (naut.)	miles (statute)	1.1516
miles (naut.)	yards	2,027.
miles (statute)	centimeters	1.609 x 10^{5}
miles (statute)	feet	5,280.
miles (statute)	inches	6.336 x 10^{4}
miles (statute)	kilometers	1.609
miles (statute)	meters	1,609.
miles (statute)	miles (naut.)	0.8684
miles (statute)	yards	1,760.
miles/hr	cms/sec	44.70
miles/hr	feet/min	88.
miles/hr	feet/sec	1.467
miles/hr	kms/hr	1.609
miles/hr	kms/min	0.02682
miles/hr	knots	0.8684
miles/hr	meters/min	26.82
miles/hr	miles/min	0.1667
miles/hr/sec	cms/sec/sec	44.70
miles/hr/sec	feet/sec/sec	1.467
miles/hr/sec	kms/hr/sec	1.609
miles/hr/sec	meters/sec/sec	0.4470
miles/min	cms/sec	2,682.
miles/min	feet/sec	88.
miles/min	kms/min	1.609
miles/min	knots/min	0.8684
miles/min	miles/hr	60.0
mil-feet	cu inches	9.425 x 10^{-6}
milliers	kilograms	1,000.
Millimicrons	meters	1 x 10^{-9}
Milligrams	grains	0.01543236
milligrams	grams	0.001
milligrams/liter	parts/million	1.0
millihenries	henries	0.001
milliliters	liters	0.001
millimeters	centimeters	0.1
millimeters	feet	3.281 x 10^{-3}
millimeters	inches	0.03937
millimeters	kilometers	10^{-6}
millimeters	meters	0.001
millimeters	miles	6.214 x 10^{-7}
millimeters	mils	39.37
millimeters	yards	1.094 x 10^{-3}
million gals/day	cu ft/sec	1.54723
mils	centimeters	2.540 x 10^{-3}
mils	feet	8.333 x 10^{-5}
mils	inches	0.001
mils	kilometers	2.540 x 10^{-8}
mils	yards	2.778 x 10^{-5}
miner's inches	cu ft/min	1.5
Minims (British)	cubic cm.	0.059192
Minims (U.S., fluid)	cubic cm.	0.061612
minutes (angles)	degrees	0.01667
minutes (angles)	quadrants	1.852 x 10^{-4}
minutes (angles)	radians	2.909 x 10^{-4}
minutes (angles)	seconds	60.0
myriagrams	kilograms	10.0
myriameters	kilometers	10.0
myriawatts	kilowatts	10.0
	N	
nepers	decibels	8.686
Newton	Dynes	1 x 105

(Continued). Alphabetical Conversion Factors

TO CONVERT	INTO	MULTIPLY BY
O		
OHM (International)	OHM (absolute)	1.0005
ohms	megohms	10^{-6}
ohms	microhms	10^{6}
ounces	drams	16.0
ounces	grains	437.5
ounces	grams	28.349527
ounces	pounds	0.0625
ounces	ounces (troy)	0.9115
ounces	tons (long)	2.790×10^{-5}
ounces	tons (metric)	2.835×10^{-5}
ounces (fluid)	cu inches	1.805
ounces (fluid)	liters	0.02957
ounces (troy)	grains	480.0
ounces (troy)	grams	31.103481
ounces (troy)	ounces (avdp.)	1.09714
ounces (troy)	pennyweights (troy)	20.0
ounces (troy)	pounds (troy)	0.08333
Ounce/sq. inch	Dynes/sq. cm.	4309
ounces/sq in.	pounds/sq in.	0.0625
P		
Parsec	Miles	19×10^{12}
Parsec	Kilometers	3.084×10^{13}
parts/million	grains/U.S. gal	0.0584
parts/million	grains/Imp. gal	0.07016
parts/million	pounds/million gal	8.345
Pecks (British)	cubic inches	554.6
Pecks (British)	liters	9.091901
Pecks (U.S.)	bushels	0.25
Pecks (U.S.)	cubic inches	537.605
Pecks (U.S.)	liters	8.809582
Pecks (U.S.)	quarts (dry)	8
pennyweights (troy)	grains	24.0
pennyweights (troy)	ounces (troy)	0.05
pennyweights (troy)	grams	1.55517
pennyweights (troy)	pounds (troy)	4.1667×10^{-3}
pints (dry)	cu inches	33.60
pints (liq.)	cu cms.	473.2
pints (liq.)	cu feet	0.01671
pints (liq.)	cu inches	28.87
pints (liq.)	cu meters	4.732×10^{-4}
pints (liq.)	cu yards	6.189×10^{-4}
pints (liq.)	gallons	0.125
pints (liq.)	liters	0.4732
pints (liq.)	quarts (liq.)	0.5
Planck's quantum	Erg – second	6.624×10^{-27}
Poise	Gram/cm. sec.	1.00
Pounds (avoirdupois)	ounces (troy)	14.5833
poundals	dynes	13,826.
poundals	grams	14.10
poundals	joules/cm	1.383×10^{-3}
poundals	joules/meter (newtons)	0.1383
poundals	kilograms	0.01410
poundals	pounds	0.03108
pounds	drams	256.
pounds	dynes	44.4823×10^{4}
pounds	grains	7,000.
pounds	grams	453.5924
pounds	joules/cm	0.04448
pounds	joules/meter (newtons)	4.448
pounds	kilograms	0.4536
pounds	ounces	16.0
pounds	ounces (troy)	14.5833
pounds	poundals	32.17
pounds	pounds (troy)	1.21528
pounds	tons (short)	0.0005
pounds (troy)	grains	5,760.
pounds (troy)	grams	373.24177
pounds (troy)	ounces (avdp.)	13.1657
pounds (troy)	ounces (troy)	12.0
pounds (troy)	pennyweights (troy)	240.0
pounds (troy)	pounds (avdp.)	0.822857
pounds (troy)	tons (long)	3.6735×10^{-4}
pounds (troy)	tons (metric)	3.7324×10^{-4}
pounds (troy)	tons (short)	4.1143×10^{-4}
pounds of water	cu feet	0.01602
pounds of water	cu inches	27.68
pounds of water	gallons	0.1198
pounds of water/min	cu ft/sec	2.670×10^{-4}
pound-feet	cm-dynes	1.356×10^{7}
pound-feet	cm-grams	13,825.
pound-feet	meter-kgs	0.1383
pounds/cu ft	grams/cu cm	0.01602
pounds/cu ft	kgs/cu meter	16.02
pounds/cu ft	pounds/cu in.	5.787×10^{-4}
pounds/cu ft	pounds/mil-foot	5.456×10^{-9}
pounds/cu in.	gms/cu cm	27.68
pounds/cu in.	kgs/cu meter	2.768×10^{4}
pounds/cu in.	pounds/cu ft	1,728.
pounds/cu in.	pounds/mil-foot	9.425×10^{-6}
pounds/ft	kgs/meter	1.488
pounds/in.	gms/cm	178.6
pounds/mil-foot	gms/cu cm	2.306×10^{6}
pounds/sq ft	atmospheres	4.725×10^{-4}
pounds/sq ft	feet of water	0.01602
pounds/sq ft	inches of mercury	0.01414
pounds/sq ft	kgs/sq meter	4.882
pounds/sq ft	pounds/sq in.	6.944×10^{-3}
pounds/sq in.	atmospheres	0.06804
pounds/sq in.	feet of water	2.307
pounds/sq in.	inches of mercury	2.036
pounds/sq in.	kgs/sq meter	703.1
pounds/sq in.	pounds/sq ft	144.0
Q		
quadrants (angle)	degrees	90.0
quadrants (angle)	minutes	5,400.0
quadrants (angle)	radians	1.571
quadrants (angle)	seconds	3.24×10^{5}
quarts (dry)	cu inches	67.20
quarts (liq.)	cu cms	946.4
quarts (liq.)	cu feet	0.03342
quarts (liq.)	cu inches	57.75
quarts (liq.)	cu meters	9.464×10^{-4}
quarts (liq.)	cu yards	1.238×10^{-3}
quarts (liq.)	gallons	0.25
quarts (liq.)	liters	0.9463
R		
radians	degrees	57.30
radians	minutes	3,438.
radians	quadrants	0.6366
radians	seconds	2.063×10^{5}
radians/sec	degrees/sec	57.30
radians/sec	revolutions/min	9.549
radians/sec	revolutions/sec	0.1592
radians/sec/sec	revs/min/min	573.0
radians/sec/sec	revs/min/sec	9.549
radians/sec/sec	revs/sec/sec	0.1592
revolutions	degrees	360.0
revolutions	quadrants	4.0
revolutions	radians	6.283
revolutions/min	degrees/sec	6.0
revolutions/min	radians/sec	0.1047
revolutions/min	revs/sec	0.01667

(Continued). Alphabetical Conversion Factors

TO CONVERT	INTO	MULTIPLY BY
revolutions/min/min	radians/sec/sec	1.745×10^{-3}
revolutions/min/min	revs/min/sec	0.01667
revolutions/min/min	revs/sec/sec	2.778×10^{-4}
revolutions/sec	degrees/sec	360.0
revolutions/sec	radians/sec	6.283
revolutions/sec	revs/min	60.0
revolutions/sec/sec	radians/sec/sec	6.283
revolutions/sec/sec	revs/min/min	3,600.0
revolutions/sec/sec	revs/min/sec	60.0
Rod	Chain (Gunters)	.25
Rod	Meters	5.029
Rods (Surveyors' meas.)	yards	5.5
rods	feet	16.5
	S	
Scruples	grains	20
seconds (angle)	degrees	2.778×10^{-4}
seconds (angle)	minutes	0.01667
seconds (angle)	quadrants	3.087×10^{-6}
seconds (angle)	radians	4.848×10^{-6}
Slug	Kilogram	14.59
Slug	Pounds	32.17
Sphere	Steradians	12.57
square centimeters	circular mils	1.973×10^{5}
square centimeters	sq feet	1.076×10^{-3}
square centimeters	sq inches	0.1550
square centimeters	sq meters	0.0001
square centimeters	sq miles	3.861×10^{-11}
square centimeters	sq millimeters	100.0
square centimeters	sq yards	1.196×10^{-4}
square feet	acres	2.296×10^{-5}
square feet	circular mils	1.833×10^{8}
square feet	sq cms	929.0
square feet	sq inches	144.0
square feet	sq meters	0.09290
square feet	sq miles	3.587×10^{-8}
square feet	sq millimeters	9.290×10^{4}
square feet	sq yards	0.1111
square inches	circular mils	1.273×10^{6}
square inches	sq cms	6.452
square inches	sq feet	6.944×10^{-3}
square inches	sq millimeters	645.2
square inches	sq mils	10^{6}
square inches	sq yards	7.716×10^{-4}
square kilometers	acres	247.1
square kilometers	sq cms	10^{10}
square kilometers	sq ft	10.76×10^{6}
square kilometers	sq inches	1.550×10^{9}
square kilometers	sq meters	10^{6}
square kilometers	sq miles	0.3861
square kilometers	sq yards	1.196×10^{6}
square meters	acres	2.471×10^{-4}
square meters	sq cms	10^{4}
square meters	sq feet	10.76
square meters	sq inches	1,550.
square meters	sq miles	3.861×10^{-7}
square meters	sq millimeters	10^{6}
square meters	sq yards	1.196
square miles	acres	640.0
square miles	sq feet	27.88×10^{6}
square miles	sq kms	2.590
square miles	sq meters	2.590×10^{6}
square miles	sq yards	3.098×10^{6}
square millimeters	circular mils	1,973.
square millimeters	sq cms	0.01
square millimeters	sq feet	1.076×10^{-5}
square millimeters	sq inches	1.550×10^{-3}
square mils	circular mils	1.273

TO CONVERT	INTO	MULTIPLY BY
square mils	sq cms	6.452×10^{-6}
square mils	sq inches	10^{-6}
square yards	acres	2.066×10^{-4}
square yards	sq cms	8,361.
square yards	sq feet	9.0
square yards	sq inches	1,296.
square yards	sq meters	0.8361
square yards	sq miles	3.228×10^{-7}
square yards	sq millimeters	8.361×10^{5}
	T	
temperature (°C) +273	absolute temperature (°C)	1.0
temperature (°C) +17.78	temperature (°F)	1.8
temperature (°F) +460	absolute temperature (°F)	1.0
temperature (°F) −32	temperature (°C)	5/9
tons (long)	kilograms	1,016.
tons (long)	pounds	2,240.
tons (long)	tons (short)	1.120
tons (metric)	kilograms	1,000.
tons (metric)	pounds	2,205.
tons (short)	kilograms	907.1848
tons (short)	ounces	32,000.
tons (short)	ounces (troy)	29,166.66
tons (short)	pounds	2,000.
tons (short)	pounds (troy)	2,430.56
tons (short)	tons (long)	0.89287
tons (short)	tons (metric)	0.9078
tons (short)/sq ft	kgs/sq meter	9,765.
tons (short)/sq ft	pounds/sq in.	2,000.
tons of water/24 hrs	pounds of water/hr	83.333
tons of water/24 hrs	gallons/min	0.16643
tons of water/24 hrs	cu ft/hr	1.3349
	V	
Volt/inch	Volt/cm.	.39370
Volt (absolute)	Statvolts	.003336
	W	
watts	Btu/hr	3.4129
watts	Btu/min	0.05688
watts	ergs/sec	107.
watts	foot-lbs/min	44.27
watts	foot-lbs/sec	0.7378
watts	horsepower	1.341×10^{-3}
watts	horsepower (metric)	1.360×10^{-3}
watts	kg-calories/min	0.01433
watts	kilowatts	0.001
Watts (Abs.)	B.T.U. (mean)/min.	0.056884
Watts (Abs.)	joules/sec.	1
watt-hours	Btu	3.413
watt-hours	ergs	3.60×10^{10}
watt-hours	foot-pounds	2,656.
watt-hours	gram-calories	859.85
watt-hours	horsepower-hrs	1.341×10^{-3}
watt-hours	kilogram-calories	0.8605
watt-hours	kilogram-meters	367.2
watt-hours	kilowatt-hrs	0.001
Watt (International)	Watt (absolute)	1.0002
webers	maxwells	10^{8}
webers	kilolines	10^{5}

Synchronous Speeds

$$\text{Synchronous Speed} = \frac{\text{Frequency} \times 120}{\text{No. of Poles}}$$

	FREQUENCY				FREQUENCY	
Poles	**60 cycle**	**50 cycle**	**25 cycle**	**Poles**	**60 cycle**	**50 cycle**
2	3600	3000	1500	42	171.4	142.9
4	1800	1500	750	44	163.6	136.4
6	1200	1000	500	46	156.5	130.4
8	900	750	375	48	150	125
10	720	600	300	50	144	120
12	600	500	250	52	138.5	115.4
14	514.3	428.6	214.3	54	133.3	111.1
16	450	375	187.5	56	128.6	107.1
18	400	333.3	166.7	58	124.1	103.5
20	360	300	150	60	120	100
22	327.2	272.7	136.4	62	116.1	96.8
24	300	250	125	64	112.5	93.7
26	276.9	230.8	115.4	66	109.1	90.9
28	257.1	214.3	107.1	68	105.9	88.2
30	240	200	100	70	102.9	85.7
32	225	187.5	93.7	72	100	83.3
34	211.8	176.5	88.2	74	97.3	81.1
36	200	166.7	83.3	76	94.7	78.9
38	189.5	157.9	78.9	78	92.3	76.9
40	180	150	75	80	90	75

Courtesy Ingersoll-Rand Co.

Temperature Conversion

NOTE: The center column of numbers in boldface refers to the temperature in degrees, either Centigrade or Fahrenheit, which it is desired to convert into the other scale. If converting from Fahrenheit to Centigrade degrees, the equivalent temperature will be found in the left column, while if converting from degrees Centigrade to degrees Fahrenhiet, the answer will be found in the column on the right.

Centigrade		Fahrenheit
−273.17	**−459.7**	
−268	**−450**	
−262	**−440**	
−257	**−430**	
−251	**−420**	
−246	**−410**	
−240	**−400**	
−234	**−390**	
−229	**−380**	
−223	**−370**	
−218	**−360**	
−212	**−350**	
−207	**−340**	
−201	**−330**	
−196	**−320**	
−190	**−310**	
−184	**−300**	
−179	**−290**	
−173	**−280**	
−169	**−273**	−459.4
−168	**−270**	−454
−162	**−260**	−436
−157	**−250**	−418
−151	**−240**	−400
−146	**−230**	−382
−140	**−220**	−364
−134	**−210**	−346
−129	**−200**	−328
−123	**−190**	−310
−118	**−180**	−292
−112	**−170**	−274
−107	**−160**	−256
−101	**−150**	−238
−96	**−140**	−220
−90	**−130**	−202
−84	**−120**	−184
−79	**−110**	−166
−73.3	**−100**	−148.0
−67.8	**−90**	−130.0
−62.2	**−80**	−112.0
−59.4	**−75**	−103.0
−56.7	**−70**	−94.0
−53.9	**−65**	−85.0
−51.1	**−60**	−76.0
−48.3	**−55**	−67.0
−45.6	**−50**	−58.0
−42.8	**−45**	−49.0
−40.0	**−40**	−40.0
−37.2	**−35**	−31.0
−34.4	**−30**	−22.0
−31.7	**−25**	−13.0
−28.9	**−20**	−4.0
−26.1	**−15**	5.0
−23.3	**−10**	.4.0

Centigrade		Fahrenheit
−20.6	**−5**	23.0
−17.8	**0**	32.0
−17.2	**1**	33.8
−16.7	**2**	35.6
−16.1	**3**	37.4
−15.6	**4**	39.2
−15.0	**5**	41.0
−14.4	**6**	42.8
−13.9	**7**	44.6
−13.3	**8**	46.4
−12.8	**9**	48.2
−12.2	**10**	50.0
−11.7	**11**	51.8
−11.1	**12**	53.6
−10.6	**13**	55.4
−10.0	**14**	57.2
−9.4	**15**	59.0
−8.9	**16**	60.8
−8.3	**17**	62.6
−7.8	**18**	64.4
−7.2	**19**	66.2
−6.7	**20**	68.0
−6.1	**21**	69.8
−5.6	**22**	71.6
−5.0	**23**	73.4
−4.4	**24**	75.2
−3.9	**25**	77.0
−3.3	**26**	78.8
−2.8	**27**	80.6
−2.2	**28**	82.4
−1.7	**29**	84.2
−1.1	**30**	86.0
−0.6	**31**	87.8
0.0	**32**	89.6
0.6	**33**	91.4
1.1	**34**	93.2
1.7	**35**	95.0
2.2	**36**	96.8
2.8	**37**	98.6
3.3	**38**	100.4
3.9	**39**	102.2
4.4	**40**	104.0
5.0	**41**	105.8
5.6	**42**	107.6
6.1	**43**	109.4
6.7	**44**	111.2
7.2	**45**	113.0
7.8	**46**	114.8
8.3	**47**	116.6
8.9	**48**	118.4
9.4	**49**	120.2
10.0	**50**	122.0
10.6	**51**	123.8

Centigrade		Fahrenheit
11.1	**52**	125.6
11.7	**53**	127.4
12.2	**54**	129.2
12.8	**55**	131.0
13.3	**56**	132.8
13.9	**57**	134.6
14.4	**58**	136.4
15.0	**59**	138.2
15.6	**60**	140.0
16.1	**61**	141.8
16.7	**62**	143.6
17.2	**63**	145.4
17.8	**64**	147.2
18.3	**65**	149.0
18.9	**66**	150.8
19.4	**67**	152.6
20.0	**68**	154.4
20.6	**69**	156.2
21.1	**70**	158.0
21.7	**71**	159.8
22.2	**72**	161.6
22.8	**73**	163.4
23.3	**74**	165.2
23.9	**75**	167.0
24.4	**76**	168.8
25.0	**77**	170.6
25.6	**78**	172.4
26.1	**79**	174.2
26.7	**80**	176.0
27.2	**81**	177.8
27.8	**82**	179.6
28.3	**83**	181.4
28.9	**84**	183.2
29.4	**85**	185.0
30.0	**86**	186.8
30.6	**87**	188.6
31.1	**88**	190.4
31.7	**89**	192.2
32.2	**90**	194.0
32.8	**91**	195.8
33.3	**92**	197.6
33.9	**93**	199.4
34.4	**94**	201.2
35.0	**95**	203.0
35.6	**96**	204.8
36.1	**97**	206.6
36.7	**98**	208.4
37.2	**99**	210.2
37.8	**100**	212.0
40.6	**105**	221
43.3	**110**	230
46.1	**115**	239
48.9	**120**	248
51.7	**125**	257

Centigrade		Fahrenheit
54.4	**130**	266
57.2	**135**	275
60.0	**140**	284
62.8	**145**	293
65.6	**150**	302
68.3	**155**	311
71.1	**160**	320
73.9	**165**	329
76.7	**170**	338
79.4	**175**	347
82.2	**180**	356
85.0	**185**	365
87.8	**190**	374
90.6	**195**	383
93.3	**200**	392
96.1	**205**	401
98.9	**210**	410
100.0	**212**	414
102	**215**	419
104	**220**	428
107	**225**	437
110	**230**	446
113	**235**	455
116	**240**	464
118	**245**	473
121	**250**	482
124	**255**	491
127	**260**	500
129	**265**	509
132	**270**	518
135	**275**	527
138	**280**	536
141	**285**	545
143	**290**	554
146	**295**	563
149	**300**	572
154	**310**	590
160	**320**	608
166	**330**	626
171	**340**	644
177	**350**	662
182	**360**	680
188	**370**	698
193	**380**	716
199	**390**	734
204	**400**	752
210	**410**	770
216	**420**	788
221	**430**	806
227	**440**	824
232	**450**	842
238	**460**	860
243	**470**	878
249	**480**	896
254	**490**	914
260	**500**	932

The formulas at the right may also be used for converting Centigrade or Fahrenheit degrees into the other scales.

Degrees Cent., $°C = \frac{5}{9}(°F + 40) - 40$

$= \frac{5}{9}(°F - 32)$

Degrees Fahr., $°F = \frac{9}{5}(°C + 40) - 40$

$= \frac{9}{5}\ C + 32$

Degrees Kelvin, $°K = °C + 273.2$

Degrees Rankine, $°R = °F + 459.7$

Courtesy Ingersoll-Rand Co.

Altitude and Atmospheric Pressures

Altitude above Sea Level			Tempera-ture**		Barometer*		Atmospheric Pressure	
Feet*	Miles	Meters*	°F	°C	Inches Hg Abs.	mm Hg Abs.	PSIA	Kg/sq cm Abs.
−5000	—	−1526	77	25	35.58	903.7	17.48	1.229
−4500	—	−1373	75	24	35.00	889.0	17.19	1.209
−4000	—	−1220	73	23	34.42	874.3	16.90	1.188
−3500	—	−1068	71	22	33.84	859.5	16.62	1.169
−3000	—	−915	70	21	33.27	845.1	16.34	1.149
−2500	—	−763	68	20	32.70	830.6	16.06	1.129
−2000	—	−610	66	19	32.14	816.4	15.78	1.109
−1500	—	−458	64	18	31.58	802.1	15.51	1.091
−1000	—	−305	63	17	31.02	787.9	15.23	1.071
−500	—	−153	61	16	30.47	773.9	14.96	1.052
0	—	0	59	15	29.92	760.0	14.696	1.0333
500	—	153	57	14	29.38	746.3	14.43	1.015
1000	—	305	55	13	28.86	733.0	14.16	.956
1500	—	458	54	12	28.33	719.6	13.91	.978
2000	—	610	52	11	27.82	706.6	13.66	.960
2500	—	763	50	10	27.32	693.9	13.41	.943
3000	—	915	48	9	26.82	681.2	13.17	.926
3500	—	1068	47	8	26.33	668.8	12.93	.909
4000	—	1220	45	7	25.84	656.3	12.69	.892
4500	—	1373	43	6	25.37	644.4	12.46	.876
5000	0.95	1526	41	5	24.90	632.5	12.23	.860
6000	1.1	1831	38	3	23.99	609.3	11.78	.828
7000	1.3	2136	34	1	23.10	586.7	11.34	.797
8000	1.5	2441	31	−1	22.23	564.6	10.91	.767
9000	1.7	2746	27	−3	21.39	543.3	10.50	.738
10,000	1.9	3050	23	−5	20.58	522.7	10.10	.710
15,000	2.8	4577	6	−14	16.89	429.0	8.29	.583
20,000	3.8	6102	−12	−24	13.76	349.5	6.76	.475
25,000	4.7	7628	−30	−34	11.12	282.4	5.46	.384
30,000	5.7	9153	−48	−44	8.903	226.1	4.37	.307
35,000	6.6	10,679	−66		7.060	179.3	3.47	.244
40,000	7.6	12,204	−70	−57	5.558	141.2	2.73	.192
45,000	8.5	13,730	−70	−57	4.375	111.1	2.15	.151
50,000	9.5	15,255	−70	−57	3.444	87.5	1.69	.119
55,000	10.4	16,781	−70	−57	2.712	68.9	1.33	.0935
60,000	11.4	18,306	−70	−57	2.135	54.2	1.05	.0738
70,000	13.3	21,357	−67	−55	1.325	33.7	.651	.0458
80,000	15.2	24,408	−62	−52	†8.273^{-1}	21.0	.406	.0285
90,000	17.1	27,459	−57	−59	5.200^{-1}	13.2	.255	.0179
100,000	18.9	30,510	−51	−46	3.290^{-1}	8.36	.162	.0114
120,000	22.8	36,612	−26	−48	1.358^{-1}	3.45	—	—
140,000	26.6	42,714	4	−16	5.947^{-2}	1.51	—	—
160,000	30.4	48,816	28	−2	2.746^{-2}	†6.97^{-1}	—	—
180,000	34.2	54,918	19	−7	1.284^{-2}	3.26^{-1}	—	—
200,000	37.9	61,020	−3	−19	5.846^{-3}	1.48^{-1}	—	—
220,000	41.7	67,122	−44	−42	2.523^{-3}	6.41^{-2}	—	—
240,000	45.5	73,224	−86	−66	9.955^{-4}	2.53^{-2}	—	—
260,000	49.3	79,326	−129	−90	3.513^{-4}	8.92^{-3}	—	—
280,000	53.1	85,428	−135	−93	1.143^{-4}	3.67^{-3}	—	—
300,000	56.9	91,530	−127	−88	3.737^{-5}	9.49^{-4}	—	—
400,000	75.9	122,040	—	—	6.3^{-7}	1.60^{-5}	—	—
500,000	94.8	152,550	—	—	1.4^{-7}	3.56^{-6}	—	—
600,000	114	183,060	—	—	5.9^{-8}	1.50^{-6}	—	—
800,000	152	244,080	—	—	1.6^{-8}	4.06^{-7}	—	—
1,000,000	189	305,100	—	—	5.1^{-9}	1.30^{-7}	—	—
1,200,000	228	366,120	—	—	2.0^{-9}	5.08^{-8}	—	—
1,400,000	266	427,140	—	—	8.2^{-10}	2.08^{-8}	—	—
1,600,000	304	488,160	—	—	3.8^{-10}	9.65^{-9}	—	—
1,800,000	342	549,180	—	—	1.8^{-10}	4.57^{-9}	—	—
2,000,000	379	610,200	—	—	9.2^{-11}	2.34^{-9}	—	—

Data from NASA Standard Atmosphere (1962).

*Temperature and barometer are approximate for negative altitudes.

**Temperatures are average existing at 40° latitude and are rounded to even numbers.

†Negative exponent shows number of spaces the decimal point must be moved to the left.

Courtesy Ingersoll-Rand Co.

Pressure Conversion Chart

BY FACTOR TO OBTAIN →

MULTIPLY GIVEN NUMBER OF ↓

GIVEN	lb/in^2	in H_2O (at +39.2°F)	cm H_2O (at +4°C)	in Hg (at +32°F)	mm Hg (Torr) (at 0°C)	dyne/cm^2 (1μ bar)	newton/m^2 (PASCAL)	kgm/cm^2	bar	atm. (A_n)	lb/ft^2	ft H_2O (at +39.2°F)
lb/in^2	1.0000	$2.7680x10^{1}$	$7.0308x10^{1}$	2.0360	$5.1715x10^{1}$	$6.8948x10^{4}$	$6.8948x10^{3}$	$7.0306x10^{-2}$	$6.8947x10^{-2}$	$6.8045x10^{-2}$	$1.4400x10^{2}$	2.3067
in H_2O (at +39.2°F)	$3.6127x10^{-2}$	1.0000	2.5400	$7.3554x10^{-2}$	1.8683	$2.4908x10^{3}$	$2.4908x10^{2}$	$2.5399x10^{-3}$	$2.4908x10^{-3}$	$2.4582x10^{-3}$	5.2022	$8.3333x10^{-2}$
cm H_2O (at +4°C)	$1.4223x10^{-2}$	0.3937	1.0000	$2.8958x10^{-2}$	0.7355	$9.8064x10^{2}$	$9.8064x10^{1}$	$9.9997x10^{-4}$	$9.8064x10^{-4}$	$9.6781x10^{-4}$	2.0481	$3.2808x10^{-2}$
in Hg (at +32°F)	$4.9116x10^{-1}$	$1.3596x10^{1}$	$3.4532x10^{1}$	1.0000	$2.5400x10^{1}$	$3.3864x10^{4}$	$3.3864x10^{3}$	$3.4532x10^{-2}$	$3.3864x10^{-2}$	$3.3421x10^{-2}$	$7.0727x10^{1}$	1.1330
mm Hg (Torr) (at 0°C)	$1.9337x10^{-2}$	$5.3525x10^{-1}$	1.3595	$3.9370x10^{-2}$	1.0000	$1.3332x10^{3}$	$1.3332x10^{2}$	$1.3595x10^{-3}$	$1.3332x10^{-3}$	$1.3158x10^{-3}$	2.7845	$4.4605x10^{-2}$
dyne/cm^2 (1μ bar)	$1.4504x10^{-5}$	$4.0147x10^{-4}$	$1.0197x10^{-3}$	$2.9530x10^{-5}$	$7.5006x10^{-4}$	1.0000	$1.0000x10^{-1}$	$1.0197x10^{-6}$	$1.0000x10^{-6}$	$9.8692x10^{-7}$	$2.0886x10^{-3}$	$3.3456x10^{-5}$
newton/m^2 (PASCAL)	$1.4504x10^{-4}$	$4.0147x10^{-3}$	$1.0197x10^{-2}$	$2.9530x10^{-4}$	$7.5006x10^{-3}$	$1.0000x10^{1}$	1.0000	$1.0197x10^{-5}$	$1.0000x10^{-5}$	$9.8692x10^{-6}$	$2.0885x10^{-2}$	$3.3456x10^{-4}$
kgm/cm^2	$1.4224x10^{1}$	$3.9371x10^{2}$	$1.00003x10^{3}$	$2.8959x10^{1}$	$7.3556x10^{2}$	$9.8060x10^{5}$	$9.8060x10^{4}$	1.0000	$9.8060x10^{-1}$	$9.678x10^{-1}$	$2.0482x10^{3}$	$3.2809x10^{1}$
bar	$1.4504x10^{1}$	$4.0147x10^{2}$	$1.0197x10^{3}$	$2.9530x10^{1}$	$7.5006x10^{2}$	$1.0000x10^{6}$	$1.0000x10^{5}$	1.0197	1.0000	$9.8692x10^{-1}$	$2.0885x10^{3}$	$3.3456x10^{1}$
atm. (A_n)	$1.4696x10^{1}$	$4.0679x10^{2}$	$1.0333x10^{3}$	$2.9921x10^{1}$	$7.6000x10^{2}$	$1.0133x10^{6}$	$1.0133x10^{5}$	1.0332	1.0133	1.0000	$2.1162x10^{3}$	$3.3900x10^{1}$
lb/ft^2	$6.9445x10^{-3}$	$1.9223x10^{-1}$	$4.882x10^{-1}$	$1.4139x10^{-2}$	$3.591x10^{-1}$	$4.7880x10^{2}$	$4.7880x10^{1}$	$4.8824x10^{-4}$	$4.7880x10^{-4}$	$4.7254x10^{-4}$	1.0000	$1.6019x10^{-2}$
ft H_2O (at +39.2°F)	$4.3352x10^{-1}$	$1.2000x10^{1}$	$3.0480x10^{1}$	$8.826x10^{-1}$	$2.2419x10^{1}$	$2.9890x10^{4}$	$2.9890x10^{3}$	$3.0479x10^{-2}$	$2.9890x10^{-2}$	$2.9499x10^{-2}$	$6.2427x10^{1}$	1.0000

Index